Computational Biology: Theory, Models and Methods

Computational Biology: Theory, Models and Methods

Editor: Ivie Nathaniel

STATES
ACADEMIC PRESS
www.statesacademicpress.com

States Academic Press,
109 South 5th Street,
Brooklyn, NY 11249, USA

Visit us on the World Wide Web at:
www.statesacademicpress.com

ISBN: 978-1-63989-121-4 (Hardback)

Cataloging-in-Publication Data

Computational biology : theory, models and methods / edited by Ivie Nathaniel.
p. cm.
Includes bibliographical references and index.
ISBN 978-1-63989-121-4
1. Computational biology. 2. Bioinformatics. 3. Systems biology. 4. Biology. I. Nathaniel, Ivie.
QH324.2 .C66 2022
570.285--dc23

Table of Contents

Preface

Computational biology uses biological data to develop algorithms or models which help in understanding biological systems and relationships. It includes many aspects of bioinformatics. There is a large amount of data which is analysed in the fields of molecular biology and genomics. Research in computational biology deals with developing analytical methods for interpreting biological information. Computational biology has been used in classifying the human genome, creating accurate models of the human brain and assisting in modeling biological systems. It makes use of concepts from the fields of biology, applied mathematics, statistics, biochemistry, chemistry, biophysics, molecular biology, genetics, genomics, computer science and evolution. The book studies, analyses and upholds the pillars of computational biology and its utmost significance in modern times. It elucidates the concepts and innovative models around prospective developments with respect to this discipline. This book, with its detailed analyses and data, will prove immensely beneficial to professionals and students involved in this area at various levels.

This book unites the global concepts and researches in an organized manner for a comprehensive understanding of the subject. It is a ripe text for all researchers, students, scientists or anyone else who is interested in acquiring a better knowledge of this dynamic field.

I extend my sincere thanks to the contributors for such eloquent research chapters. Finally, I thank my family for being a source of support and help.

Editor

1

Centrality measures for immunization of weighted networks

Mohammad Khansari, Amin Kaveh, Zainabolhoda Heshmati, Maryam Ashkpoor Motlaq
University of Tehran, Faculty of New Sciences and Technologies, Amir Abad, North Kargar Street 14395-1374, Tehran, Iran
E-mail: m.khansari@ut.ac.ir

Abstract
Effective immunization of individual communities with minimal cost in vaccination has made great discussion surrounding the realm of complex networks. Meanwhile, proper realization of relationship among people in society and applying it to social networks brings about substantial improvements in immunization. Accordingly, weighted graph in which link weights represent the intensity and intimacy of relationships is an acceptable approach. In this work we employ weighted graphs and a wide variety of weighted centrality measures to distinguish important individuals in contagion of diseases. Furthermore, we propose new centrality measures for weighted networks. Our experimental results show that Radiality-Degree centrality is satisfying for weighted BA networks. Additionally, PageRank-Degree and Radiality-Degree centralities showmoreacceptable performance in targeted immunization of weighted networks.

Keywords centrality measure; epidemic threshold; largest connected components; targeted immunization; weighted graphs.

1 Introduction

Epidemic is 'the occurrence of more cases of disease than expected in a given area or among a specific group of people over a particular period of time' (Parthasarathy, 2013). Nowadays, nearly 13 million people die annually by infectious diseases which imposes high costs to society. Between 2009 and 2010, 14,000 people lost their lives as a result of an influenza epidemic (Jain et al., 2009). Since 1981, AIDS is known as an epidemic disease and according to estimates by WHO and UNAIDS only in 2013, it was the cause of death of 1.5 million people in the world. As a result, prediction and control of epidemics among populations has attracted many researchers.

Immunization prevents diseases from spreading and saves many people from suffering and death (Parthasarathy, 2013). Vaccination is one of the immunization techniques that protects the vaccinated person as well as prevents the spread of disease, simultaneously (Cornforth et al., 2011). In traditional methods (mass vaccination) the entire population will be vaccinated, which is not affordable due to high costs (Gallos et al.,

2007; Vidondo et al., 2012). A more effective method is targeted vaccination in which people are clustered by some known criteria. In these categories people with more influence are identified and vaccinated (Chen et al., 2008; Cornforth et al., 2011; Eames et al., 2009; Hartvigsen et al., 2007; Miller and Hyman, 2007; Peng et al., 2010; Schneider et al., 2012; Shams and Khansari, 2014; Vidondo et al., 2012).The purpose of targeted vaccination is that people are selected in such a way that unsafe clusters are as small as possible (Schneider et al., 2012). In other words, the work is done by immunization with intercommunal individuals.

Complex networks has become a suitable tool for modeling, representing and describing the characteristics of many natural phenomena (Zhgang and Zhan, 2011; Zhang, 2012). Among these networks, social networks can be outlined in which each node represents an individual while communication between a pair of individuals is represented as a link. As a result, many studies have been done on disease distributions and immunization within these networks (Cornforth et al., 2011; Hartvigsen et al., 2007; Shams and Khansari, 2014). In these studies different centralities have been defined and nodes with high centrality have been vaccinated and the result of the vaccination has been analyzed.

Although representation of social networks with simple graphs leads to suitable models, adopting weighted networks can be more appropriate to include detailed information. In weighted social networks, the quality of communication or relationship between a pair of nodes is considered as the weight of the edge between them. Some works which have studied centrality measures in weighted networks are (Abbasi and Hossain, 2013; Abbasi et al., 2011; Wang et al., 2008; Zhai et al., 2013).

Shams and Khansari have proposed an evaluation framework for targeted immunization algorithms in which some centrality measures are utilized (Shams and Khansari, 2014). These centrality measures have been used to distinguish effective nodes, though other meaningful and powerful ones are not considered.

To our knowledge, a complete study in capability of various centrality measures in detection of effective nodes in weighted networks has not been done. Thus, we conduct a comprehensive study to determine the effectiveness of different centrality measures in detecting important nodes within weighted networks. In this study we propose new centrality measures while comparing them with the current existing measures. All the evaluations are done on real and artificial networks.

The rest of this paper is organized as follows: section 2 presents current existing centrality measures for weighted networks, following four new centrality measures which are introduced in section 3. Model evaluation methods and analysis are introduced in section 4 while experimental results are presented in the following section 5. Finally, Section 6 concludes this work.

2 Centrality Measures for Weighted Networks

Node centrality is one of the most analyzed concepts which determines the node effect on network flows. Centrality of a node represents its prominence based on the definition of centrality. In a social network node centrality exhibits the individuals' communication pattern and his/her position in the network.

Leavitt defined this concept for the first time in 1951 (Leavitt, 1951)and Freeman defined three important and applicable centralities: *degree*, *closeness* and *betweenness* in 1978 (Freeman, 1978). In the following subsections current existing centrality measures are reviewed with emphasis on weighted networks.

2.1 Degree centrality

Degree centrality is a simple local centrality which is defined based on neighborhood concept. In a social network it exhibits a node popularity and in our examinations it shows the node influence in spreading and suppression of diseases. In weighted networks, a node degree centrality is the sum of the weights of the edges attached to that node which is also nominated as the node *strength*. This measure represents the whole involvement of a node in the network (Opsahl et al., 2010):

$$C_D^W(i) = \sum_{j=1}^{N} W(i,j) \tag{1}$$

where, W is the weighted adjacency matrix and $w(i,j)$ is the weight of edge which ties node $i$to j.

2.2 Distance-based Centralities

Other centralities are distance-based measures and have a close relationship with the concept of "path"in weighted networks. In contrast to binary networks, different values can be assigned to different links in weighted networks. Edge weight can represent cost (Dijkstra, 1959) as well as strength of connection (Newman, 2001). In social networks edge weight represents strength of physical connection or sense of intimacy between individuals. Hence, distance between two adjacent nodes is the inverse of the weight of their corresponding link. Thus to calculate the geodesic path we use the method proposed in (Newman, 2001). Distance-based centralities are as follows:

2.2.1 Closeness centrality

Closeness centrality is a global centrality which represents the independence of a node in the network (Freeman, 1978). The most central node has the minimum sum of the 'geodesic' distances to all other nodes in the network:

$$C_C^W(i) = \left(\sum_j d^w(i,j)\right)^{-1} \tag{2}$$

where, $d^w(i,j)$ is the weighted geodesic path between nodes$i$and j.

2.2.2 Betweenness centrality

Betweenness centrality represents the node's ability to control the data flow in the network (Freeman, 1978). This measure is the proportion of number of geodesic paths that pass through the given node to number of geodesic paths between any pair of nodes in the network:

$$C_B^W(i) = \frac{g_{jk}^w(i)}{g_{jk}^w} \tag{3}$$

where, $g_{jk}^w(i)$is the number of weighted geodesic paths which pass through node i and g_{jk}^w is the number of weighted geodesic paths between two nodes.

2.2.3 Radiality centrality

Radiality represents the extent of access into the network privided by the node's neighbors (Valente and Foreman, 1998). High radiality means that it takes less time for the infectious node to reach others in the network:

$$C_{Rad}^W(i) = \frac{\sum_{j \in V}(\Delta_G^W + 1 - d^w(i,j))}{g_{jk}^w} \tag{4}$$

where, Δ_G^W is the diameter of graph G, V is the set of nodes in graph G, $d^w(i,j)$ is the weighted geodesic path between nodes$i$ and j and $|V| = n$.

2.2.4 Katz centrality

Katz centrality is the generalization of degree centrality. In other words, a node's degree centrality is the number of its neighbors while Katz centrality is the number of nodes which are accessible through a specific path. However, contribution of distant nodes is reduced:

$$C_{Katz}^{W}(i) = \alpha \sum_j W(i,j) \,.\, C_{Katz}^{W}(j) + \beta \tag{5}$$

where, $W(i,j)$ is the element of row i and column j of weighted adjacency matrix.α, β are positive constants where, α is an attenuation factor,$0 \leq \alpha \leq 1$(Katz, 1953). By $\beta > 0$, nodes with zero in-degree (in directed networks) get positive centrality.

2.2.5 Subgraph centrality

This measure is based on spectral properties and characterizes the contribution of a node in different possible subgraphs. Subgraph centrality of a node is the sum of closed paths with different lengths which launch from and terminate in that node. Shorter path lengths have more influence in the calculations (Estrada and Rodriguez, 2005):

$$C_{SG}^{W}(i) = \sum_{k=0}^{\infty} \frac{\mu_k^W(i)}{k!} \tag{6}$$

where, $\mu_k^W(i)$ is the number of closed walks of cost k starting and ending at node i.

2.2.6 Eccentricity centrality

A node's eccentricity centrality is the inverse of the longest geodesic path from that node to other nodes. Hence, a node with the smallest longest geodesic path is the most important node in network. In social networks this centrality exhibits how distant, at most, is each person from others.

$$C_{Ecc}^{W}(i) = max\{d^W(i,j)\}^{-1} \tag{7}$$

where, $d^W(i,j)$ is the weighted shortest path between nodes$i$and j.

2.2.7 Communication centrality

Communication centrality represents the capability of a node to communicate with other nodes. This measure depends on node's degree, communication ability of neighbors and node's edges weight. Communication centrality of a node is defined based on multiplying the *h-degree* of a neighbor in the weight of the edge connecting the node to that neighbor.

Zhai et al. defined the communication centrality of node x as "the largest integer k such that the node x has at least k neighbor nodes satisfying the product of each node's *h-degree* and the weight of the edge linked with node x is no fewer than k" (Zhai et al., 2013). If the *h-degree* of node's neighbors are marked as $d_h(n_1)$, $d_h(n_2)$,..., $d_h(n_M)$ and the node's edge weight are w_1, w_2, ..., w_M then their product sequence is:

$$w_1 d_h(n_1), w_2 d_h(n_2), \ldots, w_M d_h(n_M)$$

If we suppose:

$$w_1 d_h(n_1) \geq w_2 d_h(n_2) \geq \ldots \geq w_M d_h(n_M)$$

Then,

$$C_{Com}(i) = \max\{k : w_k d_h(n_k) \geq k\} \tag{8}$$

2.3 Spectral centralities

Spectral centrality measures are those that their definition and calculation relies on the eigenvalues and the eigenvectors of adjacency and Laplacian matrices of the graph. The most well-known two are:

2.3.1 Laplacian centrality

Laplacian centrality is based on the Laplacian matrix. This matrix contains practical information about dynamics and geometry of the network (Pauls and Remondini, 2012). Laplacian matrix is defined as:

$$L(G) = X(G) - W(G)$$

where, $X(G)$ is the graph's degree matrix and $W(G)$ is the graph's weighted adjacency matrix. If we nominate eigenvalues of $L(G)$as λ_1, λ_2,..., λ_n the Laplacian energy of graph G is:

$$E_L(G) = \sum_{i=1}^{n} \lambda_i^2$$

A node's *Laplacian* centrality is the difference between Laplacian energy of network with and without that node (Qi et al., 2012):

$$E_L(v_i, G) = \frac{E_L(G) - E_L(G_i)}{E_L(G)} \qquad (9)$$

2.3.2 PageRank centrality

This centrality measure represents a node's relative importance in the network (Sarma et al., 2011). The node's importance depends on the importance of its neighbors. *PageRank* centrality is based on links analysis:

$$C_{PR}^{w}(i) = (I - d.\overline{W})^{-1} + (1 - d)v_i \qquad (10)$$

where d is known as damping factor that is usually set to $0.85.\overline{W}$is the normalized weighted adjacency matrix and v is the preference network.

It is worth noting that *Subgraph Centrality* can be identified as spectral centrality.

2.4 Hybrid centralities

One of the major problems in pure centralities such as *degree*, *betweenness* and *closeness* centralities is that they are not a convenient measure to estimating node availability. In other words, the degree centrality of a node with n neighbors with edge weight of 1 is equal to the degree centrality of a node with just 1 neighbor with edge weight of$n$. However, by removing an edge from each node the former is available and the latter is not. Hence, hybrid centrality measures are proposed for better understanding of importance and influence of nodes which are applicable to both weighted and binary networks.

2.4.1 Degree-Degree centrality

This measure highlights the nodes which have more connections to more important nodes. In other words, the node degree as well as the degree of node's neighbors are considered in calculation of this centrality measure.

$$DD^{w}(i) = \sum_{j=1}^{n} (w(i,j).C_D(j)) \qquad (11)$$

where, $C_D(j)$ is the degree centrality of node j and $w(i,j)$ is the weight of edge between nodes $i$and j.

2.4.2 Closeness-Degree centrality

This measure not only represents the node's influence in control of data flow, but also represents the node's performance in communication with other nodes in the network.

$$DC^{w}(i) = \sum_{j=1}^{n} (w(i,j).C_C(j)) \qquad (12)$$

where, $C_C(j)$ is the closeness centrality of node j and $w(i,j)$ is the weight of edge between nodes $i$and j.

2.4.3 Betweenness-Degree centrality

This centrality represents popularity of a node as well as its influence in control of data flows in the network.

$$DB^{w}(i) = \sum_{j=1}^{n}(w(i,j).C_B(j)) \tag{13}$$

where, $C_B(j)$ is the betweenness centrality of node j and $w(i,j)$ is the weight of edge between nodes $i$and j.

3 The Proposed Centrality Measures

To detect the most effective nodes in the network and study vaccination of these nodes in the immunization network, we propose four new hybrid centrality measures regarding the pattern proposed in (Abbasi et al., 2011):

3.1 PageRank-Degree centrality

As aforementioned, PageRank centrality measure represents the importance of nodes based on their adjacent nodes. We Introduce PageRank-Degree centrality as a combination of PageRank measure with degree centrality. The following formula represents this new centrality measure:

$$C_{DP}(i) = \sum_{j=1}^{n} w(i,j).((1-d) + d\sum_{v\in B(j)}\frac{C_{PR}(v)}{N_v}) \tag{14}$$

where, $w(i,j)$ is the weight of the edge between nodes i and j, $C_{PR}(v)$ is the PageRank centrality of node v, d is damping factor with value of 0.85, $B(j)$ is the set of nodes which have edge with node j and N_v is number of outlinks of node v.

3.2 Radiality-Degree centrality

Radiality centrality exhibits the degree of a node's access to other nodes in the network through its neighbors. Thus, if a node has more neighbors with high Radiality centrality, its Radiality centrality is high as well. We define Radiality-Degree centrality measure as:

$$C_{DR}(i) = \sum_{j=1}^{n} w(i,j).\sum_{u\in V}\frac{\Delta_G+1-d^{w}(j,u)}{n-1} \tag{15}$$

where, $w(i,j)$ is the weight of edge between nodes $i$and j, Δ_G is the diameter of graph G, $V$is the set of nodes of graph G, n is the number of nodes in the graph and $d^{w}(j,u)$ is the geodesic path between nodes$j$ and u.

3.3 Subgraph-Degree centrality

We define subgraph-degree centrality as:

$$C_{DS}(i) = \sum_{j=1}^{n} w(i,j).\sum_{k=0}^{\infty}\frac{\mu_k(j)}{k!} \tag{16}$$

where, $w(i,j)$ is the weight of edge between nodes $i$and j, $\mu_k(j)$ is the number closed walks of cost k starting and ending on node j.

3.4 Katz-Degree centrality

Hybrid Katz-degree centrality is defined as follows:

$$C_{DK}(i) = \sum_{j=1}^{n} w(i,j).\alpha\sum_{j} A_{ju}x_{ju} + \beta \tag{17}$$

where, $w(i,j)$ is the weight of edge between nodes $i$and j. α, β are positive constants where, α is attenuation factor and $0 \leq \alpha \leq 1$.

4 Evaluation Method

The purpose of this study is to vaccinate the nodes with highest effect on disease distribution and remove them from the network. Effective nodes will be selected by centrality measures which were discussed in the previous section. By vaccinating these nodes the links in the network will be decreased therefore the network is assorted into smaller elements. The framework of this model is the same as in (Shams and Khansari, 2014). We use two criteria to estimate immunization algorithms: largest connected component (LCC) and epidemic threshold.

4.1 Increment of epidemic threshold of network

Epidemic threshold of network means the minimum number of people who have to be infected to reach an epidemic level. Higher epidemic threshold indicates lower probability to reach epidemic (Chakrabarti et al., 2008; Kitchovitch and Lio, 2011; Masuda, 2009; Peng et al., 2010; Shams and Khansari, 2014). The epidemic threshold of a network is the inverse of largest eigenvalue of network adjacency matrix (Chakrabarti et al., 2008; Kitchovitch and Lio, 2011; Shams and Khansari, 2014). We study the influence of removal of a specified node in reducing the largest eigenvalue of the network adjacency matrix. The purpose of this study is to assess the capability of considered centrality measures to detect and immunize the most effective nodes. As described in (Shams and Khansari, 2014), we calculate $E_1 = \frac{\lambda_1'}{\lambda_1}$ where λ_1 is the largest eigenvalue of the original network and λ_1' is the largest eigenvalue of the vaccinated network.

4.2 Decrement of Largest-Connected-Component size

A Connected-Component is a subgraph in which there is at least one path between every two nodes. Thus, the largest epidemic size of a network is its largest connected component (LCC) size (Chen et al., 2008; Gallos et al., 2007; Schneider et al., 2012; Shams and Khansari, 2014). Hence, we calculate $E_2 = \frac{LCC'}{LCC}$ where LCC is the largest connected component of the original network and LCC' is the largest connected component of the immunized network.

5 Experimental Results

In this section we investigate the performance of the predefined and proposed centrality measures in detecting and immunizing influential nodes in the network. We calculate the centrality measures in three artificial networks and in one real network and investigate their performance. To generate the artificial networks and to calculate the different centrality measures we use R.3.1.1 software.

5.1 Datasets

To generate the artificial networks we use three most famous models: Scale-free, Erdős-Rényi model (ER) and Small-World. Scale-free network is generated based on (Albert and Barabási, 2002) with 500 nodes and parameter $m = 3$ which leads to 1500 edges. ER network is generated with 500 nodes and 1500 edges which is greater than the threshold of connectedness of a random graph (Erdős and Rényi, 1961). Small-World network is produced with 500 nodes and 3 initial neighbors in each side which result in 1500 edges and rewiring probability is $p = 0.1$ (Watts and Strogatz, 1998). Edge weight is a random number between 1 and 20. For each network and for each centrality measure we generate five datasets and calculate the average of desired parameters. Then we use them in evaluation of immunization performance.

The real network includes Facebook-like (FBL) (Opsahl and Panzarasa, 2009) network which is frequently used in immunization literature and consists of 1899 nodes and 13838 edges. The weight of the edges are uniformly distributed between 1 and 20.

5.2 Diagnose and vaccination

Two methods are proposed to diagnose and vaccinate individuals based on their centrality in networks: the first is initial method in which centrality measure is calculated in initial network and most central nodes will be immunized regarding number of available vaccination resources. In the second method, adaptive method, in each iteration the most central node will be immunized and centrality measure will be recalculated (Schneider et al., 2012; Shams and Khansari, 2014). We determine and vaccinate individuals based on the initial method. Our artificial networks consist of 500 nodes and in each iteration we immunize (remove) 10 nodes. The utilized real network consist of 1890 nodes and in each iteration 40 individuals are vaccinated.

5.3 LCC in network models

Fig. 1 illustrates the efficiency of targeted immunization using LCC size in case of scale-free (BA), Small-world (SW) and Random (ER) networks, regarding different centrality measures. We plotted the proportion of LCC size to network size versus number of vaccinated (removed) nodes.Obviously, for BA model which is presented by ● symbol, by vaccinating 50 as well as 100 individuals *Communication* centrality outperforms others. *Radiality-Degree*, *PageRank* centralities represent the next best performance with almost same level. On the other hand, *Betweenness-Degree* centrality is the worst strategy to immunize the weighted BA network. *Subgraph-Degree* and *Eccentricity* are the next worst metrics. Our experiments show that in all centrality measures except *Betweenness-Degree*, the worst expected epidemic size (LCC) decreases significantly in the first 20 iterations (immunization of 40% of society). By using the *Communication* centrality, the proportion of worst expected epidemic size (LCC) to network size is 0.04 when %20 individuals of network are immunized which has the best result.

Considering different centrality measures to determine and immunize most effective nodes in weighted ER networks is illustrated by ■ symbol. The worst expected epidemic size reduces considerably by using *Subgraph* centrality. *Communication* and *Radiality-Degree* centralities have the next best performance. On the contrary, *Eccentricity* and *Degree-Degree* Centralities exhibit the worst results.

Proportion of LCC size to network size versus vaccinated nodes regarding different centrality measures for Wattz-Strogatz networks are represented by ▲ symbol. The *Betweenness* centrality outperforms others in reducing LCC size. By using *Betweenness* centrality to detect and immunize the most important nodes, the worst expected epidemic size will be reduced to one fifth of network size by vaccinating 40% of society. Table 1 summarizes our experiments in network models.*Closeness-Degree* and *Radiality-Degree* centralities are the next best measures to immunize small-world networks. Take, for example, in order to diminish the size of LCC to one fifth of network, we have to immunize 40%, 44% and 48% of the society by using *Betweenness*, *Closeness-Degree* and *Radiality-Degree* respectively. *Katz-Degree*, *Katz* and *Eccentricity* centralities show the worst results.

5.4 Epidemic threshold in network models

Effectiveness of different centrality measures to reduce the largest eigenvalue of network adjacency matrix versus number of vaccinated individuals are illustrated in Fig. 2 in the case of scale-free (BA), Small-world (SW) and Random (ER) networks.

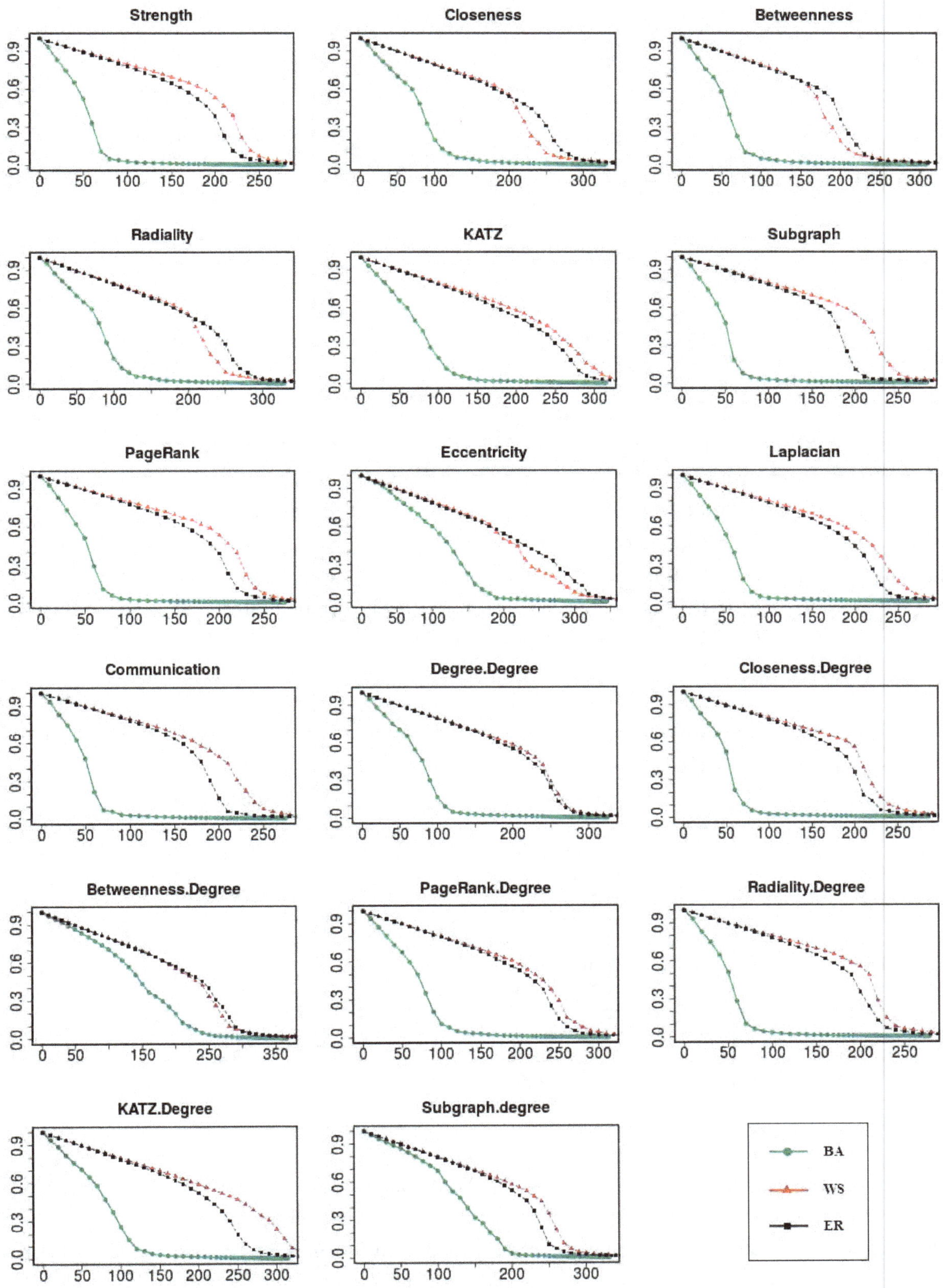

Fig. 1 Proportion of LCC size to inital network size versus number of vaccinated nodes in BA, ER and WS networks.

Table 1 Centrality measures which have the best and worst performance in decreasing LCC size in network models.

Network Model	Outperform	Underperform
Barabási-Albert	Communication Radiality-Degree PageRank	Betweenness-Degree Subgraph-Degree Eccentricity
Erdős-Rényi	Subgraph Communication Radiality-Degree	Eccentricity Degree-Degree
Watts-Strogatz	Betweeness Closeness-Degree Radiality-Degree	Katz-Degree Katz Eccentricity

Proportion of largest eigenvalue of adjacency matrix of immunized network to largest eigenvalue of original weighted BA network (λ_1) is illustrated by ● symbol. *Radiality-Degree*, *PageRank-Degree* and *Strength* centralities represent the best performance respectively. On the other hand, *Eccentricity*, *Subgraph-Degree* and *Betweenness-Degree* exhibit the worst.

Reducing the largest eigenvalue of immunized network versus immunized nodes is depicted by ■ symbol. Obviously, *Closeness-Degree*, *Degree-Degree* and *PageRank-Degree* centralities have the most performance while *Eccentricity* and *Subgraph* centralities underperform others.

Table 2 Centrality measures which have the best as well as the worst performance in increasing epidemic threshold in network models.

Network Model	Outperform	Underperform
Barabási-Albert	Radiality-Degree PageRank-Degree Strength	Eccentricity Subgraph-Degree Betweenness-Degree
Erdős-Rényi	Closeness-Degree Degree-Degree PageRank-Degree	Eccentricity Subgraph
Watts-Strogatz	Strength Degree-Degree PageRank-Degree	Betweenness Radiality Eccentricity

The same process has been done for Small-World (Wattz-Strogatz) networks which is illustrated by ▲ symbol. Our experiments show that by vaccination of 47% of society, the proportion of largest eigenvalue of network to largest eigenvalue of initial network would be 0.3 by using *Strength* centrality which is the best result. *Degree-Degree* and *PageRank-Degree* centralities exhibit the next best performance. *Betweenness*, *Radiality* and *Eccentricity* show the worst results respectively. Table 2 summarizes our experiments in increasing epidemic threshold in different network models.

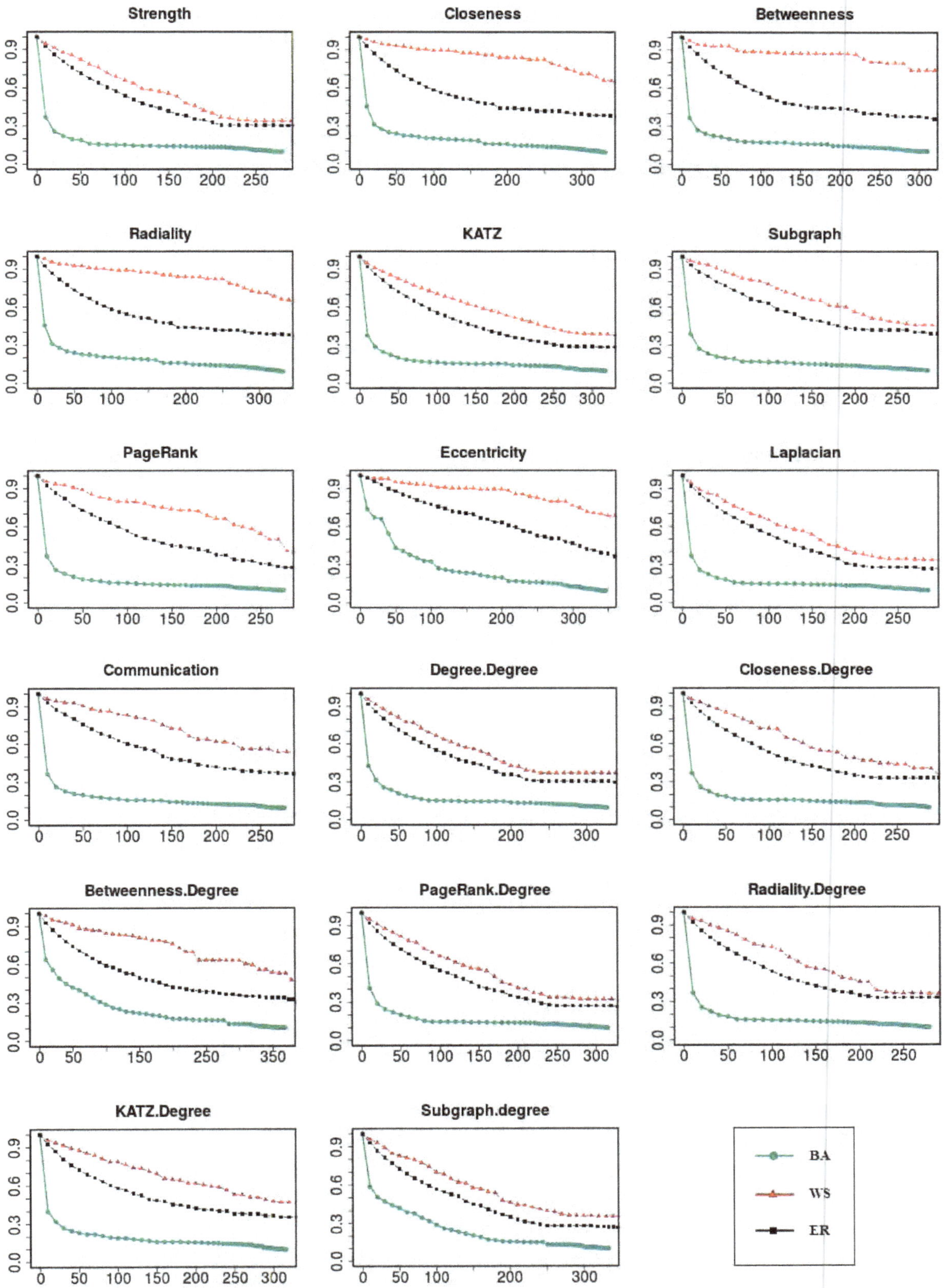

Fig. 2 Proportion of largest eigenvalue of immunized network to largest eigenvalue of initial network versus number of vaccinated nodes in BA, ER and WS networks

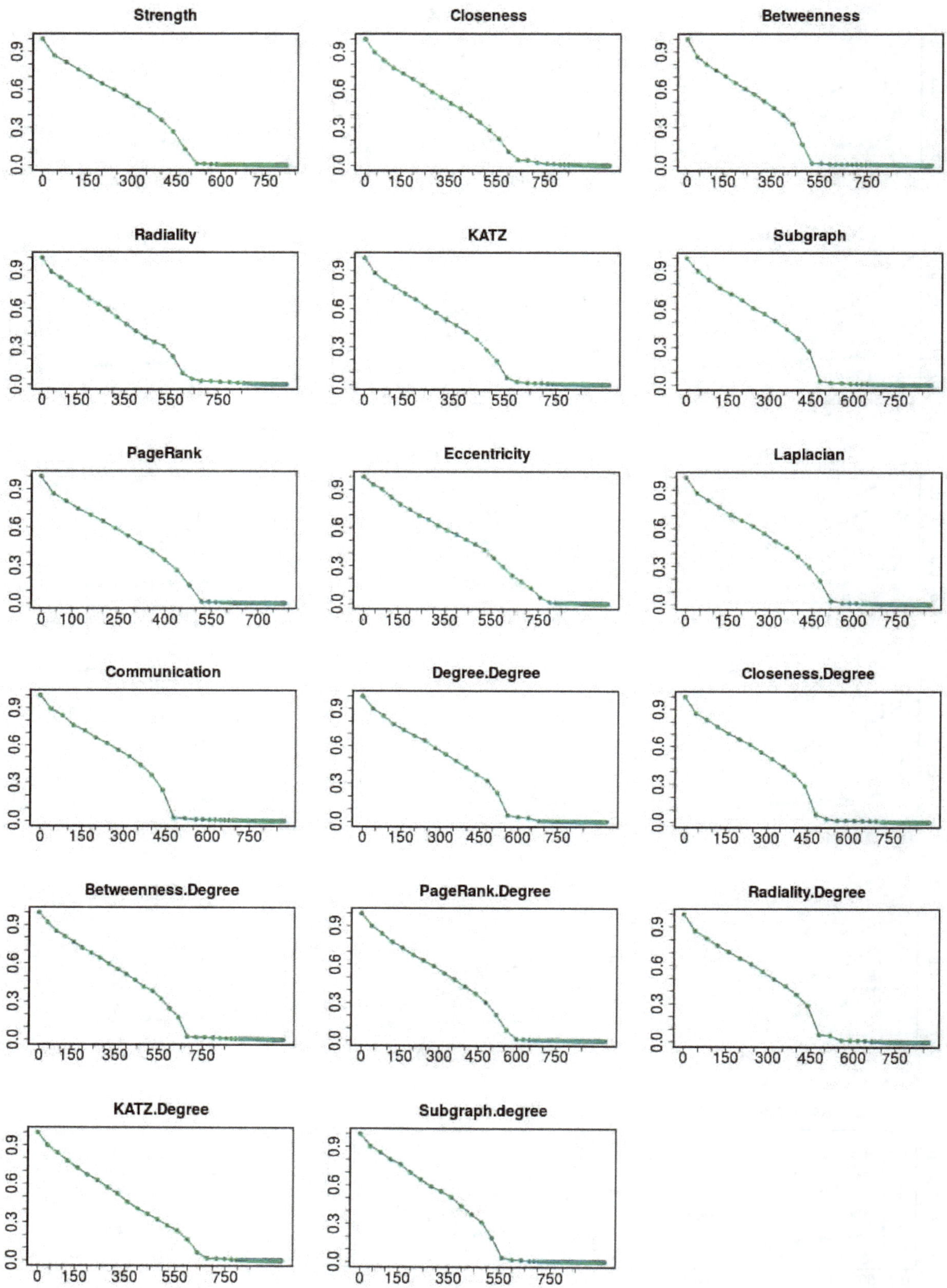

Fig. 3 Proportion of LCC size to initial network size versus number of vaccinated nodes in real network (FBL).

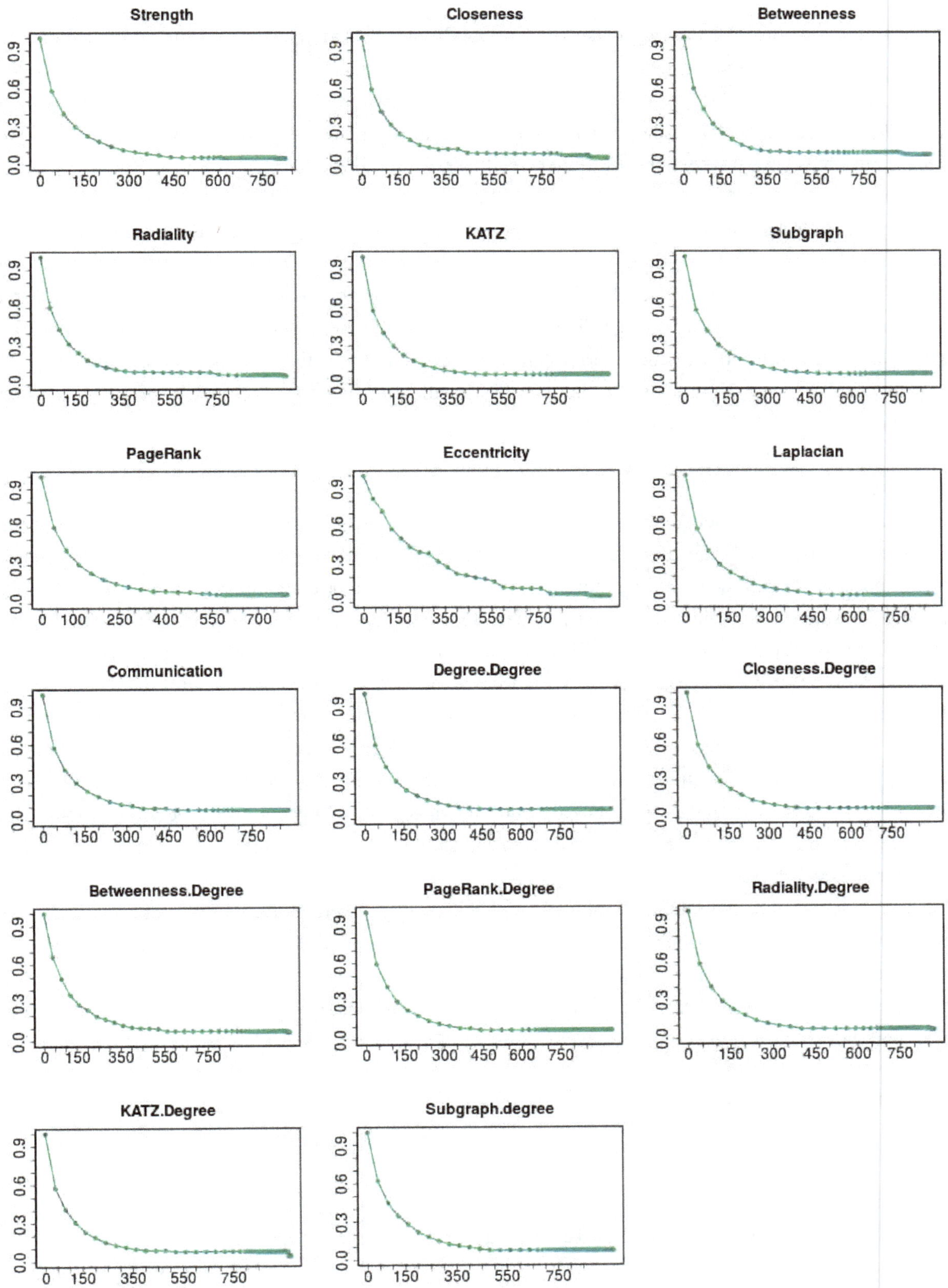

Fig. 4 Proportion of largest eigenvalue of immunized network to largest eigenvalue of initial network versus number of vaccinated nodes in real network (FBL).

5.5 Real network

As aforementioned we use the Facebook-like (FBL) (Opsahl and Panzarasa, 2009) as the real network. The major drawback of this dataset is that it has a cybernetic essence and does not reflect individuals physical contacts. However, due to lack of information in contact networks we used this dataset. Our experiments show that *Radiality-Degree* centrality is the best measure to detect most influenced nodes. *Communication* and *PageRank* centralities have the next best fulfillment as illustrated in Fig. 3.

To reduce the worst epidemic size to one fifth of the network size, we have to vaccinate %23, %23.7 and 24.3% of most important nodes by using *Radiality-Degree*, *Communication* and *PageRank* centralities, respectively.

Fig. 4 illustrates the impact of targeted vaccination on reducing the largest eigenvalue of epidemic threshold of proposed real network regarding different centrality measures. Our experiments show that *Closeness-Degree* centrality has the best performance in increasing the network epidemic threshold. *Degree-Degree* and *Katz-Degree* centralities show the next best results. By contrast, *Eccentricity* and *Betweenness-Degree* centralities exhibit the worst results respectively.

6 Conclusions

Due to accuracy of weighted networks in representing physical contacts in society, we employed them in this work. Hence, we took advantage of 13 predefined centrality measures for weighted networks to determine prominent nodes as central nodes. Furthermore, we proposed 4 new hybrid centrality measures which weight of links are considered in their computations. These 17 centrality measures were considered in targeted immunization realm. We made a comparison between these centrality measures for eminent network models: BA, ER and WS, as well as a real network (FBL) based on two metrics: *epidemic threshold* and *largest connected component* of network.In case of increasing size of largest connected component, *Radiality-Degree centrality* represented satisfying results as well as *PageRank-Degree* used to be applicable in increasing epidemic threshold in all three models: BA, ER and WS. On the other hand, *Subgraph-Degree centrality* exhibited poor performance among 4 proposed centrality measures. Moreover, *Katz-Degree* demonstrated unacceptable outcomes in almost all networks although exhibited valuable results in increasing epidemic threshold in real network (FBL).

Concisely, among pure centralities, Communication and Strength were applicable in decreasing LCC and increasing epidemic threshold respectively. Moreover, *Eccentricity* presented entirely unacceptable results. Among predefined hybrid centrality measures, (*Degree-Degree*, *Betweenness-Degree* and *Closeness-Degree*) we concluded that *Betweenness-Degree* centrality is impractical in decreasing LCC. Results to the other two measures were not so clear. Last but not least, for weighted BA network, *Radiality-Degree* and *Betweenness-Degree* centralities represented best and worst performance respectively. No such unique conclusion was inferred for the other two network models, i.e. Erdős-Rényi and Wattz-Strogatz.

Acknowledgment

We are thankfull to partiallysupportfrom Iran Telecommunication Research Center (ITRC).

References

Abbasi A, Hossain L. 2013. Hybrid centrality measures for binary and weighted networks. In Complex networks. 1-7, Springer Berlin Heidelberg, Germany

Abbasi A, Altmann J, Hossain L. 2011. Identifying the effects of co-authorship networks on the performance of scholars: A correlation and regression analysis of performance measures and social network analysis measures. Journal of Informetrics, 5(4): 594-607

Albert R, Barabási AL. 2002. Statistical mechanics of complex networks. Reviews of Modern Physics, 74(1): 47

Chakrabarti D, Wang Y, Wang C, Leskovec J, Faloutsos C. 2008. Epidemic thresholds in real networks. ACM Transactions on Information and System Security, 10: 1-26

Chen Y, Paul G, Havlin S, et al. 2008. Finding a Better Immunization Strategy. Physical Review Letters, 101: 2-5

Cornforth DM, Reluga TC, Shim E, et al. 2011. Erratic flu vaccination emerges from short-sighted behavior in contact networks. PLoS Computational Biology, 7: e1001062

Dijkstra EW. 1959. A note on two problems in connexion with graphs. Numerischemathematik, 1(1): 269-271

Eames KTD, Read JM, Edmunds WJ. 2009. Epidemic prediction and control in weighted networks. Epidemics, 1: 70-76

Erdős P, Rényi A. 1961. On the strength of connectedness of a random graph. Acta Mathematica Hungarica, 12(1-2): 261-267

Estrada E. Rodriguez-Velazquez JA. 2005. Subgraph centrality in complex networks. Physical Review E, 71(5): 056103

Freeman LC. 1978. Centrality in social networks conceptual clarification. Social Networks, 1: 215-239

Gallos L, Liljeros F, Argyrakis P, et al. 2007. Improving immunization strategies. Physical Review E, 75: 1-4

Hartvigsen G, Dresch JM, Zielinski AL, Macula AJ, et al. 2007. Network structure, and vaccination strategy and effort interact to affect the dynamics of influenza epidemics. Journal of Theoretical Biology, 246: 205-213

Jain S, Kamimoto L, Bramley AM, Schmitz AM, Benoit SR, Louie J, et al. 2009. Hospitalized patients with 2009 H1N1 influenza in the United States, April–June 2009. New England Journal of Medicine, 361(20): 1935-1944

Katz L. 1953. A new status index derived from sociometric analysis. Psychometrika, 18(1): 39-43

Kitchovitch S, Lio P. 2011. Community structure in social networks: applications for epidemiological modelling. PloS one, 6: e22220.

Leavitt HJ. 1951. Some effects of certain communication patterns on group performance. The Journal of Abnormal and Social Psychology, 46(1): 127-134

Masuda N. 2009. Immunization of networks with community structure. New Journal of Physics, 11: 123018.

Miller JC, Hyman JM. 2007. Effective vaccination strategies for realistic social networks. Physica A: Statistical Mechanics and Its Applications, 386: 780-785

Newman ME. 2001. Scientific collaboration networks. II. Shortest paths, weighted networks, and centrality. Physical Review E, 64(1): 016132

Opsahl T, Panzarasa P. 2009. Clustering in weighted networks. Social Networks, 31: 155-163

Opsahl T, Agneessens F, Skvoretz J. 2010. Node centrality in weighted networks: Generalizing degree and shortest paths. Social Networks, 32(3): 245-251

Parthasarathy A. 2013. Textbook of Pediatric Infectious Diseases. JP Medical Ltd, London, UK

Pauls SD, Remondini D. 2012. Measures of centrality based on the spectrum of the Laplacian. Physical Review E, 85(6): 066127

Peng C, Jin X, Shi M. 2010. Epidemic threshold and immunization on generalized networks. Physica A: Statistical Mechanics and Its Applications, 389: 549-560

Qi X, Fuller E, Wu Q, Wu Y, Zhang CQ. 2012. Laplacian centrality: A new centrality measure for weighted networks. Information Sciences, 194: 240-253

Sarma AD, Gollapudi S, Panigrahy R. 2011. Estimating pagerank on graph streams. Journal of the ACM (JACM), 58(3): 13

Schneider CM, Mihaljev T, Herrmann HJ. 2012. Inverse targeting —An effective immunization strategy. Europhysics Letters, 98: 46002

Shams B, Khansari M. 2014. Using network properties to evaluate targeted immunization algorithms. Network Biology, 74: 21

Valente TW, Foreman RK. 1998. Integration and radiality: measuring the extent of an individual's connectedness and reachability in a network. Social networks, 20(1): 89-105

Vidondo B, Schwehm M, Bühlmann A, et al. 2012. Finding and removing highly connected individuals using suboptimal vaccines. BMC Infectious Diseases, 12: 51

Wang H, Martin Hernandez J, Van Mieghem P. 2008. Betweenness centrality in a weighted network. Physical Review E, 77: 046105

Watts D J, Strogatz S H. 1998. Collective dynamics of 'small-world' networks.nature, 393(6684): 440-442

Zhai L, Yan X, Zhang G. 2013. A centrality measure for communication ability in weighted network. Physica A: Statistical Mechanics and its Applications, 392(23): 6107-6117

Zhang WJ. 2012. Computational Ecology: Graphs, Networks and Agent-based Modeling. World Scientific, Singapore

Zhang WJ, Zhan CY. 2011. An algorithm for calculation of degree distribution and detection of network type: with application in food webs. Network Biology, 1(3-4): 159-170

2

Controlling ecological and biological networks via evolutionary modelling

Alessandro Ferrarini
Department of Evolutionary and Functional Biology, University of Parma, Via G. Saragat 4, I-43100 Parma, Italy
E-mail: sgtpm@libero.it, alessandro.ferrarini@unipr.it

Abstract
The controllability of network-like systems is a topical key-issue in many disciplines, including ecology and biology. It relies on the ability to guide a system's behaviour towards the desired state through the appropriate handling of a few input variables. To date, controllability of networks is based on the identification of the set of driver nodes that can guide the system's dynamics. I introduce here a new framework for the controllability of both network nodes and edges based on the use of evolutionary modelling, and provide an exemplification of its application.

Keywords edges control; genetic algorithms; network control; nodes control.

1 Introduction

Network controllability is the ability to guide a system's behaviour towards the desired state through the appropriate handling of a few input variables (Luenberger, 1979; Slotine and Li, 1991). The controllability of network-like systems is becoming a trendy key-issue in many disciplines, including ecology (Caldarelli, 2007) and biology (Dorogovtsev and Mendes, 2003; Kim and Motter, 2009; Marucci et al., 2009).

In Nature, Liu et al. (2011) proposed analytical tools for the controllability of complex networks, identifying the set of driver nodes that can guide the system's dynamics. Their way to the controllability of networks is based on the identification of the set of driver nodes that can guide the system's dynamics, in other words on the choice of a subset of nodes that are selected to be permanently controlled. This assumption seems motivated by real-world networks observation, where a decentralized control action is often applied only to part of the nodes.

In 2011, I remarked the need for five key-improvements to this usual approach in networks controllability (Ferrarini, 2011). In particular, I noticed that several tools, like genetic optimization, could reasonably lower the seemingly intractable problem of edges control.

Hence, I introduce here a new framework for network controllability based on the use of evolutionary modelling, and provide an exemplification of its application.

2 Mathematical Formulation

An ecological (or biological) dynamical system of n interacting species is as follows:

$$\frac{d\vec{S}}{dt} = \varphi(\vec{S}, t) \tag{1}$$

where S_i is the number of individuals (or the total biomass) of the generic i-th species. If we also consider inputs (e.g. species reintroductions) and outputs (e.g. hunting or road mortality) from outer universe, we must write:

$$\frac{d\vec{S}}{dt} = \varphi(\vec{S}, t) + \vec{I}(t) + \vec{O}(t) \tag{2}$$

As noted by numerous authors (e.g., Liu et al., 2011; Slotine and Li, 1991) most real systems' dynamics can be modelled and simulated using a system of canonical, linear equations, as follows:

$$\begin{cases} \frac{dS_1}{dt} = a_{11}S_1 + \ldots + a_{1n}S_n + I_1 + O_1 \\ \ldots \\ \frac{dS_n}{dt} = a_{n1}S_1 + \ldots + a_{nn}S_n + I_n + O_n \end{cases} \tag{3}$$

that can be written in the compact form:

$$\frac{d\vec{S}}{dt} = A * \vec{S} + \vec{I} + \vec{O} \tag{4}$$

with initial values

$$\vec{S}_0 = \langle S_1(0), S_2(0) \ldots S_n(0) \rangle \tag{5}$$

and co-domain limits

$$\begin{cases} \mathrm{S}_{1\min} \le S_1(t) \le \mathrm{S}_{1\max} \\ \ldots \\ \mathrm{S}_{n\min} \le S_n(t) \le \mathrm{S}_{n\max} \end{cases} \tag{6}$$

and where

$$A = \begin{pmatrix} a_{11} & \ldots & a_{1n} \\ \ldots & \ldots & \ldots \\ a_{n1} & \ldots & a_{nn} \end{pmatrix} \tag{7}$$

is the matrix of the per unit-of-time effect on S_i due to unitary S_j.

3 Network Control via Evolutionary Modelling

Genetic algorithms (GAs) consist of optimization procedures based on principles inspired by natural selection. GAs involves 'chromosomal' representations of proposed problem solutions which undergo genetic operations such as selection, crossover and mutation (Holland, 1975; Goldberg, 1989, Parolo et al., 2009; Ferrarini, 2012). Under genetic optimization, equation (3) becomes:

$$\begin{cases} \left(\frac{dS_1}{dt}\right)_{OPT} = a_{11*}S_1^* + ... + a_{1n*}S_n^* + I_{1*} + O_{1*} \\ ... \\ \left(\frac{dS_n}{dt}\right)_{OPT} = a_{n1*}S_1^* + ... + a_{nn*}S_n^* + I_{n*} + O_{n*} \end{cases} \quad (8)$$

where asterisks stand for the optimization of edges (i.e., coefficients of interaction among variables) or nodes (i.e., initial stocks), that is the modification of their values at the beginning of the network dynamics in order to get a certain goal (e.g., maximization of the final value of a certain variable).

4 An Applicative Example

Fig. 1 depicts an ecological network borrowed with modification from Ferrarini (2012b). Greenish nodes represent positive actors or events for the goal of the network control, i.e. the increase of individuals of the target species (centre of the network). Reddish nodes represent ecological actors or events with negative impact on the target species. Blueish nodes represent resources needed by the target species. The goal is to preserve target species' occurrence in the study area. Stocks stand for the actual amounts of individuals or biomass. Updates stand for yearly internal dynamics (i.e., intraspecific gains due to births and/or immigration rates minus losses due to deaths and/or emigration rates). Minimum and maximum values stand for lowest and highest values of stock values. For the sake of simplicity, the maximum possible value for each actor (in italic hereafter) has been set to 100.The percent value associated to links represent the percentage of the receiver that is yearly consumed by the transmitter at the beginning of the network simulation. Road mortality and re-introductions account for 18 and 10 individuals per year respectively.

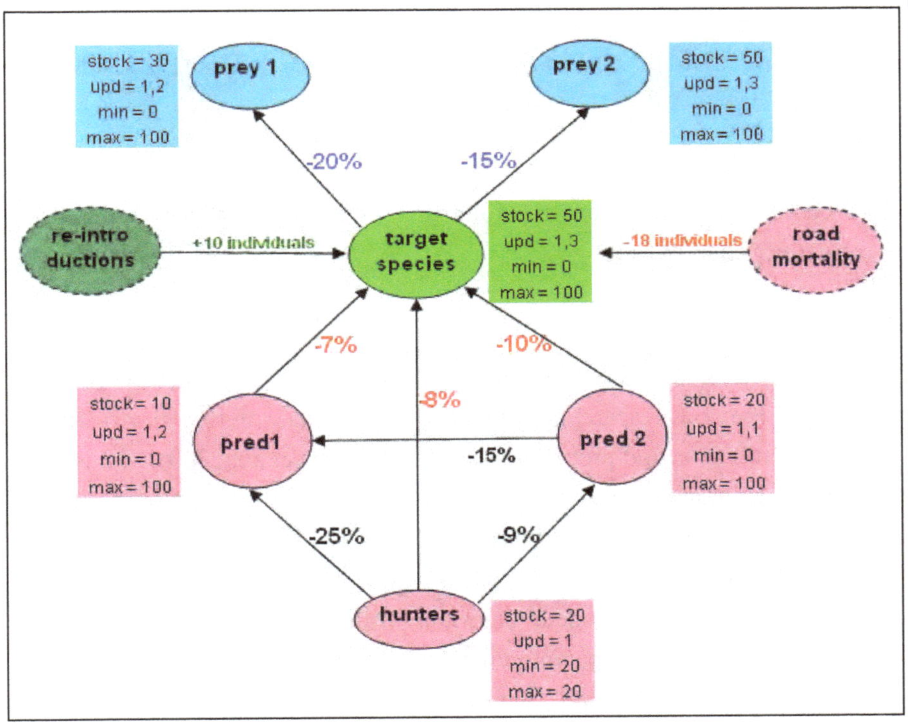

Fig. 1 The ecological network on which evolutionary control has been applied.

Since data are yearly-based, I express equation (8) using a system of difference recurrent equations, instead of differential ones:

$$\begin{cases} (S_1)_{t+1\,OPT} = a_{11*}(S_1)_t + \ldots + a_{1n*}(S_n)_t + (I_1)_{t*} + (O_1)_{t*} \\ \ldots \\ (S_n)_{t+1\,OPT} = a_{n1*}(S_1)_t + \ldots + a_{nn*}(S_n)_t + (I_n)_{t*} + (O_n)_{t*} \end{cases} \tag{9}$$

For instance, *prey1*'s dynamic (Fig. 2) is given by:

$$\frac{d(prey1)}{dt} = +0.2 * prey1 + a * target\_species \tag{10}$$

$$\text{where } a = \frac{-20}{100} * \frac{30}{50} = -0.12 \tag{11}$$

or, expressed using a difference equation:

$$\begin{cases} prey1_{t+1} = +0.2 * prey1_t - 0.12 * target\_species_t \\ prey1(0) = 30 \end{cases} \tag{12}$$

INTERACTIONS MATRIX (per-cycle changes to receivers per-unit of transmitters)											
		Receivers									
		target	prey1	prey2	pred1	pred2	hunters	var G	var H	var I	var J
Transmitters	target	1.3	-0.12	-0.15	0	0	0	0	0	0	0
	prey1	0	1.2	0	0	0	0	0	0	0	0
	prey2	0	0	1.3	0	0	0	0	0	0	0
	pred1	-0.35	0	0	1.2	0	0	0	0	0	0
	pred2	-0.2	0	0	-0.075	1.1	0	0	0	0	0
	hunters	-0.2	0	0	-0.125	-0.09	1	0	0	0	0
	var G	0	0	0	0	0	0	0	0	0	0
	var H	0	0	0	0	0	0	0	0	0	0
	var I	0	0	0	0	0	0	0	0	0	0
	var J	0	0	0	0	0	0	0	0	0	0

Fig. 2 The interaction matrix relative to the ecological network of Fig. 1. The matrix has been calculated using Quant-Lab (Ferrarini, in preparation).

The previous ecological network has the following inertial dynamics (Fig. 3), with the *target species* (green line) going extinct after 7 years.

5 Network Control Using GAs

Since GAs are based on random searching for solutions, I performed iterative processes for determining appropriate parameters. Previous research revealed that the optimal solution may be to search at a high rate of crossover, a low rate of mutation and proper population size (Kuo et al., 2000). In this study, crossover was set at a probability of 60% while mutations occur with a probability of 5%. This low setting helps to avoid getting trapped local optima during the search (D'heygere et al., 2006). The initial population consisted of 500 chromosomes that were evolved over minimal 10,000 generations. These parameters were set after preliminary experiments. I applied a steady-state genetic algorithm with a one-point crossover operator (Wall, 1996) to accomplish crossover. In this case the parent genome strings are cut at some random position to produce two "head" and two "tail" segments. The "tail" segments are swapped to produce two new genomes. For parent selection the roulette wheel selection method was used (Goldberg, 1989), where the likelihood of selection is proportionate to the fitness score given by the performance criterion. After crossover and mutation, the individuals with the lowest fitness scores were removed. Hence, I was able to find a set of feasible solutions to the goal of preserving the *target species*.

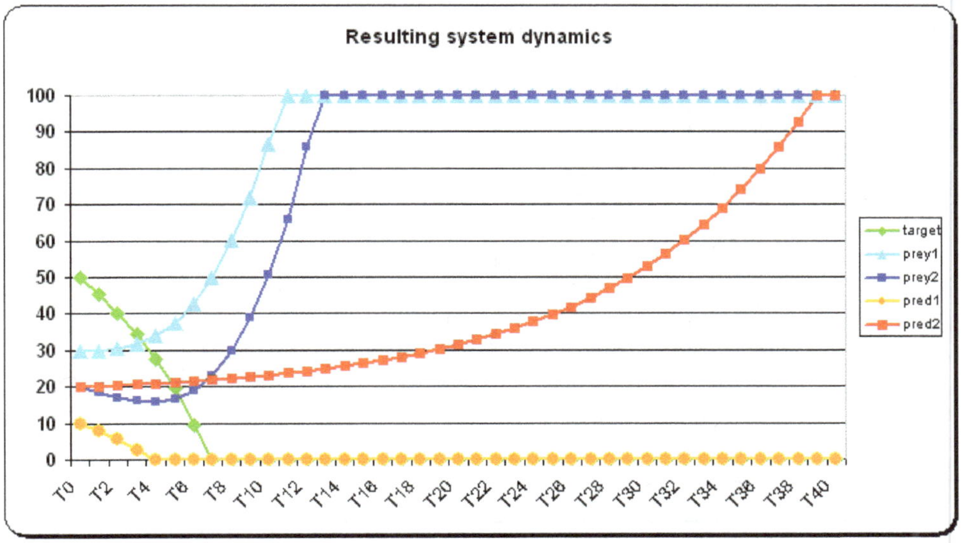

Fig. 3 Resulting dynamics for the network of Fig. 1. X-axis measures time in years. Dynamics have been calculated using Quant-Lab (Ferrarini, in preparation).

Being all other simulation parameters equal, by imposing *pred2*'s initial stock to be <8 (instead of 20), I achieved the following dynamics (Fig. 4). Both *target species* and *prey1* are stably safe.

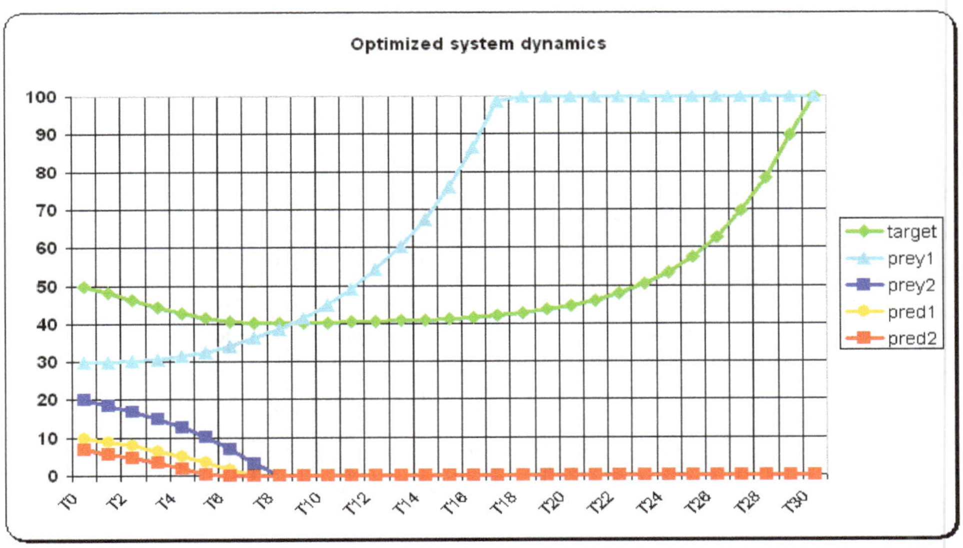

Fig. 4 Solution 1: acting on *pred2*'s initial stock. It is set here to 7 instead of 20. Dynamics have been calculated using Quant-Lab (Ferrarini, in preparation).

Interestingly, with *pred2*'s initial stock ≤5 the stocks of both preys (*prey1* and *prey2*) would go to 0 (Fig. 5), thus only solutions with 6 ≤ *prey2*'s stock <8 are feasible for long-term conservation of the *target species*. Otherwise, a successive prey's re-stocking would be needed.

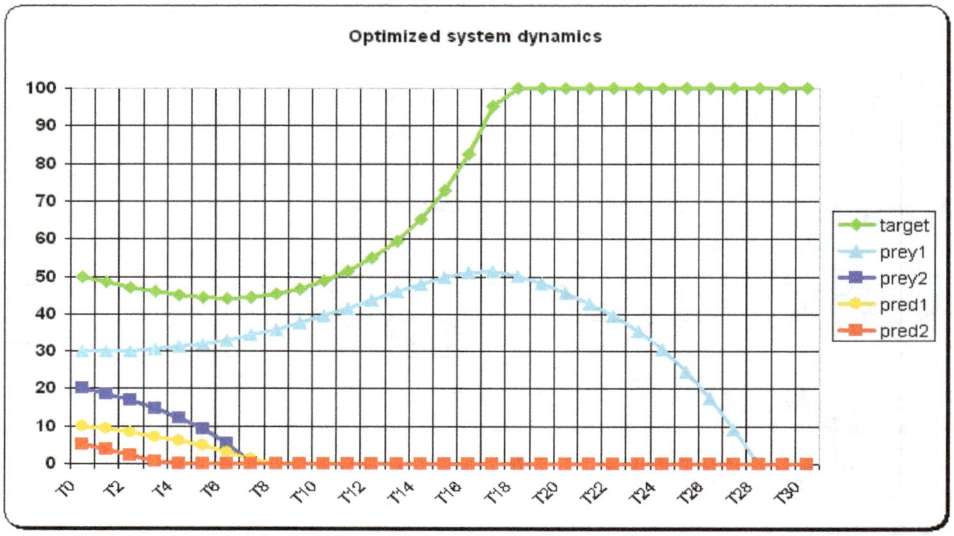

Fig. 5 Resulting dynamics with *prey2*'s initial stock set to 5.

Another solution is possible by increasing *target species*' intraspecific coefficient above the value 1.36 (instead of the actual value 1.3; Fig. 6). This would also require *prey1*'s (or *prey2*) re-stocking after 28 years in order to preserve some prey for the *target species*.

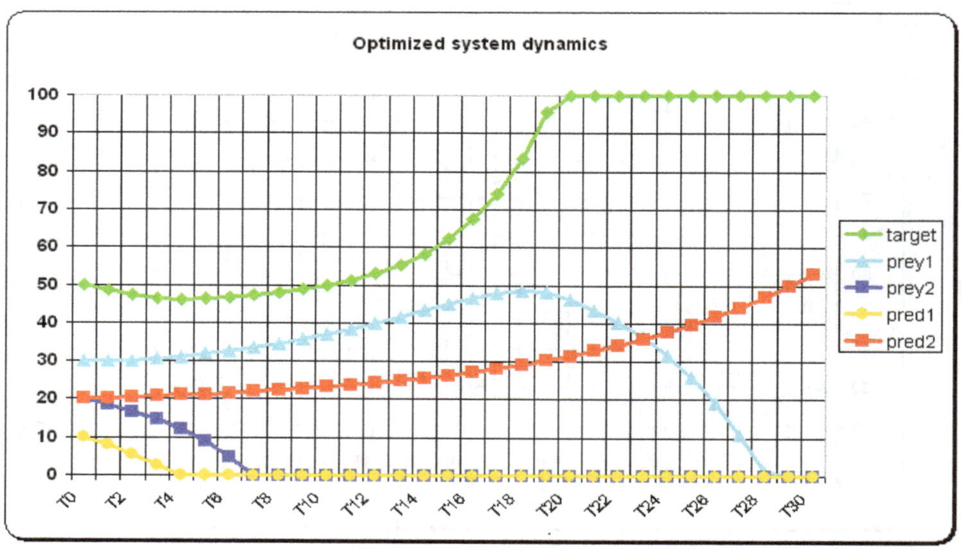

Fig. 6 Solution 2. Resulting dynamics with *target species*' intraspecific coefficient set to 1.361.

A further solution is possible by increasing input-output balance (*reintroductions* and *road mortality*). Actually it's equal to -8 in the study area (10 reintroductions and 18 road casualties per year), but if it were >-6 it would determine the following dynamics (Fig. 7). This could be achieved by increasing reintroduction activities, or by limiting road casualties with proper road signals. *Prey1*'s restocking would be also required after 21 years.

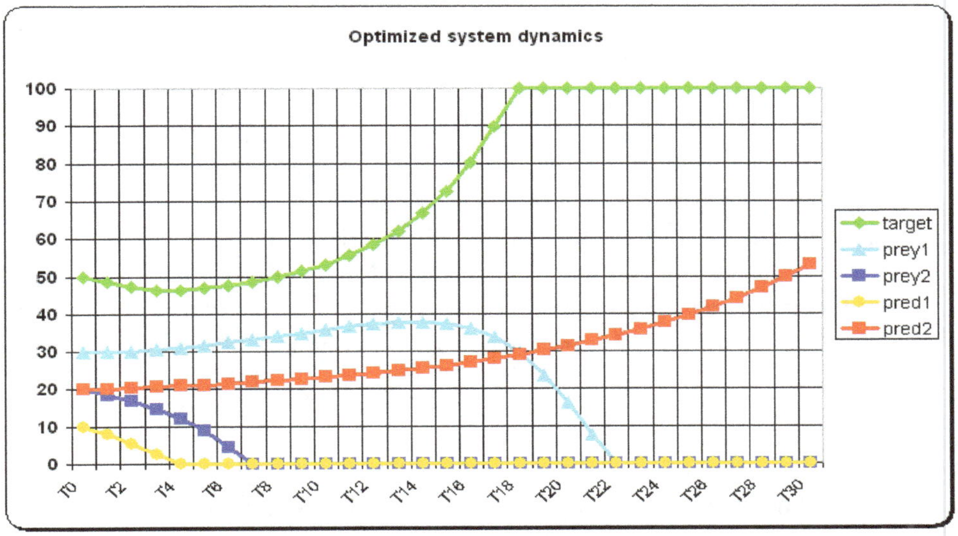

Fig. 7 Solution 3. Resulting dynamics with input-output balance set to -5.

Another solution is given by yearly subtraction of *pred1* and *pred2's* individuals. At least 5 individuals of both species must be yearly removed in order to preserve the *target species* (Fig. 8).

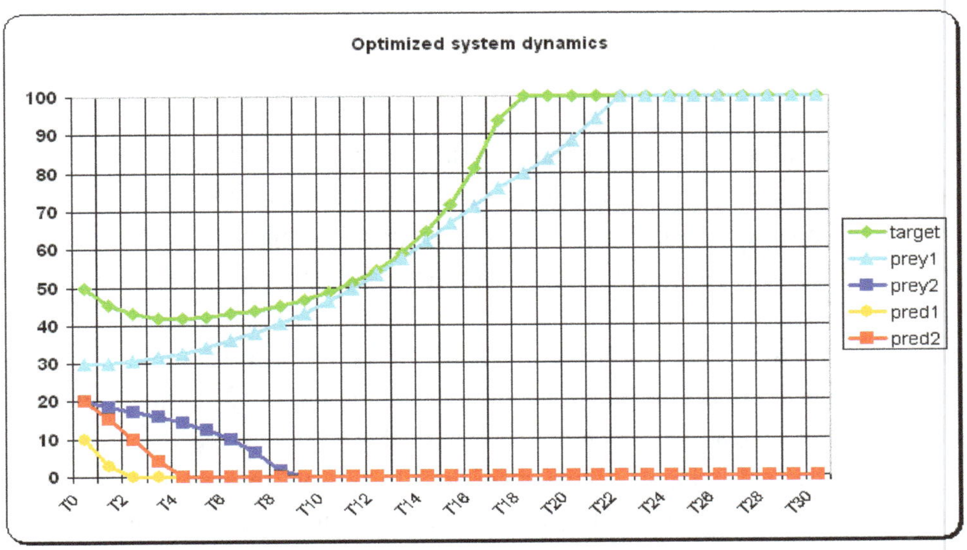

Fig. 8 Solution 4. Resulting dynamics with yearly removal of *pred1* and *pred2's* individuals (5 individuals of both species).

Interestingly, the removal of *hunters* (Fig. 9) at the beginning of the simulation (i.e., *hunters*(0)=0) doesn't allow to achieve the proposed goal. In fact, although in the study area hunters kill *target species*' individuals, they also subtract individuals to the two species of predators.

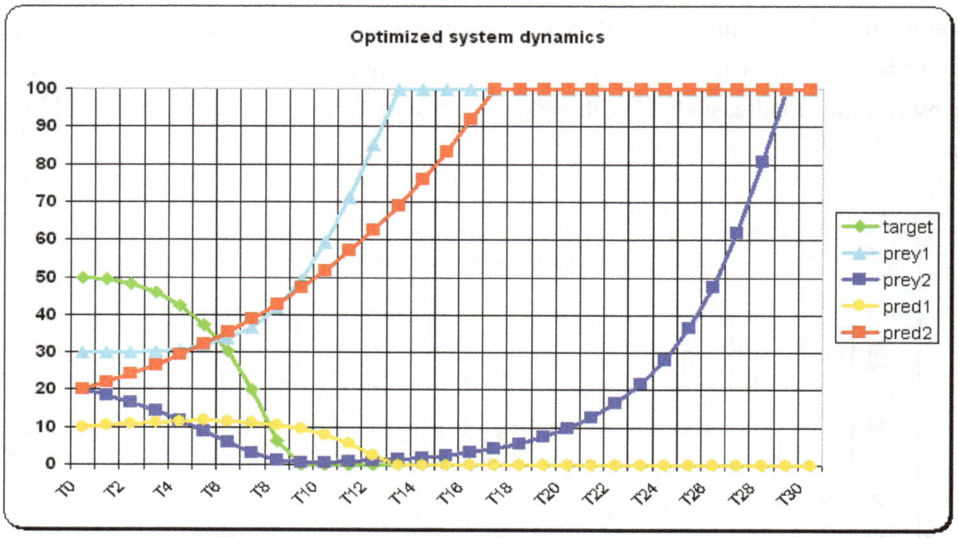

Fig. 9 Resulting dynamics achieved by removing hunters at T=0 (i.e., year 2012).

Since many alternative solutions can be found by applying evolutionary modelling to the control of ecological and biological networks, I also propose the following decision framework:

- goal setting (e.g. *target species* maximization);
- network setup;
- application of GAs to network dynamics;
- multiple solutions detection via evolutionary modelling;
- setup of cost-benefit ratio (CBR) for each detected solution;
- choice of the action (or actions) that minimize CBR.

A step forward could be given by *a priori* incorporating CBRs into goal setting, i.e. the goal of the network control would be to minimize CBR while maximizing *target species*' stock.

6 Conclusions

To date, the controllability of ecological and biological networks is based on the identification of the set of driver nodes that can guide the system's dynamics, in other words on the choice of a subset of nodes that are selected to be permanently controlled. In addition, it is commonly supposed that, even if all network weights are known, the control based on a brute-force algorithm would be computationally prohibitive for large networks.

Although a *n*-node network could bear up to $n*(n-1)$ links (or $n*n$ if we also consider self-links) among nodes, this could be conceived as a better chance of network control, not just like a computational difficulty. The higher the number of switches on which one can act, the higher the chance to commute the actual network into the desired one. The application of genetic algorithms to network dynamics allows to act on the highest number of switches, while maintaining the computational effort to tractable levels. Genetic algorithms can reasonably lower the seemingly intractable problem of edge control. In addition I showed here that they allowed to control multiple nodes and edges at the same time.

The framework proposed here might also be applied to semi-quantitative and qualitative networks (Ferrarini 2011b; Ferrarini 2011c). In addition, I suggest that it could be applied to find out how to counteract the negative impacts of climate change on biodiversity. A full understanding of the effects of climate-induced scenarios on biotic communities necessarily necessitates the consideration of the whole interaction network among species (Ferrarini, 2012c). This requires a $n \times n$ "climate-induced" interaction matrix whose off-diagonal elements w_{ij} take into account the inter-specific effects of species i upon species j as primed by a climate change scenario, while elements w_{jj} take into account intra-specific effects (on growth, phenology and reproduction) on single species. Outcomes of such network of "climate-induced" interactions could predict undesirable effects on biodiversity that can be counterbalanced by management actions that can be properly decided by applying evolutionary modelling to such dynamics.

References

Caldarelli G. 2007. Scale-Free Networks: Complex Webs in Nature and Technology. Oxford University Press, UK

D'heygere T, Goethals PLM, Niels De Pauw N. 2006. Genetic algorithms for optimisation of predictive ecosystems models based on decision trees and neural networks. Ecological Modelling, 195: 20–29

Dorogovtsev SN, Mendes JFF. 2003. Evolution of Networks: From Biological Nets to the Internet and WWW. Oxford University Press, UK

Ferrarini A. 2011. Some thoughts on the controllability of network systems. Network Biology, 1: 186-188

Ferrarini A. 2011b. Fuzzy cognitive maps outmatch loop analysis in dynamic modeling of ecological systems. Computational Ecology and Software, 1(1): 55-59

Ferrarini A. 2011c. Some steps forward in semi-quantitative network modelling. Network Biology, 1(1): 72-78

Ferrarini A. 2012. Betterments to biodiversity optimal sampling. Proceedings of the International Academy of Ecology and Environmental Sciences, 2(4): 246-250

Ferrarini A. 2012b. Network di biodiversità in alcune aree della regione Emilia-Romagna. Technical relation (in Italian)

Ferrarini A. 2012c. I think different: models of climate change impact on plant species are unrealistic. Environmental Skeptics and Critics, 1(2): 30-33

Ferrarini A. 201X. Quant-Lab: a software for solving Quantitative Ecological Networks. In preparation

Goldberg D.E. 1989. Genetic Algorithms in Search Optimization and Machine Learning. Addison-Wesley, Reading, USA

Holland J.H. 1975. Adaptation in Natural and Artificial Systems: An Introductory Analysis with Applications to Biology, Control and Artificial Intelligence. University of Michigan Press, Ann Arbor, USA

Kim DH, Motter AE. 2009. Slave nodes and the controllability of metabolic networks. New Journal of Physics, 11: 113047

Kuo SF, Merkley GP, Liu, C.W. 2000. Decision support for irrigation project planning using a genetic algorithm. Agriculture and Water Management, 45: 243–266

Liu YY, Slotine JJ, Barabasi AL. 2011. Controllability of complex networks. Nature, 473:167-173

Luenberger DG. 1979. Introduction to Dynamic Systems: Theory, Models, & Applications. Wiley, USA

Marucci L, Santini S, di Bernardo M. 2009. How to turn a genetic circuit into a synthetic tunable oscillator, or a bistable switch. PLoS ONE, 4: e8083

Parolo G, Ferrarini A, Rossi G. 2009. Optimization of tourism impacts within protected areas by means of genetic algorithms. Ecological Modelling, 220: 1138-1147

Slotine JJ, Li W. 1991. Applied Nonlinear Control. Prentice-Hall, USA

Wall M. 1996. GAlib: A C++ Library of Genetic Algorithm Components, version 2.4. Massachusetts Institute of Technology, USA

3

Cryptic successors unrevealed even by network analysis: A comparative study of two paper wasp species

Anindita Bhadra[1], Ferenc Jordán[2]

[1]Behaviour and Ecology Lab, Department of Biological Sciences, Indian Institute of Science Education and Research - Kolkata, Mohanpur, India

[2]The Microsoft Research – University of Trento Centre for Computational and Systems Biology, Rovereto, Italy

E-mail: jordan.ferenc@gmail.com

Abstract

Understanding queen succession could be a key contribution to the better understanding of the origins and evolution of eusociality. In order to investigate the nature of organizational changes during queen succession, we analyzed two closely related paper wasp species (*Ropalidia cyathiformis* and *Ropalidia marginata*). We compared the effects of in vivo and in silico queen removal on the structure of their interaction networks (the former resulting in queenless colonies with potential queens). We studied several structural measures. There is no major structural difference between full (queenright) and in silico queen-removed colonies but there are major differences between queenless and in silico queen-removed ones. This suggests that queen succession is accompanied by a major reorganization of the society, in Rm but not so much in Rc. We also analysed the centrality ranks of potential queens and found that their positional importance changes a lot during queen succession in *R. marginata*, as they are processed in the colony. In the queenright colonies of *R. marginata*, the direction of links is a better predictor of the identity of the potential queen than the strength of links.

Keywords centralization; small world; queen determination; behavioral caste.

1 Introduction

A major challenge in the study of primitive insect societies is to understand queen succession. In these species, individuals are not pre-destined to be queens or workers by genetics or development, as in highly eusocial species. Instead, social interactions play a major role in determining whether an individual can become a queen. When the queen is lost from a colony, one of the workers becomes the next queen. In most species, the identity of the next queen can be predicted, based on age, body size, aggression levels, ovarian condition or a combination of these factors. In *Polistes dominulus*, for example, the queen is always succeeded by the beta individual in the dominance hierarchy, while in the Stenogastrine wasps the oldest individual in the colony at the time of queen loss becomes the next queen. Thus the queen's successor can be predicted even in the presence of the existing queen (Pardi, 1948; West-Eberhard, 1969; Cronin and Field, 2007; Cant et al., 2006; Röseler et al., 1986; Heinze et al., 1996; Ito and Higashi, 1991; Deshpande et al., 2006; Turillazzi, 1991).

This is the case also in *Ropalidia cyathiformis*, a primitively eusocial wasp where the queen's successor can be predicted, because the beta individual in the dominance hierarchy always succeeds the queen, who occupies the alpha position in the hierarchy, similar to *P. dominulus* (Deshpande et al., 2006). However, in *Ropalidia marginata* (a congeneric, sympatric species), the queen is a docile, non-interactive individual, who does not use aggression to maintain her reproductive monopoly. When the queen is lost from the colony, one of the workers becomes extremely aggressive within minutes, and develops her ovaries within a few days, to become the next queen. We designate this individual as the potential queen (PQ; Premnath et al., 1996; Kardile and Gadagkar, 2002; Sumana and Gadagkar, 2001; Gadagkar, 2001). Unlike in the case of *R. cyathiformis*, we cannot predict the queen's successor in *R. marginata*, because she seems to be an unspecialized individual with respect to her behaviour, age and ovarian condition (Deshpande et al., 2006; Chandrashekara and Gadagkar, 1992). But, like in *R. cyathiformis*, the potential queen is the most aggressive individual in the colony after queen removal. However, it is intriguing that the queen's successor, though cryptic to us in the presence of the queen, is "known" to other individuals in the colony (Bhadra and Gadagkar, 2008). In fact there is a pre-determined queue of at least five successors to the queen. Age is a significant but not perfect predictor of an individual's position in this queue (Bang and Gadagkar, 2012).

In an earlier study (Bhadra et al., 2009), we used data from behavioural experiments to characterize the social networks of *R. marginata* and *R. cyathiformis* based on dominance-subordinate interactions among individuals. Our preliminary analysis of the social networks of the two species corroborated our earlier behavioural results on both species. Experimental queen removal in both species leads to the emergence of a potential queen, and the resulting queen-less network (QL) is significantly different from the initial, queen-right one (QR). However, the queens in the two species are very different behaviourally, and also from the perspective of their position in the interaction network. While the queen in *R. cyathiformis* is a central node in the network, the node representing the queen in *R. marginata* is more peripheral (or isolated in 8 out of 12 colonies). Nevertheless, in all the 12 colonies, the interaction network underwent a drastic change after queen removal and the emergence of the potential queen. It is surprising from a structural perspective that removal of such a peripheral node (i.e. a non-significant individual) from the network can lead to such a drastic change in network structure. These results make it obvious that the queen in *R. marginata* is an important player in the colony, even though she does not stand out in the network of dominant-subordinate interactions (Bhadra et al., 2009). We note that these interactions represent only one, although evidently important type of interactions (considering networks representing several kinds of interactions is in preparation). Here we study whether (and how) we can predict the identity of the cryptic successor in *R. marginata* using network analysis. We explore several measures for network position and compare indices to see which one is the best predictor of the PQ's identity.

2 Methods

2.1 Data

We used data from queen removal experiments each in *R. marginata* (Rm) and *R. cyathiformis* (Rc) collected for other experiments. Each data set consisted of the identifiers of the initiator and recipient individuals of pair-wise dominant-subordinate interactions (directed graph), and the number of interactions between pairs of individuals (weighted graph). Data was available on the queen-right and queen-less conditions of each colony (we had 48 networks in total). Network analysis was performed using the software UCINET 6.0 (Borgatti et al., 2002), and statistical analysis was done using STATISTICA 7.0 and StatisticXL.

2.2 Network analysis: global properties

Social networks contain information on the relative position of individuals in networks, characterising part-to-whole (individual-to-group) relationships (for methodical background, see Wasserman and Faust, 1994). We have considered a few basic, structural properties of social networks (Croft et al., 2008, Wey et al., 2008).

We calculated the degree–based *network centralisation index* (NCI^D, Wasserman and Faust, 1994; Zhang, 2012b) for the unweighted and undirected versions of our networks, where *degree* (D) is the number of adjacent nodes to a graph node (number of neighbours). Note that network centrality, expressed in percentages, is maximal (100%) if a central node is directly connected to all other nodes that are only linked to the central one (a perfect star), and minimal (0%) in a regular graph, where the positional importance of all nodes are equal (a lattice).

The *clustering coefficient* (CL) of a graph node quantifies the frequency with which its neighbours are connected to each other. It can be averaged over the whole graph if one wants to provide a global network measure (we used its version weighted by the size of neighbourhood of each node).

The *average path length* (d) of a graph is defined as the mean distance between pairs of nodes. For nodes i and j, distance is defined as the length of the shortest path between them (length meaning the number of edges connecting the two nodes). The average characterises the whole network (low average path length meaning easy and fast communication).

In regular graphs (where each node has exactly n neighbours, like on a lattice), both clustering coefficient (CL) and average distance (d) are high. In random networks, both measures are low. Small world networks are characterized as having mixed properties of the two, i.e. relatively high clustering but relatively low average distance (Zhang, 2012a). Thus, a high CL / d ratio (denoted by SW) indicates that the given network is more "small world –like" (sensu Watts and Strogatz, 1998).

2.3 Network analysis: local properties

Beyond the global properties of the networks, we were also interested in the local indices characterising the particular positions of graph nodes in the networks. Positions of nodes in the social networks were quantified by three indices. Degree (D) provides the simplest, unweighted and undirected measure of neighbourhood (Zhang, 2012b). Weighted degree (wD) considers also the frequency of undirected interactions among individuals. Net status (ns) is a directed but unweighted measure of positional importance in hierarchies (Harary, 1959): it is the summed distance from a given node i to all subordinates minus the same value calculated for the graph with reversed directions. Since these two indices consider either the direction or the weight of links, their comparison (to each other and to the unweighted and undirected case) may provide a key to understanding the mechanisms in the background of queen succession.

Based on these three measures, we calculated the ranks of all individuals in each colony. Then, we derived the relative ranks of the PQs (their actual rank over the number of individuals in the network): If the rank of individual i is k in a network of n individuals, its $(k/n)_i$ relative rank tells how close it is to the leading alpha position (clearly, the rank k is not enough to characterise this: it does make a difference whether the third individual, with k=3, is the third in a colony of four or fifty wasps). If $(k/n)_i$ is smaller, the individual is more central in the colony network, and in general, the alpha individual has a relative rank of $1/n$. We have analysed (1) which index gives the highest ranks to PQs (on average) and (2) which index gives the most consistent estimation for ranks of PQs (described by the coefficient of variation of the k/n values).

2.4 *In silico* queen removal experiments

We carried out *in silico* queen removal experiments in the 24 queen-right networks: we removed the queen with all her interactions (both initiated and received) from the data set and constructed the "*in silico* queen removed" network. We also removed the PQ (identified from the experimental queen-less condition) and a random worker from the QR data and generated "*in silico* PQ-removed" and "*in silico* worker-removed" networks, respectively. The five types of networks (QR, QL and the three removals) were then compared using a Mann-Whitney U test for global network properties (NCI^{D}, *d*, *SW*). We investigated also the local indices for queens and PQs in the *in silico* queen removed networks of *R. marginata*.

In an earlier experiment, we tested the cryptic successor hypothesis in *R. marginata* (Bhadra and Gadagkar, 2008). In five of the eight colonies used, we had two potential queens, one of which was the successor (PQ2), and the other we designated here as the loser (PQ1). In the remaining three experiments, we had only the successor (PQ1). We used data on dominant-subordinate interactions from the queen-right colony to construct a degree-based hierarchy for the colony. We also constructed a similar hierarchy based on all paired interactions observed in the QR colony. In both kinds of hierarchies, we looked for the ranks of the eight successors (5 PQ2's and 3 PQ1's) and the losers (5 PQ1's).

3 Results

We have shown previously that the *R. marginata* QR and QL networks are at the two extremes of centrality, with some QL *R. marginata* networks achieving 100% centrality (i.e. a perfect "star" topology). The *R. cyathiformis* QR networks are more centralized than the *R. marginata* ones, while the QL networks of *R. cyathiformis* are also highly but much less centralized than the QL networks of *R. marginata* (Bhadra et al., 2009). We removed particular individuals, *in silico*, from the QR networks; (we could remove the queen *in silico* only from four of the 12 *R. marginata* networks; in the remaining eight, queens were isolated network nodes). We could remove the PQ in silico from 11 of the 12 networks in *R. marginata*. Both the queen and the PQ could be removed similarly from all the *R. cyathiformis* networks. Fig. 1 shows the QR, QL and queen-removed (Qrem) networks of a typical colony in both species. The Qrem networks of both species were similar to their own QR networks, and they were significantly different from the QL networks in their centrality measures. The queen-removed, PQ-removed and worker-removed networks were similar within a species (Fig. 2). All the comparisons were made by Mann-Whitney U test, with alpha set to 0.007, after Bonferroni correction.

(a)

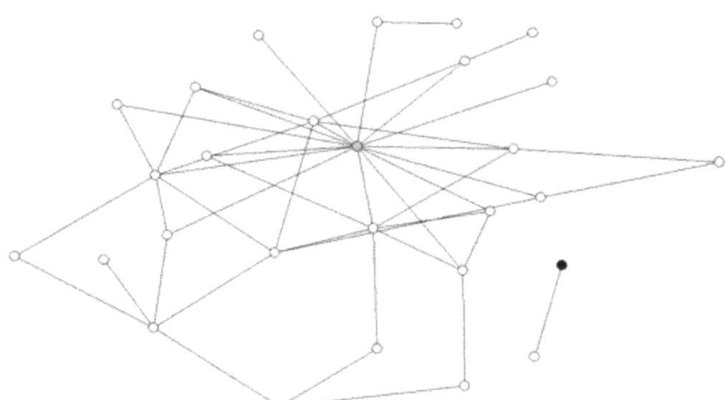

(b)

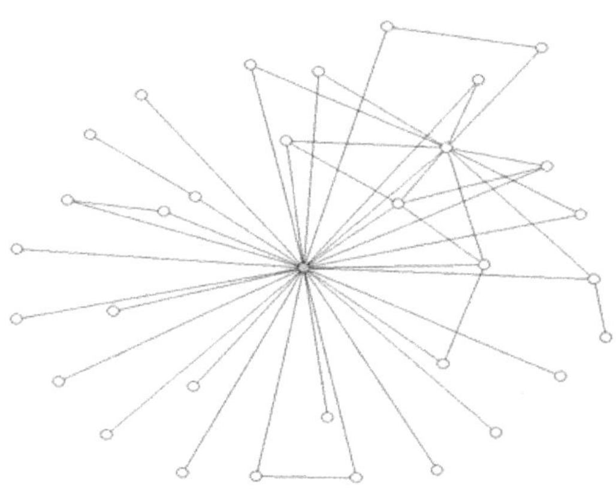

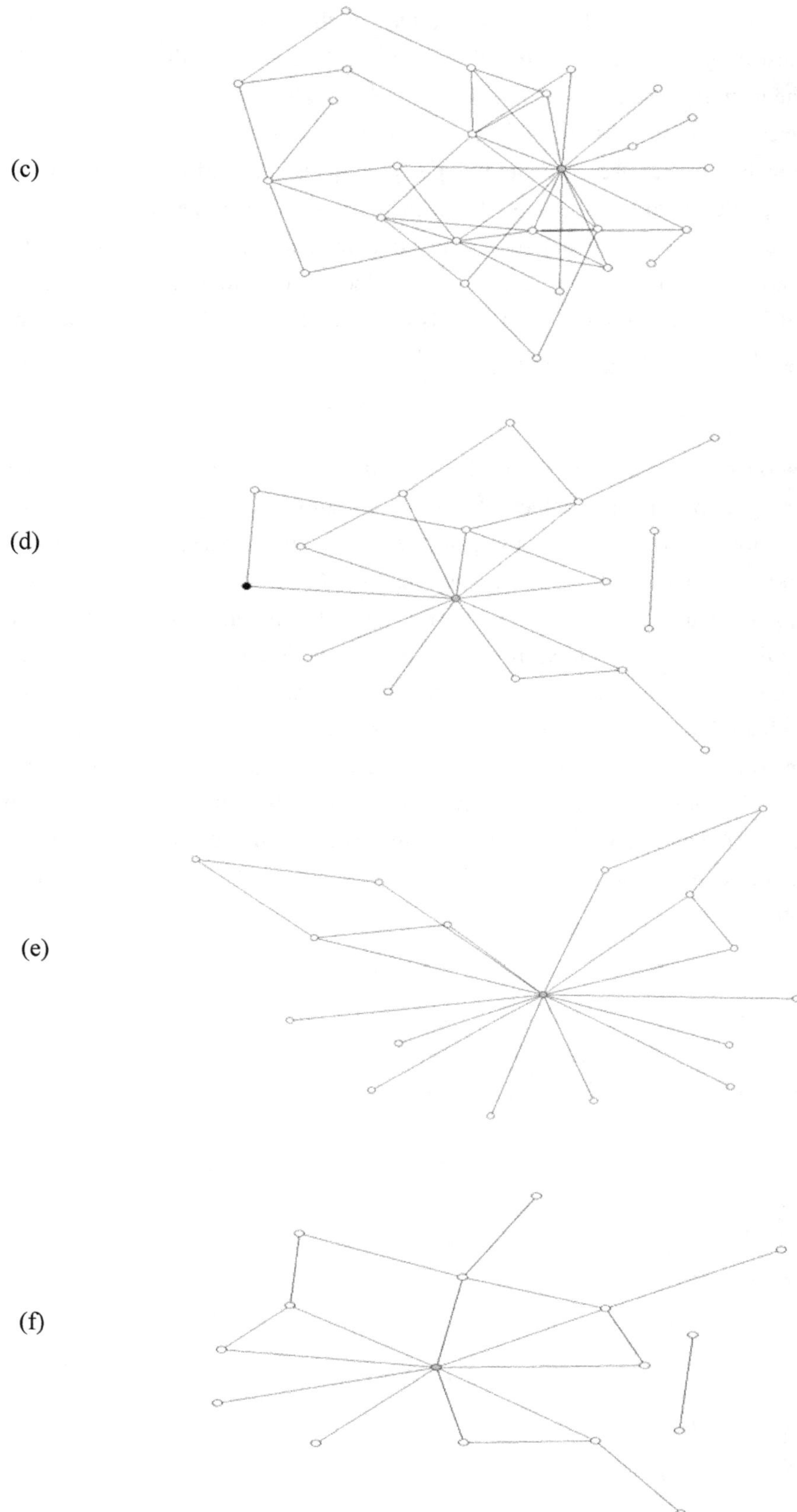

Fig. 1 The queen-right (a: V269 QR), queen-less (b: V269 QL) and *in silico* queen removed (c: V269 Qrem) networks of a *R. marginata* colony constructed using dominance-subordinate interactions. The similarly constructed queen-right (d: C08 QR), queen-less (e: C08 QL) and *in silico* queen removed (f: C08 Qrem) networks of a *R. cyathiformis* colony. Black, gray and white

nodes show the queen (Q), the potential queen (PQ) and the workers, respectively.

In *R. marginata*, the Qrem networks were more small-world-like than all the other networks. In *R. cyathiformis*, the PQrem networks were the least small-world-like, and were significantly different only from the Qrem networks (Fig. 3). All the comparisons were made by Mann-Whitney U test, with alpha set to 0.007, after Bonferroni correction.

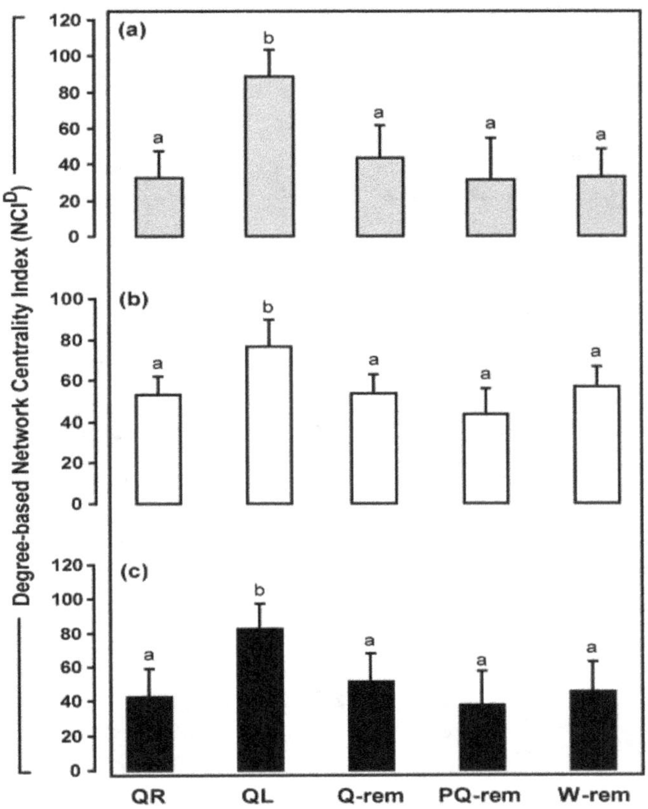

Fig. 2 Mean and standard deviation of the degree-based network centrality index (NCI^D) for the queen-right, queen-less, queen-removed, PQ-removed and worker-removed networks in (a) *R. marginata* and (b) *R. cyathiformis* and (c) for pooled data. Bars carrying different letters are significantly different from each other (Mann Whitney U-test, alpha set to 0.007 after Bonferroni correction).

It is an important finding that the difference between the QL and Qrem versions of the same networks is larger in *R. marginata* than in *R. cyathiformis*. This suggests that the loss of the queen is followed by serious network reorganisation especially in *R. marginata* (Fig. 4).

For illustration, we present the actual ranks of relevant individuals in the colonies (based on their degree values), while relative ranks will be used below for comparing indices to each other. *R. cyathiformis* potential queens were the top-ranking nodes in both the QL and the Qrem networks, and both the queens and the potential queens had high ranks in the QR networks (Table 1). However, the potential queens did not rank high in the Qrem networks of *R. marginata*, though they always had rank 1 in the QL networks (Table 2). In the QR networks of *R. marginata*, neither the queens nor the potential queens had high ranks (Table 2).

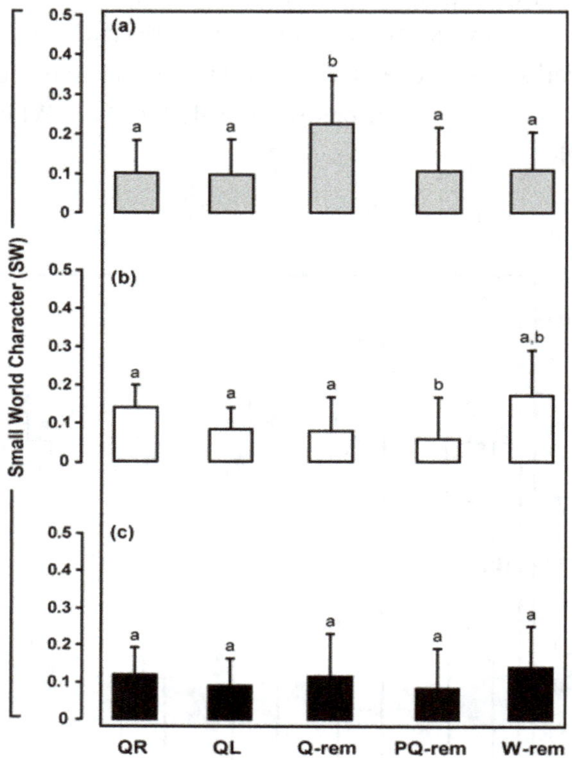

Fig. 3 Mean and standard deviation of the small world character (*SW*) for the queen-right, queen-less, queen-removed, PQ-removed and worker-removed networks in (a) *R. marginata* and (b) *R. cyathiformis* and (c) for pooled data. Bars carrying different letters are significantly different from each other (Mann Whitney U-test, alpha set to 0.007 after Bonferroni correction).

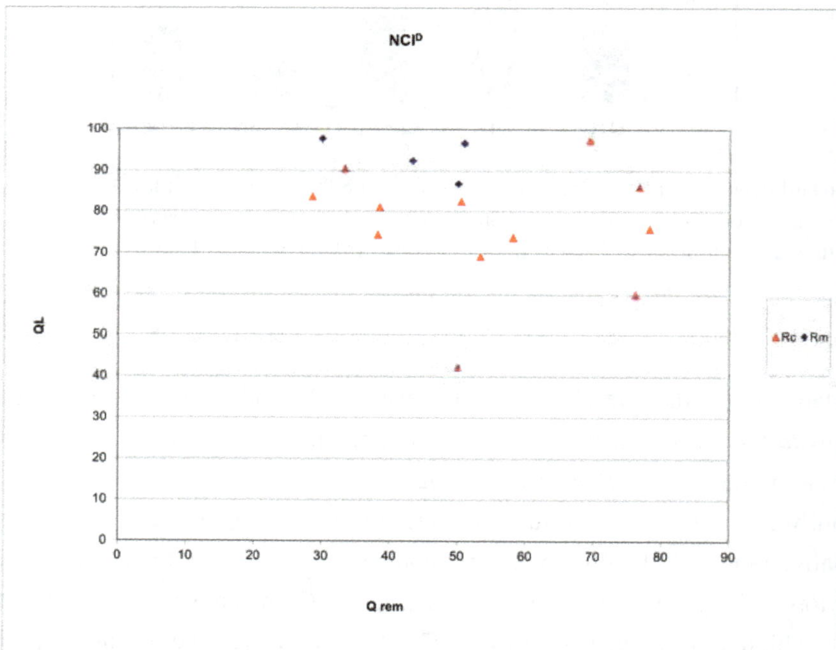

Fig. 4 The plots show the pairs of NCI^D values for QL and Qrem versions of the same network. Values are calculate for the 12 Rc and 4 Rm colonies where *in silico* queen removal was possible. If the loss of the queen was not more than deleting a node from the network, NCI^D values would equal for a pair of networks. Thus, the larger the difference between the pairs, the more important the reorganisation of the colony network after queen loss is.

Table 1 Degree-based ranks of the queens in the queen-right, as well as PQs in the queen-right, QL and queen removed networks of *R. cyathiformis*. The PQ's are ranked 1 in most of the QL and queen removed networks. Note that there is a single network (C86) where neither the queen nor the PQ are most central in QR.

	QR		QL	Qrem
Nest	**Q**	**PQ**	**PQ**	**PQ**
C01	2	1	1	1
C03	4	1.5	1	1.5
C07	2	1	1	1
C08	8.5	1	1	1
C33	1	2	1	1
C34	2	1	1	1
C38	1	2	1	1
C50	2.5	1	1	1
C58	5	1	1	1
C59	1	2.5	11	1
C60	4	1	1	1
C86	2.5	2.5	1	2

Table 2 Degree-based ranks of the queens in the QR, as well as PQ's in the QR, QL and Qrem networks of *R. marginata*. The PQ's do not have unique ranks in the queen removed networks, but they are ranked 1 in all QL networks.

	QR		QL	Qrem
Nest	**Q**	**PQ**	**PQ**	**PQ**
V262	6	7	1	6
V269	25	1	1	1
V272	8	4	1	6.5
V277	12.5	2	1	2
V519	0	4.5	1	1.5
V532	12	1	1	0
V614	33.5	25	1	2
V638	15	1.5	1	10.5
V657	18	2	1	1
V800	2.5	16.5	1	9

It is not easy to predict the potential queens on the basis on their degree-based rank, and the choice of the centrality index to be used is a key question: whether the unweighted and undirected measure of degree is suitable, or whether it should be a weighted or directed index. For this comparison, we have used the k/n relative ranks. Our findings suggest that weighted degree (wD) gives better ranks for PQs than the directed net status index (ns), but ns may provide more consistent prediction on the identity of the PQ (Table 3, see also

Fig. 5). We suggest that an index consistently predicting a lower rank has higher predictive power than another one predicting higher rank but in a less consistent way.

Table 3 The relative ranks for the PQ's in *Rm* networks, based on three indices (*wD*, *D*, and *ns*). We also show the number of individuals in the colonies (*N*). On the right, the average (*av*), standard deviation (*std*) and coefficient of variation (*CV*) of relative ranks are shown.

Colony	v262r	v269r	v272r	v277r	V519	V532	V586	V614	V638	V621	V657	V800	av	std	CV
N	10	28	23	13	17	8	17	28	31	40	16	44			
wD (PQ)	0.15	0.04	0.17	0.19	0.06	0.56	0.06	0.05	0.06	0.03	0.38	0.33	0.17	0.17	0.97
D (PQ)	0.15	0.04	0.17	0.15	0.09	0.56	0.06	0.07	0.06	0.03	0.28	0.23	0.16	0.15	0.95
ns (PQ)	0.20	0.25	0.09	0.15	0.76	0.31	0.06	0.75	0.34	0.40	0.34	0.31	0.33	0.23	0.68

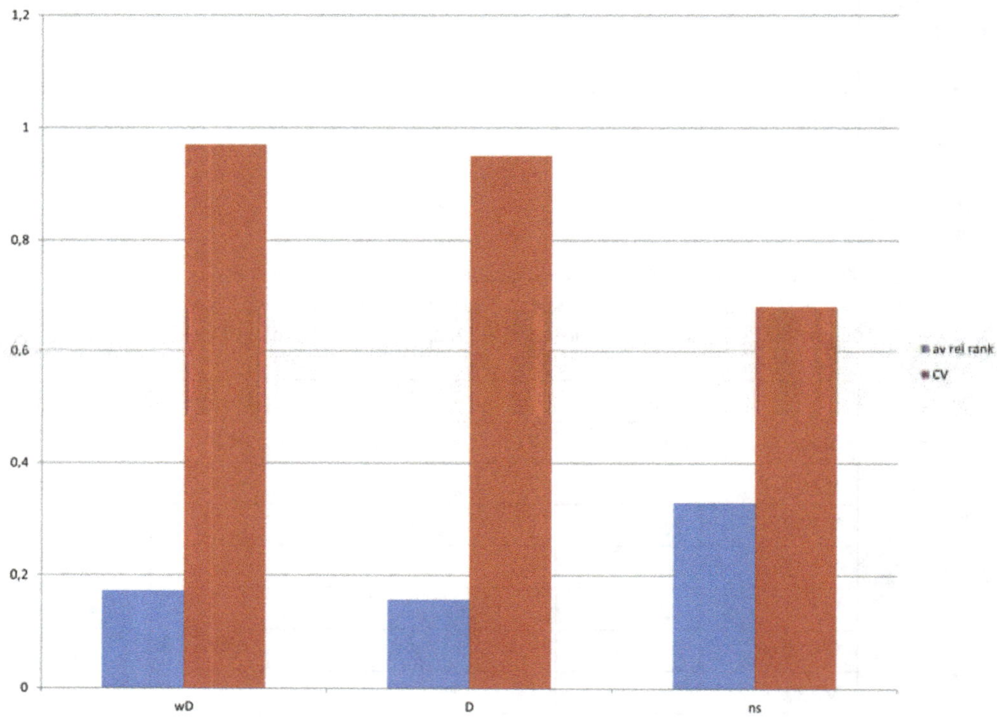

Fig. 5 The average relative rank and its coefficient of variation (*CV*) for PQ individuals in the Rm QR, based on three network metrics, degree (*D*), weighted degree (*wD*) and directed net status (*ns*).

In the experiments used to test the cryptic successor hypothesis, neither the cryptic successors nor the losers had high ranks in the hierarchies of their colonies, and there was no consistent pattern observed in the rank order of the queen and the two potential queens in a colony (Table 4). Since we know that the queen in *R. marginata* is non-aggressive, as opposed to the queen of *R. cyathiformis*, and that we cannot identify the queen's successor in the QR colony of *R. marginata* by her position in the dominance hierarchy, we considered it prudent to consider the network of all paired interactions that take place in the QR colony to attempt prediction of the queen's successor. However, even considering all interactions we could not predict the identity of the queen's successor in the QR colony (Table 5).

Table 4 Ranks of the queens, cryptic successors and losers in the QR networks of *R. marginata* based on all dominance-subordinate interactions. The PQs do not have unique ranks in the networks.

Nest	Q	CS	LO	n
V519	15	1.5		17
V532	4.5	0	4.5	8
V586	0	4.5	1	17
V614	18	2		28
V621	0	18	18	40
V638	25	11	6.5	25
V657	12	1	3	20
V800	2.5	16.5		47

Table 5 Ranks of the queens, cryptic successors and losers in the QR networks of *R. marginata* based on all paired interactions. The PQs do not have unique ranks in the networks.

Nest	Q	CS	LO	n
V519	12.5	2		20
V532	5	7	10	19
V586	21	10	1	34
V614	14	2		50
V621	50.5	9.5	9.5	56
V638	33.5	25	8	34
V657	12	1	3	20
V800	2.5	16.5		47

4 Discussion

R. marginata and *R. cyathiformis* together present a highly exciting scenario for studying the evolution of eusociality. While *R. marginata* is known to display certain characteristics of highly evolved societies, *R. cyathiformis* is a typical primitively eusocial species. This dichotomy between the two closely related species is most obvious in the behaviour of their queens. While on the one hand, the queen of *R. marginata* is a non-aggressive, non-interactive individual who seems to use pheromones rather than aggression to maintain reproductive monopoly in her colony, the queen of *R. cyathiformis* occupies the alpha position in the dominance hierarchy of her colony, and acts as the central "pacemaker" of the colony (Kardile and Gadagkar, 2003). The behavioural contrast between these species is also reflected in their social networks.

Based on the number of neighbours in the QR network, predicting the identity of the new queen (PQ) is quite easy in *R. cyathiformis* but is much harder in *R. marginata*. Even when we use weighted networks (i.e. information on the frequency of interactions, measured by *wD*), the situation does not alter. However, if we consider the direction of the links (*ns*), the average relative rank of PQs is larger (further away from the alpha-position) but the results are more consistent: the PQ is around the one-third mark of the individual ranks. Though there is a lot of variation even in this, of the three indices used, ns has the least CV (0.68), which is still quite high on an absolute scale. Thus we have to accept that network analysis does not take us any closer to predicting the identity of the PQ in the queen-right colony of *R. marginata* (Deshpande et al., 2006; Chandrashekara and Gadagkar, 1992; Bhadra and Gadagkar, 2008).

The *in silico* removals of the potential queen and a random worker affected network centrality in none of these species. This was surprising for *R. cyathiformis*, because the potential queen is in quite an important position even in the QR network (perhaps the presence of the queen reduced the effect of removing the potential queen). It was not surprising that these targeted removals did not affect the *R. marginata* colonies. From the biological perspective, the fact that the removal of the potential queen from the queen-right network does not affect the colony is important as it means that though there is a designated successor to the queen, the colony also has a back-up mechanism. If the queen dies, the potential queen would take up the responsibility of the queen, but if the designated successor is missing, there would be another to take up her place. This is consistent with the fact that there is actually a queue of successors present in the colony at any time, at least in *R. marginata* (Bang and Gadagkar, 2012). However, the result that the centrality measure does not change when the queen is removed *in silico* from *R. cyathiformis* colonies is very surprising. A closer look at the *R. cyathiformis* QR networks reveals that the queen usually interacts directly with the potential queen and a few other individuals, and the network is dense. Hence, the removal of the queen might not cause great alterations in the centrality of the network. Interestingly, *in silico* queen removed networks of *R. marginata* are more small-world-like than the QR or QL networks. This is probably because the queens show very few interactions, and the removal of these nodes increases the clustering coefficient of the networks. Generally, larger difference in a network measure between the QL and Qrem versions of the same colony indicates more important reorganisation after queen loss (Fig. 4). This result is also interesting in the light of recent work carried out on these two species with respect to the mechanism of the evolution of complexity in the social wasps. It has been demonstrated that by tweaking with the queen and worker interaction strategies in the relatively more primitive *R. cyathiformis*, a social organization like that of the more socially complex *R. marginata* can be easily achieved theoretically (Nandi et al., 2013).

We have also demonstrated that neither the potential queens in the dominance-subordinate interaction networks from the queen removal experiments, nor the cryptic successors and losers in both the dominance-subordinate and all interactions networks from the Q-PQ exchange experiment ranked high in the degree-based hierarchies. These results help us to conclude with more confidence that the queen's successor cannot be identified in *R. marginata* as long as the queen is present, using all the social network analysis methods by which we can identify her in *R. cyathiformis*. Since *R. marginata* seems to have acquired certain characteristics of more evolved societies, the presence of a designated successor to the queen who is acceptable to all other individuals in the colony, but cannot be perceived by behavioural or apparent anatomical and physiological differences is another aspect of the more evolved eusociality of *R. marginata*. Since the queen of *R. marginata* is known to produce a pheromone that she applies on the nest material, which the workers can perceive just by being on the nest (Bhadra et al., 2010, Mitra et al., 2011), it is possible that they use some similar subtle cues to identify their successor in the presence of the queen. Such perception would not necessarily involve active interactions between the potential queen and other workers and this might be the reason why she does not rank high in the centrality rank of the dominance network, unlike in the QL condition, when the PQ directs aggression actively towards her nestmates in order to step up the process of her own ovarian development (Lamba et al., 2007).

Acknowledgements

The work on wasps was supported by grants from the Department of Science and Technology, Department of Biotechnology and the Ministry of Environment and Forests, Government of India (to RG) and a CSIR Senior Research Fellowship to AB. We are thankful to Dr. Annagiri Sumana and Dr. Sujata A. Deshpande for allowing us to use their data on R. marginata and R. cyathiformis respectively for the analysis of this paper. Behavioural observations on wasps for the cryptic successor experiment were carried out by AB. Data analysis was performed by AB and FJ. We gratefully acknowledge the inputs provided on this work by Prof. Raghavendra Gadagkar, which have enriched the manuscript greatly. AB and FJ co-wrote the paper. Our experiments comply with regulations for animal care in India.

References

Bang A, Gadagkar R. 2012. Reproductive queue without overt conflict in the primitively eusocial wasp *Ropalidia marginata*. Proceedings of the National Academy of Sciences of USA, 109: 14494-14499

Bhadra A, Gadagkar R. 2008. We know that the wasps "know": Cryptic successors to the queen in *Ropalidia marginata*. Biology Letters, 4: 634-637

Bhadra A, Jordán F, Sumana A, et al. 2009. A comparative social network analysis of wasp colonies and classrooms: linking network structure to functioning. Ecological Complexity, 6: 48-55

Bhadra A, Mitra M, Deshpande S, et al. 2010. Regulation of reproduction in the primitively eusocial wasp *Ropalidia marginata*: On the trail of the Queen pheromone. Journal of Chemical Ecology, 36: 424-431

Borgatti SP, Everett MG, Freeman LC. 2002. Ucinet for Windows: Software for Social Network Analysis. Analytic Technologies, Harford, USA

Cant MA, Llop JB, Field J. 2006. Individual variation in social aggression and the probability of inheritance: Theory and a field test. American Naturalist, 167: 837-852

Chandrashekara K, Gadagkar R. 1992. Queen succession in the primitively eusocial tropical wasp Ropalidia Marginata (lep.) (Hymenoptera: Vespidae). Journal of Insect Behavior, 5: 193-209

Croft DP, James R, Krause J. 2008. Exploring Animal Social Networks. Princeton University Press, Chicago, USA

Cronin AL, Field J. 2007. Social aggression in an age-dependent dominance hierarchy. Behaviour, 144: 753-765

Deshapande SA, Sumana A, Surbeck M, et al. 2006. Wasp who would be queen: A comparative study of two primitively eusocial species. Current Science, 91: 332-336

Gadagkar R. 2001. The Social Biology of *Ropalidia marginata*. Harvard University Press, Cambridge, MA, USA

Harary F. 1959. Status and contrastatus. Sociometry, 22: 23-43

Heinze J, Trunzer B, Oliveira PS, et al. 1996. Regulation of reproduction in the ponerine ant, *Pachycondyla villosa*. Journal of Insect Behavior, 9: 441-450

Ito F, Higashi S. 1991. A linear dominance hierarchy regulating reproduction and polyethism of the queenless ant Pachycondyla sublaevis. Naturwissenschaften, 78: 80-82

Kardile S, Gadagkar R. 2002. Docile sitters and active fighters in paper wasps: a tale of two queens. Naturwissenschaften, 89: 176-179

Lamba S, Kazi YC, Deshpande S, et al. 2007. A possible novel function of dominance behaviour in queen-less colonies of the primitively eusocial wasp *Ropalidia marginata*. Behavioural Processes, 74: 351-356

Mitra A, Saha P, Chaoulideer ME, et al. 2011. Chemical communication in *Ropalidia marginata*: Dufour's gland contains queen signal that is perceived across colonies and does not contain colony signal. Journal of Insect Physiology, 57: 280-284

Nandi AK, Bhadra A, Sumana A, et al. 2013. The evolution of complexity in social organization – a model using dominance-subordinate behaviour in two social wasp species. Journal of Theoretical Biology (in press)

Pardi L. 1948. Dominance order in Polistes wasps. Physiological Zoology, 21: 1-13

Premnath S, Sinha A, Gadagkar R. 1996. Dominance relationships in the establishment of reproductive division of labour in a primitively eusocial wasp (*Ropalidia marginata)*. Behavioral Ecology and Sociobiology, 39: 125-132

Röseler PF, Röseler I, Strambi A. 1986. Studies of the dominance hierarchy in the paper wasp, *Polistes gallicus* (l.). Monitore Zoologico Italino, 20: 283-290

Sumana A, Gadagkar R. 2001. The structure of dominance hierarchies in the primitively eusocial wasp *Ropalidia marginata*. Ethology, Ecology and Evolution, 13: 273-281

Turillazzi S. 1991. The Stenogastrinae. In: The Social Biology of Wasps (Ross KG, Matthews RW, eds). Comstock, Ithaca, New York, USA

Wasserman S, Faust K. 1994. Social Network Analysis: Methods and Applications. Cambridge University Press, USA

Watts DJ, Strogatz SH. 1998. Collective dynamics of 'small-world' networks. Nature, 393: 440-442

West-Eberhard MJ. 1969. The social biology of polistine wasps. Miscellaneous Publications Museum of Zoology University of Michigan, 140: 1-101

Wey T, Blumstein D, Shen W, et al. 2008. Social network analysis of animal behaviour: a promising tool for the study of sociality. Animal Behaviour, 75: 333-344

Zhang WJ. 2012a. Computational Ecology: Graphs, Networks and Agent-based Modeling. World Scientific, Singapore

Zhang WJ. 2012b. Several mathematical methods for identifying crucial nodes in networks. Network Biology, 2(4): 121-126

4

Dynamics of fractional order modified Morris-Lecar neural model

Ranjit Kumar Upadhyay, Argha Mondal
Department of Applied Mathematics, Indian School of Mines, Dhanbad-826004, India
E-mail: ranjit_ism@yahoo.com, arghamondalb1@gmail.com

Abstract
Most of the beautiful biological functions in neural systems are expected to happen considering the system with memory effect. Fractional differential equations are very useful to investigate long-range interacting systems or systems with memory effect. In this paper, a fractional order nonlinear three dimensional modified Morris-Lecar neural system (M-L system) has been studied. The fractional order M-L system is a generalization of the integer order M-L system. The paper presents an approximate analytical solution of the fractional order M-L system, using Homotopy Perturbation Method (HPM) and Variational Iteration Method (VIM). The fractional derivatives are described in the Caputo sense. We have used the above methods as they show very efficient result for very small time region. Solutions are obtained in the form of rapidly convergent infinite series and only a few iterations are needed to obtain the approximate solutions. Comparison of both HPM and VIM reveals that the two present methods of solution are elegant and powerful for solving the nonlinear fractional order biological as well as neural systems.

Keywords Morris-Lecar neural model; fractional order; homotopy perturbation method; variational iteration method.

1 Introduction

Human brain is the most complicated part of the body that can display behavior which is periodic or chaotic of varying dimensionality (Destexhe, 1994; Freeman, 1994). Neurons are basic function units in nervous systems. Billions of neurons are connected in a complex network. They are the key elements in signal processing. Neurons respond to stimuli by generating sequences of brief electrical pulses, referred to as action potentials (Tateno and Pakdaman, 2004). The form of action potentials varies little, so that information concerning the stimulus cannot be readily conveyed by their shape. Conversely, the timing of these electrical discharges is stimulus dependent (Tateno and Pakdaman, 2004). The information is encoded, transmitted and decoded through firing activity of neurons, which is characterized by relaxation oscillation process of producing and transmitting action potential (Alexander et al., 2011). Bifurcation mechanisms involved in the generation of action potentials by neurons were reviewed by Izhikevich (2000).

Many neuron models have been proposed to understand a real nervous system that exhibits complicated dynamics. The following are some of the popular models: (i) Hodgkin-Huxley (H-H) model (Hodgkin and Huxley, 1952), (ii) FitzHugh-Nagumo (F-H-N) model (FitzHugh, 1955; Aqil et al., 2012), (iii) Wilson-Cowan (W-C) model (Wilson and Cowan, 1972) and (iv) Hindmarsh-Rose (H-R) model (Hindmarsh and Rose, 1984). The Morris-Lecar (M-L) model (1981) belongs to the family of conductance based membrane models of which H-H model is a well-known model. It represents an electrical circuit equivalent to a cellular membrane crossed by three different transmembraner currents referred to the voltage gated Ca^{+2} current, delayed rectifier K^{+} current and the leakage current (Morris and Lecar, 1981). Experiments on Barnacle muscle fiber with extra current show that such neuron model could produce much complex oscillation behavior (Gutkin and Ermentrout, 1998; Xie et al., 2004; Hodgkin and Rushton, 1946), but the generation mechanism of such behaviors is different from H-H model. The M-L system supports only the two types of dynamics, resting and sustained spiking activity and does not exhibit chaotic oscillations. Larter et al. (1999) developed a coupled ODE lattice model for the CA3 region of the Hippocampus for the simulation of epileptic seizures. The model consists of a hexagonal lattice of nodes, each describing a subnetwork consisting of a group of prototypical excitatory pyramidal cells with membrane potential V and a group of prototypical inhibitory interneurons with membrane potential Z interconnected via on/off excitatory and inhibitory synapses and both feed by current from the excitatory pathway (see figure 1 of Larter et al., 1999). These two types of neurons are included in the subnetwork which forms the basis of the model. The growth of mean membrane potential of excitatory principal cells is controlled by that of the inhibitory interneurons. The dynamical behavior of this subnetwork was described by a system of three differential equations based on a two variable reduction of the H-H model. To model the behavior of the entire subnetwork Larter et al. (1999) have added an equation to the original M-L model system to simulate the effect of a population of inhibitory interneurons synapsing on pyramidal cells. Rai et al. (2013) reported the existence of phase coupled oscillations in electrical activity of the neuronal cells that carry together amplitude, phase and time information for cellular signaling in the model developed by Larter et al. (1999). Nadar and Rai (2012) numerically simulated this model and commented that chaos is vital for functioning of a healthy brain and synchronization of the neural system occurs when all its regions are in transient periodicity represented by chaotic saddles in state space. This is how the intermittent pathology of epileptic seizure is created.

In this paper, we have modified the model developed by Larter et al. (1999) and studied by Nadar and Rai (2012) and Rai et al. (2013). We have considered a three dimensional fractional order nonlinear modified M-L neuron model. It is a generalization of the integer order modified M-L model with fast – slow variables, where the fractional order derivative is used for the memory effect and power law of neural cell membranes. The V-W subsystem has faster time scale than that of variable Z. Interesting behavior of the model system resulted from an interaction between the faster subsystem and a slower one with the population of inhibitory interneurons. The excitatory neurons carry information flow (long range projections) and inhibitory neurons are responsible for regulating the activation of excitatory neurons. Recently, Shi and Wang (2014) proposed a fractional order M-L neuron model and different types of bursting patterns were investigated using the bifurcation theory of fast-slow dynamical systems. It helps us in understanding the neuronal activities, efficient information processing and stimulus anticipation as well as in frequency independent phase shifts of oscillatory neuronal firing. Lundstrom et al. (2008) found that the dynamics, underlined spike rate adaptation to stimulus steps in single cortical neurons, functionally approximated fractional differentiation. This provides a general model for the firing- rate response to time-varying stimulus statistics or envelope coding. Fractional order dynamics have been observed in the vestibular-ocular systems (Paulin et al., 2004; Anastasio, 1994) and in the fly motion sensitive neuron H1 (Fairhall et al., 2001). A range of mechanisms which contribute to fractional order dynamics and power law includes synaptic mechanisms (Fusi et al., 2005), geometrical properties of cells (Thorson and Biederman-Thorson, 1974), circuits (Anastasio, 1998) and dendrites (Anastasio, 1994).

In recent years, fractional order differential equations have been used to solve physical, biological and neural systems. Since there are no methods available to obtain exact solution of fractional order differential equations, approximate analytical techniques are used to solve these equations. Several analytical and numerical methods have been proposed to solve fractional order differential equations. The most commonly used are Adomian Decomposition Method (ADM) (Adomian, 1998), Homotopy perturbation method (HPM) (He, 1999), Variational iteration method (VIM) (He, 1999a), Fractional Difference Method (FDM) (Podlubny, 1999) etc. To find an approximate analytical solution of the three dimensional nonlinear fractional order modified M-L neural system, we have applied the Homotopy perturbation method (HPM) (He, 2000; He, 2005; He, 2005a; Momani and Odibat, 2007; Abdulaziz et al., 2008; Chakraverty and Tapaswini, 2013) and Variational iteration method (VIM) (Momani and Odibat, 2007; Odibat and Momani, 2008). HPM is an approach for finding the approximate solution of linear or nonlinear differential and integral equations and for fractional order differential equations (Abdulaziz et al., 2008; Chakraverty and Tapaswini, 2013). This method was first proposed by He (1999) and was successfully applied to solve nonlinear wave equations (He, 2005; He, 2005a). The method which is a coupling of the traditional perturbation method and homotopy in topology deforms continuously to simple problems which can easily be solved. The method does not require small parameters in the equation which overcomes the limitations of the traditional perturbation method. The VIM, which was first applied by He (1999), is relatively a new approach to provide analytical approximate solutions to nonlinear problems and it provides visible symbolic terms of analytical solutions to fractional differential equations. The principles of VIM and its applicability for various kinds of differential equations are given by Momani and Odibat (2007) and Odibat and Momani (2008).

The organization of the paper is as follows: Section 2 describes the details about the modified Morris-Lecar neural model and the proposed fractional order modified M-L model based on its integer order model. Section 3 presents the analytical solutions using HPM and VIM. Numerical results are presented in Section 4. Finally, in Section 5, discussions and conclusions are presented.

2 The Modified Morris- Lecar Neuron Model

Consider a population of neurons as a dynamical system (Speelman, 1997) in which the neurons interact with each other. The differential equations that represent the dynamical system were derived by considering the behavior of prototypical single neurons (M-L neurons). The interactions between the populations of excitatory cells and inhibitory interneurons are governed by the system of nonlinear differential equations (Larter et al., 1999; Rai et al., 2013; Nadar and Rai, 2012)

$$\frac{dV}{dt} = -g_{Ca} m_\infty(V)(V-1) - g_K W(V-V^K) - g_L(V-V^L) + I - \alpha_{inh}(Z)Z, \tag{1a}$$

$$\frac{dW}{dt} = \phi(W_\infty - W)/\tau_W, \tag{1b}$$

$$\frac{dZ}{dt} = b'(c' I + \alpha_{exc}(V)V), \tag{1c}$$

where $\tau_W = 1/\cosh((V-V_3)/2V_4)$, $m_\infty(V) = 0.5(1+\tanh((V-V_1)/V_2))$, $W_\infty(V) = 0.5(1+\tanh((V-V_3)/V_4))$,

$$\alpha_{exc}(V) = \alpha_{exc}(1+\tanh((V-V_5)/V_6)),\ \alpha_{inh}(Z) = \alpha_{inh}(1+\tanh((Z-V_7)/V_6)). \tag{2}$$

The variables V and Z are mean membrane potentials for excitatory and inhibitory cells respectively. The variable W is the fraction of open potassium channels at any point of time. The detailed meanings of the parameters, their values and other details are given by Larter et al. (1999).

Fractional order Modified Morris-Lecar neural model

Resistor-capacitor circuit theory is applied to the modified M-L system (1). Since the total volume of all ionic currents Ca^{+2}, K^{+} and leaky currents varies, the mean membrane potential of excitatory neuron V changes reflecting the firing activity of the neuron. It is governed by the equation (Larter et al., 1999)

$$C\frac{dV}{dt} = I_{ion} + I_{ext} - \alpha_{inh}(Z)Z = I_{Ca} + I_K + I_L + I - \alpha_{inh}(Z)Z, \tag{3}$$

where $I_{Ca} = -g_{Ca}m_{\infty}(V)(V-1), I_K = -g_K W(V-V^K), I_L = -g_L(V-V^L)$, I and α_{inh} denote the applied current and dimensionless synaptic strength of inhibitory interneurons respectively. From electrical circuit theory, we have

$$I = C\,dV/dt, \tag{4}$$

where the capacitance C is a constant, $I = dQ/dt$ and $Q = CV$.

Normal capacitive behavior of real dielectrics and insulators exhibit the fractional differentiation relationship (Westerlund and Ekstam, 1994)

$$I = C\,d^{\alpha}V/dt^{\alpha}, \quad 0<\alpha<1. \tag{5}$$

Assume that a step voltage $V_0 u(t)$ is applied at $t = 0$. Using Laplace transforms, solutions of (4) and (5) under this initial conditions are given by (Westerlund and Ekstam, 1994; Magin, 2004)

$$I(t) = CV_0\delta(t), \tag{6a}$$

$$I(t) = \frac{CV_0}{\Gamma(1-\alpha)}t^{-\alpha}, \quad 0<\alpha<1, \tag{6b}$$

where $\delta(t)$ is the Dirac delta function. From (6b), observe that the current follows a power law and decays at a rate of $t^{-\alpha}$ with time. In the limiting case as $\alpha \to 1$, (6b) is consistent with (6a) as $1/\Gamma(1-\alpha) \to 0$ as $\alpha \to 1$.

From (4) and (6a), we get $V' = V_0\delta(t)$. According to the definition of Caputo's fractional order derivative (Podlubny, 1999)

$$\frac{d^{\alpha}V}{dt^{\alpha}} = \frac{1}{\Gamma(1-\alpha)}\int_0^t (t-\tau)^{-\alpha}V'(\tau)d\tau,$$

$$= \frac{V_0}{\Gamma(1-\alpha)}\int_0^t (t-\tau)^{-\alpha}\delta(\tau)d\tau = \frac{V_0}{\Gamma(1-\alpha)}t^{-\alpha}.$$

Now, $I = C\dfrac{d^{\alpha}V}{dt^{\alpha}} = \dfrac{CV_0}{\Gamma(1-\alpha)}t^{-\alpha}$.

Hence, it is reasonable to use a fractional order capacitor theory with order $0<\alpha<1$ in modeling neural dynamics. For computations, we have assumed $C = 1$ and $V(0) = 0.01, W(0) = 0.1, Z(0) = 0.01$ at $t = 0$ (Larter et al., 1999; Nadar and Rai, 2012; Rai et al., 2013) for HPM and VIM solution. As a result, a fractional order modified M-L system with order $0 < \alpha \le 1$, in the sense of Caputo derivative can be proposed as

$$D_t^{\alpha} V(t) = -g_{Ca} m_{\infty}(V)(V-1) - g_K W(V - V^K) - g_L(V - V^L) + I - \alpha_{inh}(Z)Z, \tag{7a}$$

$$D_t^{\beta} W(t) = \phi(W_{\infty} - W)/\tau_W, \tag{7b}$$

$$D_t^{\gamma} Z(t) = b'(c' I + \alpha_{exc}(V)V), \tag{7c}$$

where $D_t^{\alpha}, D_t^{\beta}$ and D_t^{γ} are Caputo derivatives of orders $0 < \alpha, \beta, \gamma \le 1$. These fractional equations are obtained from the classical equations by replacing the first order time derivatives by fractional derivatives of orders α, β, γ, $0 < \alpha \le 1, 0 < \beta \le 1, 0 < \gamma \le 1$. The advantage of fractional order systems is that they allow greater degrees of freedom in the model.

The following definitions and properties of the fractional calculus (Podlubny, 1999; Miller and Ross, 1993) are used in the derivation.

Definition 1 A real function $f(t), t > 0$, is said to be in the space $C_{\mu}, \mu \in R$, if there exists a real number $p > \mu$, such that $f(t) = t^p f_1(t)$, where $f_1(t) \in C(0, \infty)$ and it is said to be in the space C_{μ}^n if and only if $f^n \in C_{\mu}, n \in N$.

Definition 2 The Riemann-Liouville fractional integral operator J^{α} of order $\alpha \ge 0$, of a function $f \in C_{\mu}, \mu \ge -1$, is defined as

$$J^{\alpha} f(t) = \frac{1}{\Gamma(\alpha)} \int_0^t (t-\tau)^{\alpha-1} f(\tau) d\tau, \quad (\alpha > 0), \qquad J^0 f(t) = f(t).$$

Some of the properties of the operator J^{α} which are used are the following:

(i) $J^{\alpha} J^{\beta} f(t) = J^{\alpha+\beta} f(t)$, (ii) $J^{\alpha} J^{\beta} f(t) = J^{\beta} J^{\alpha} f(t)$, (iii) $J^{\alpha} t^{\gamma} = \dfrac{\Gamma(\gamma+1)}{\Gamma(\alpha+\gamma+1)} t^{\alpha+\gamma}$.

Definition 3 Fractional derivative $D^{\alpha} f(t) = d^{\alpha} f(t)/dt^{\alpha}$ in the Caputo sense is defined as

$$D^{\alpha} f(t) = \frac{1}{\Gamma(n-\alpha)} \int_0^t (t-\tau)^{n-\alpha-1} f^n(\tau) d\tau, \text{for } n-1 < \alpha \le n, \;\; n \in N, \; t > 0, \; f \in C_{-1}^n.$$

3 Analytical Solution of the Fractional Order Modified M-L neural System

3.1 HPM solution for fractional order modified Morris-Lecar equations

Applying HPM, we construct the following homotopy of the model system (7) as

$$D_t^{\alpha} V(t) = p\{-g_{Ca} m_{\infty}(V)(V-1) - g_K W(V - V^K) - g_L(V - V^L) + I - \alpha_{inh}(Z)Z\}, \tag{8a}$$

$$D_t^{\beta} W(t) = p\{\phi(W_{\infty} - W)/\tau_W\}, \tag{8b}$$

$$D_t^{\gamma} Z(t) = p\{b'(c' I + \alpha_{exc}(V)V)\}, \tag{8c}$$

where $p, 0 \le p \le 1$ is the homotopy parameter.

$$D_t^{\gamma} Z(t) = p\{b'(c' I + \alpha_{exc}(V)V)\}, \tag{8c}$$

where $p, 0 \le p \le 1$ is the homotopy parameter.

Substituting the values of $m_{\infty}(V),\ \alpha_{inh}(Z),\ W_{\infty}(V),\ \tau_W$ and $\alpha_{exc}(V)$ in (8a) to (8c), we obtain

$$D_t^{\alpha} V(t) = p\{-0.5 g_{Ca}[1 + \tanh((V - V_1)/V_2)](V - 1) - g_K W(V - V^K) - g_L(V - V^L) + I \\ - \alpha_{inh}[1 + \tanh((Z - V_7)/V_6)]Z\}, \tag{9a}$$

$$D_t^{\beta} W(t) = p\{\phi(0.5[1 + \tanh((V - V_3)/V_4)] - W)\cosh((V - V_3)/2V_4)\}, \tag{9b}$$

$$D_t^{\gamma} Z(t) = p\{b'(c' I + \alpha_{exc}[1 + \tanh((V - V_5)/V_6)]V)\}. \tag{9c}$$

Linearizing (9a)-(9c) (take Maclaurin's series of hyperbolic functions and ignore second and higher order terms), we obtain

$$D_t^{\alpha} V(t) = p\{-0.5 g_{Ca}[1 + ((V - V_1)/V_2)](V - 1) - g_K W(V - V^K) - g_L(V - V^L) + I \\ - \alpha_{inh}[1 + ((Z - V_7)/V_6)]Z\}, \tag{10a}$$

$$D_t^{\beta} W(t) = p\{\phi(0.5[1 + ((V - V_3)/V_4)] - W)\}, \tag{10b}$$

$$D_t^{\gamma} Z(t) = p\{b'(c' I + \alpha_{exc}[1 + ((V - V_5)/V_6)]V)\}, \tag{10c}$$

where

$$V(t) = V_0' + pV_1' + p^2 V_2' + p^3 V_3' + \cdots, \tag{11a}$$

$$W(t) = W_0' + pW_1' + p^2 W_2' + p^3 W_3' + \cdots, \tag{11b}$$

$$Z(t) = Z_0' + pZ_1' + p^2 Z_2' + p^3 Z_3' + \cdots. \tag{11c}$$

Insert the expressions from (11a)-(11c) into (10a)-(10c) and use the set of parameters values (Nadar and Rai, 2012) and the initial conditions as

$V_1 = -0.01, V_2 = 0.15, V_3 = 0.03, V_4 = 0.3, V_5 = 0.0, V_6 = 0.4, V_7 = 0.05, g_{Ca} = 1.1, g_K = 2.0, g_L = 1.0, V^K = -0.7,$
$V^L = -0.5, I = 0.3, b' = 0.15, \alpha_{exc} = 1, \alpha_{inh} = 1,\ \phi = 0.4,\ c' = 0.238$ and $V(0) = 0.01, W(0) = 0.1$ and $Z(0) = 0.01$.

Taking $V_0' = a = 0.01, W_0' = b = 0.1, Z_0' = c = 0.01$, we obtain the following set of differential equations for $V_0', W_0', Z_0', \ldots$

$$p^0 : D_t^{\alpha} V_0' = 0, \tag{12a}$$

$$D_t^{\beta} W_0' = 0, \tag{12b}$$

$$D_t^{\gamma} Z_0' = 0. \tag{12c}$$

$$p^1: D_t^\alpha V_1' = 2.08a - 3.6685a^2 + 0.3869 - 2ab - 1.4b - 0.875c - 2.5c^2, \tag{13a}$$

$$D_t^\beta W_1' = 0.18 + 0.666a - 0.4b, \tag{13b}$$

$$D_t^\gamma Z_1' = 0.0107 + 0.15a + 0.375a^2. \tag{13c}$$

$$p^2: D_t^\alpha V_2' = V_1'(2.08165 - 7.3315a - 2b) - 2W_1'(a+0.7) - Z_1'(0.875 + 5c), \tag{14a}$$

$$D_t^\beta W_2' = 0.67V_1' - 0.4W_1', \tag{14b}$$

$$D_t^\gamma Z_2' = 0.15V_1'(1+5a). \tag{14c}$$

$$p^3: D_t^\alpha V_3' = V_2'(-7.337a + 2.08165 - 2b) - 3.6685V_1'^2 - 2(a+0.7)W_2' - 2V_1'W_1' - Z_2'(5c+0.875) - 2.5Z_1'^2, \tag{15a}$$

$$D_t^\beta W_3' = 0.67V_2' - 0.4W_2', \tag{15b}$$

$$D_t^\gamma Z_3' = 0.15(1+5a)V_2' + 0.375V_1'^2, \tag{15c}$$

and so on. The method is based on applying the operators $J_t^\alpha, J_t^\beta, J_t^\gamma$ respectively (the inverse operators of the Caputo derivatives $D_t^\alpha, D_t^\beta, D_t^\gamma$ respectively) on both sides of (12a) to (15c).

Finally, we obtain the approximate solutions for $V(t)$, $W(t)$ and $Z(t)$ as

$$V(t) = \lim_{N\to\infty} \delta_N(t),\ W(t) = \lim_{N\to\infty} \lambda_N(t),\ Z(t) = \lim_{N\to\infty} \psi_N(t),$$

where $\delta_N(t) = \sum_{n=0}^{N-1} V_n'(t),\ \lambda_N(t) = \sum_{n=0}^{N-1} W_n'(t),\ \psi_N(t) = \sum_{n=0}^{N-1} Z_n'(t),\ N \ge 1.$

The above three series converge very rapidly. The rapid convergence means that only few terms are required to get the approximate solutions. Solving (12a) to (15c), we obtain

$$V_0' = 0.01,\ W_0' = 0.1,\ Z_0' = 0.01,$$

$$V_1' = P(a,b,c)\frac{t^\alpha}{\Gamma(\alpha+1)},\ W_1' = P(a,b)\frac{t^\beta}{\Gamma(\beta+1)},\ Z_1' = P(a)\frac{t^\gamma}{\Gamma(\gamma+1)},$$

$$V_2' = P(a,b,c)P_1(a,\ b)\frac{t^{2\alpha}}{\Gamma(2\alpha+1)} - 2P(a,b)(a+0.7)\frac{t^{(\alpha+\beta)}}{\Gamma(\alpha+\beta+1)} - P(a)P(c)\frac{t^{(\alpha+\gamma)}}{\Gamma(\alpha+\gamma+1)},$$

$$W_2' = 0.67P(a,b,c)\frac{t^{(\alpha+\beta)}}{\Gamma(\alpha+\beta+1)} - 0.4P(a,b)\frac{t^{2\beta}}{\Gamma(2\beta+1)},$$

$$Z_2' = 0.15(1+5a)P(a,b,c)\frac{t^{(\alpha+\gamma)}}{\Gamma(\alpha+\gamma+1)},$$

$$V_3' = P_1(a,b)(P(a,b,c)P_1(a,b)\frac{t^{3\alpha}}{\Gamma(3\alpha+1)} - 2(a+0.7)P(a,b)\frac{t^{(2\alpha+\beta)}}{\Gamma(2\alpha+\beta+1)})$$

$$-P_1(a,b)P(a)P(c)\frac{t^{(2\alpha+\gamma)}}{\Gamma(2\alpha+\gamma+1)}-3.6685(P(a,b,c))^2\frac{t^{3\alpha}\Gamma(2\alpha+1)}{(\Gamma(\alpha+1))^2\Gamma(3\alpha+1)}$$
$$-2(a+0.7)(0.67P(a,b,c)\frac{t^{(2\alpha+\beta)}}{\Gamma(2\alpha+\beta+1)}-0.4P(a,b)\frac{t^{(\alpha+2\beta)}}{\Gamma(\alpha+2\beta+1)})$$
$$-2P(a,b,c)P(a,b)\frac{t^{(2\alpha+\beta)}\Gamma(\alpha+\beta+1)}{\Gamma(\alpha+1)\Gamma(\beta+1)\Gamma(2\alpha+\beta+1)}$$
$$-(0.55(1+5a)P(a,b,c))P(c)\frac{t^{(2\alpha+\gamma)}}{\Gamma(2\alpha+\gamma+1)}-2.5(P(a))^2\frac{t^{(\alpha+2\gamma)}\Gamma(2\gamma+1)}{(\Gamma(\gamma+1))^2\Gamma(\alpha+2\gamma+1)},$$

$$W_3'=0.67(P(a,b,c)P_1(a,b)\frac{t^{(2\alpha+\beta)}}{\Gamma(2\alpha+\beta+1)}-2(a+0.7)P(a,b)\frac{t^{(\alpha+2\beta)}}{\Gamma(\alpha+2\beta+1)})$$
$$-0.67P(a)P(c)\frac{t^{(\alpha+\beta+\gamma)}}{\Gamma(\alpha+\beta+\gamma+1)}-0.4(0.67P(a,b,c)\frac{t^{(\alpha+2\beta)}}{\Gamma(\alpha+2\beta+1)}-0.4P(a,b)\frac{t^{3\beta}}{\Gamma(3\beta+1)}),$$

$$Z_3'=0.15(1+5a)(P(a,b,c)P_1(a,b)\frac{t^{(2\alpha+\gamma)}}{\Gamma(2\alpha+\gamma+1)}-2(a+0.7)P(a,b)\frac{t^{(\alpha+\beta+\gamma)}}{\Gamma(\alpha+\beta+\gamma+1)}$$
$$-P(a)P(c)\frac{t^{(\alpha+2\gamma)}}{\Gamma(\alpha+2\gamma+1)})+0.375(P(a,b,c))^2\frac{t^{(2\alpha+\gamma)}\Gamma(2\alpha+1)}{(\Gamma(\alpha+1))^2\Gamma(2\alpha+\gamma+1)},$$

where

$$P(a,b,c)=(2.08a-3.6685a^2+0.3869-2ab-1.4b-0.875c-2.5c^2),P(a,b)=(0.18+0.666a-0.4b),$$

$$P_1(a,b)=(2.08165-7.3315a-2b),\quad P(a)=(0.0107+0.15a+0.375a^2),\ P(c)=(0.875+5c).$$

In a similar manner, the rest of the components can be obtained, and hence the approximate solutions for *V*(*t*), *W*(*t*) and *Z*(*t*).

3.2 VIM solution for fractional order modified Morris-Lecar equations

The variational iteration method is also valid for solving fractional differential equations. In this paper, we extend the application of the variational iteration method to solve the fractional order modified M-L neural model as follows: The standard VIM formula for (7a) - (7c) leads to the following iteration formula (after inserting the values of $g_{Ca}, g_K, g_L, I, \alpha_{exc}, \alpha_{inh}, \phi, b', c'$)

$$V'_{n+1}=V_n'-J_t^\alpha[D_t^\alpha V_n'+0.55\{1+(V_n'-V_1)/V_2\}(V_n'-1)+2W_n'(V_n'-V^K)+(V_n'-V^L)-0.3+\{1+(Z_n'-V_7)/V_6\}Z_n'], \tag{16a}$$

$$W'_{n+1}=W_n'-J_t^\beta[D_t^\beta W_n'-0.2\{1+(V_n'-V_3)/V_4\}+0.4W_n'], \tag{16b}$$

$$Z'_{n+1}=Z_n'-J_t^\gamma[D_t^\gamma Z_n'-0.01071-0.15\{1+(V_n'-V_5)/V_6\}V_n']. \tag{16c}$$

Assume $V_0'=a=0.01, W_0'=b=0.1, Z_0'=c=0.01$. The values of the parameters are taken as in Nadar and Rai (2012). For $n=0,1,2,3$, we obtain

$$V_1'=a-A\frac{t^\alpha}{\Gamma(\alpha+1)},\ W_1'=b-B\frac{t^\beta}{\Gamma(\beta+1)},\ Z_1'=c-G\frac{t^\gamma}{\Gamma(\gamma+1)},$$

where

$A=(3.6685a^2-2.08165a-0.38685+2ab+1.4b+2.5c^2+0.875c),$
$B=(-0.18-0.666a+0.4b),\ G=(-0.01071-0.15a-0.375a^2),$

$$V_2' = 2.0817(a\frac{t^{\alpha}}{\Gamma(\alpha+1)} - A\frac{t^{2\alpha}}{\Gamma(2\alpha+1)}) - 3.6685(a^2\frac{t^{\alpha}}{\Gamma(\alpha+1)} + A^2\frac{t^{3\alpha}\Gamma(2\alpha+1)}{(\Gamma(\alpha+1))^2\Gamma(3\alpha+1)})$$
$$+7.337aA\frac{t^{2\alpha}}{\Gamma(2\alpha+1)} - 2(ab\frac{t^{\alpha}}{\Gamma(\alpha+1)} - Ab\frac{t^{2\alpha}}{\Gamma(2\alpha+1)} - aB\frac{t^{\alpha+\beta}}{\Gamma(\alpha+\beta+1)})$$
$$+2AB\frac{t^{2\alpha+\beta}\Gamma(\alpha+\beta+1)}{\Gamma(\alpha+1)\Gamma(\beta+1)\Gamma(2\alpha+\beta+1)} - 1.4(b\frac{t^{\alpha}}{\Gamma(\alpha+1)} - B\frac{t^{\alpha+\beta}}{\Gamma(\alpha+\beta+1)}) + 0.3869\frac{t^{\alpha}}{\Gamma(\alpha+1)}$$
$$-0.875(c\frac{t^{\alpha}}{\Gamma(\alpha+1)} - G\frac{t^{\alpha+\gamma}}{\Gamma(\alpha+\gamma+1)}) - 2.5(ac\frac{t^{\alpha}}{\Gamma(\alpha+1)} - cB\frac{t^{\alpha+\beta}}{\Gamma(\alpha+\beta+1)} - bG\frac{t^{\alpha+\gamma}}{\Gamma(\alpha+\gamma+1)})$$
$$-2.5BG\frac{t^{2\alpha+\gamma}\Gamma(\alpha+\gamma+1)}{\Gamma(\alpha+1)\Gamma(\gamma+1)\Gamma(2\alpha+\gamma+1)}.$$

Simplifying V_2', we have

$$V_2' = H\frac{t^{\alpha}}{\Gamma(\alpha+1)} + K\frac{t^{2\alpha}}{\Gamma(2\alpha+1)} - 3.6685A^2\frac{t^{3\alpha}\Gamma(2\alpha+1)}{(\Gamma(\alpha+1))^2\Gamma(3\alpha+1)} + J\frac{t^{\alpha+\beta}}{\Gamma(\alpha+\beta+1)}$$
$$-2AB\frac{t^{2\alpha+\beta}\Gamma(\alpha+\beta+1)}{\Gamma(\alpha+1)\Gamma(\beta+1)\Gamma(2\alpha+\beta+1)} + P\frac{t^{\alpha+\gamma}}{\Gamma(\alpha+\gamma+1)} - 2.5BG\frac{t^{2\alpha+\gamma}\Gamma(\alpha+\gamma+1)}{\Gamma(\alpha+1)\Gamma(\gamma+1)\Gamma(2\alpha+\gamma+1)},$$

$$W_2' = L\frac{t^{\beta}}{\Gamma(\beta+1)} - 0.066A\frac{t^{\alpha+\beta}}{\Gamma(\alpha+\beta+1)} + 0.4B\frac{t^{2\beta}}{\Gamma(2\beta+1)},$$

$$Z_2' = M\frac{t^{\gamma}}{\Gamma(\gamma+1)} + N\frac{t^{\alpha+\gamma}}{\Gamma(\alpha+\gamma+1)} + 0.375A^2\frac{t^{2\alpha+\gamma}\Gamma(2\alpha+1)}{(\Gamma(\alpha+1))^2\Gamma(2\alpha+\gamma+1)},$$

where

$H = (2.0817a - 3.6685a^2 - 2ab - 1.4b + 0.3869 - 0.875c - 2.5ac)$, $K = (-2.0817A + 7.337aA + 2Ab)$,
$J = (2aB + 1.4B + 2.5cB)$, $P = (0.875G + 2.5bG)$, $L = (0.18 + 0.066a - 0.4b)$,
$M = (0.01071 + 0.15a + 0.375a^2)$, $N = (-0.15A - 0.75aA)$.

For $n = 2$, we obtain the next iterated values V_3', W_3' and Z_3' as

$$V_3' = 2.0817J_t^{\alpha}V_2' - 3.6685J_t^{\alpha}V_2'^2 + 0.38685\frac{t^{\alpha}}{\Gamma(\alpha+1)} - 2J_t^{\alpha}V_2'W_2' - 1.4J_t^{\alpha}W_2' - 0.875J_t^{\alpha}Z_2' - 2.5J_t^{\alpha}Z_2'^2,$$

$$W_3' = 0.18\frac{t^{\beta}}{\Gamma(\beta+1)} + 0.666J_t^{\beta}V_2' - 0.4J_t^{\beta}W_2',$$

$$Z_3' = 0.01071\frac{t^{\gamma}}{\Gamma(\gamma+1)} + 0.15J_t^{\gamma}V_2' + 3.75J_t^{\gamma}V_2'^2.$$

Inserting the values of V_2', W_2' and Z_2', we can simplify the above expressions. In a similar manner, the rest of the components can be obtained. Finally, we get the approximate solutions for $V(t), W(t)$ and $Z(t)$.

Detailed expressions for V_3', W_3' and Z_3' are given in the Appendix A1.

4 Numerical Simulation Results

In this section, numerical results of the modified Morris-Lecar equations given by (1a) to (1c), for different fractional order derivatives $\alpha = \beta = \gamma = 1/3, 1/2, 2/3$ and for the standard derivative $\alpha = \beta = \gamma = 1$ are calculated using the methods HPM and VIM for $t \in [0,1]$. Numerical results are depicted graphically by varying $t \in [0,1]$ and keeping γ constant and by taking $\gamma = 0.5$ and 1 for different values of fractional orders $\alpha = \beta = 1/3, \alpha = \beta = 1/2, \alpha = \beta = 2/3$ and for integer order $\alpha = \beta = 1$. Next, fixing $t = 0.1$ and varying

$\gamma \in [0,1]$, the obtained results are shown graphically for various order fractional derivatives $\alpha = \beta = 1/3, \alpha = \beta = 1/2, \alpha = \beta = 2/3,$ and $\alpha = \beta = 1.$ Varying both $t, \gamma \in [0,1]$ for $\alpha = \beta = 1/3, \alpha = \beta = 1/2, \alpha = \beta = 2/3,$ and $\alpha = \beta = 1,$ the results are depicted graphically using the above two methods.

In the two methods, it was found that only four terms are sufficient to obtain the approximate solutions. The accuracy of the results can be improved by introducing more terms.

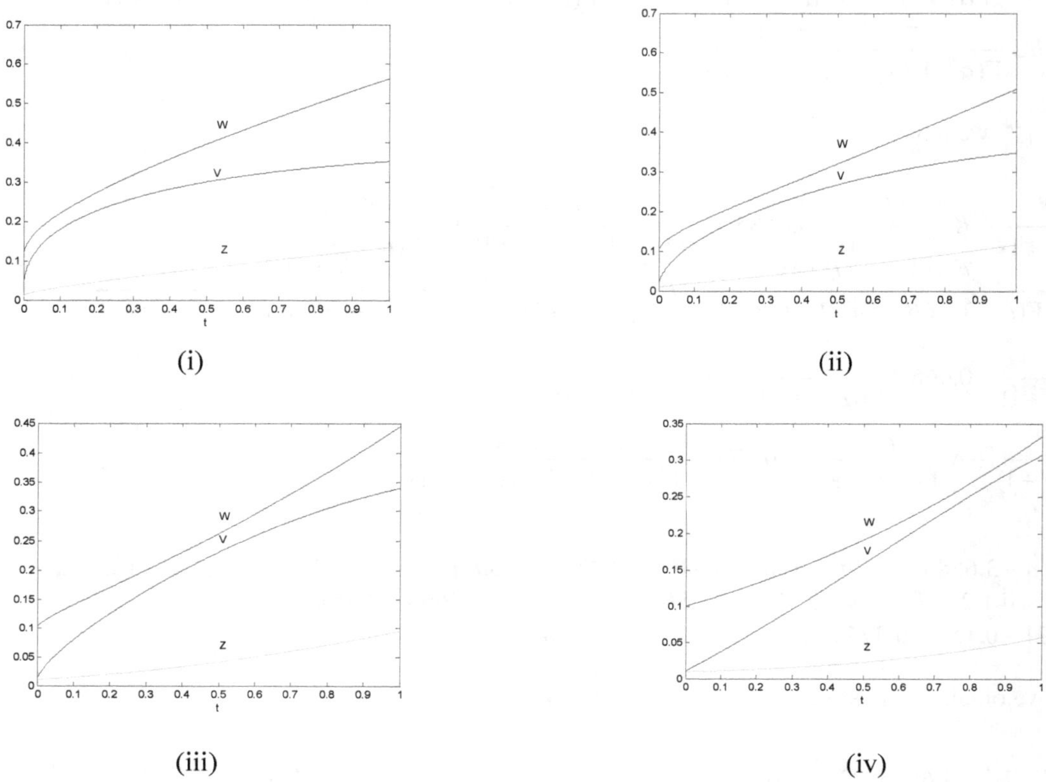

Fig. 1 Fractional order HPM solutions for $V(t), W(t), Z(t)$ for $t \in [0,1]$ and (i) $\alpha = \beta = \gamma = 1/3$, (ii) $\alpha = \beta = \gamma = 1/2$, (iii) $\alpha = \beta = \gamma = 2/3$, (iv) $\alpha = \beta = \gamma = 1$.

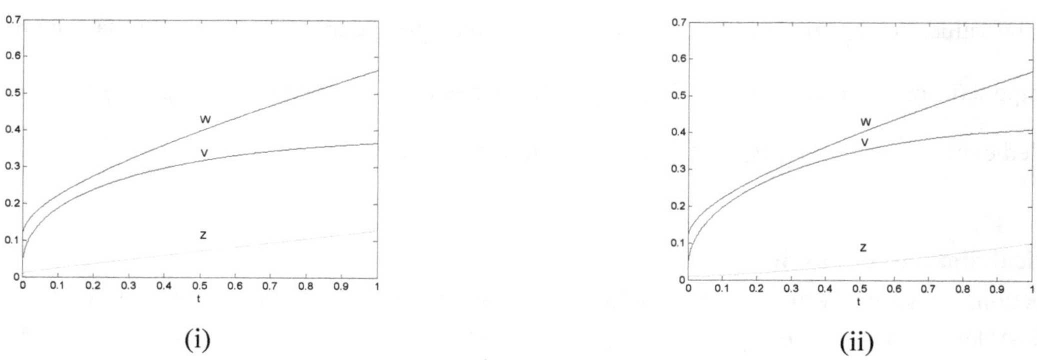

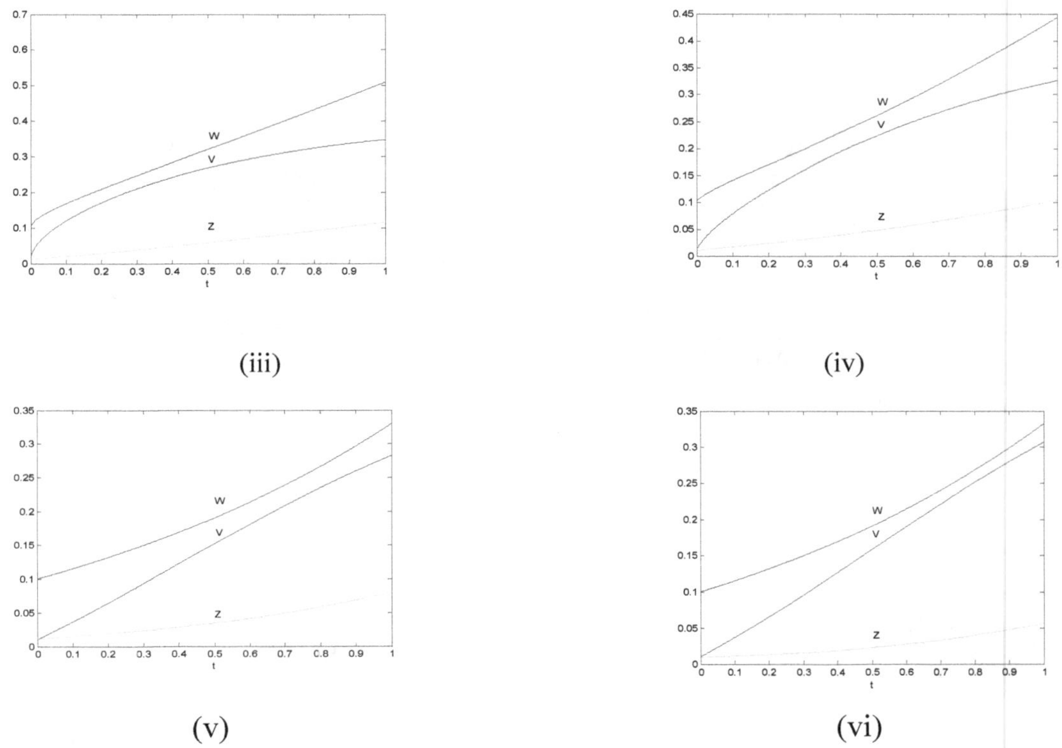

Fig. 2 Fractional order HPM solutions for $V(t), W(t), Z(t)$ for $t \in [0,1]$ and (i) $\alpha = \beta = 1/3, \gamma = 0.5$, (ii) $\alpha = \beta = 1/3, \gamma = 1$, (iii) $\alpha = \beta = 1/2, \gamma = 0.5$, (iv) $\alpha = \beta = 2/3, \gamma = 0.5$, (v) $\alpha = \beta = 1, \gamma = 0.5$, (vi) $\alpha = \beta = 1, \gamma = 1$.

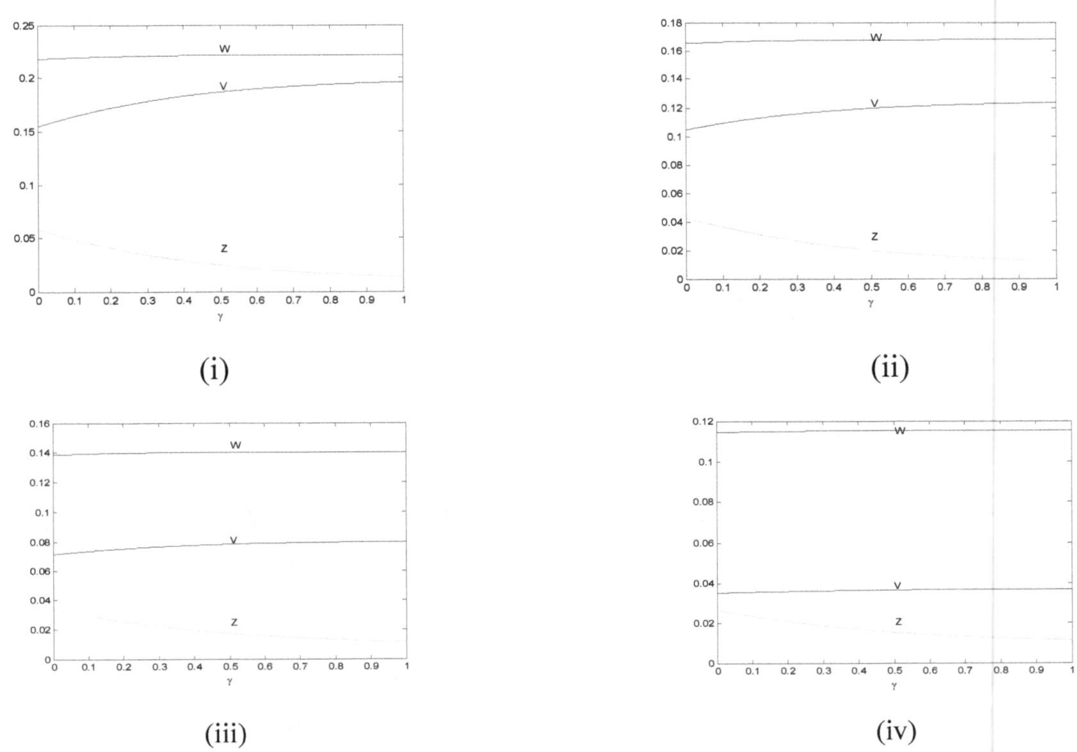

Fig. 3 Fractional order HPM solutions for $V(t), W(t), Z(t)$ at $t = 0.1$ and $\gamma \in [0,1]$ (i) $\alpha = \beta = 1/3$, (ii) $\alpha = \beta = 1/2$,

(iii) $\alpha = \beta = 2/3$, (iv) $\alpha = \beta = 1$.

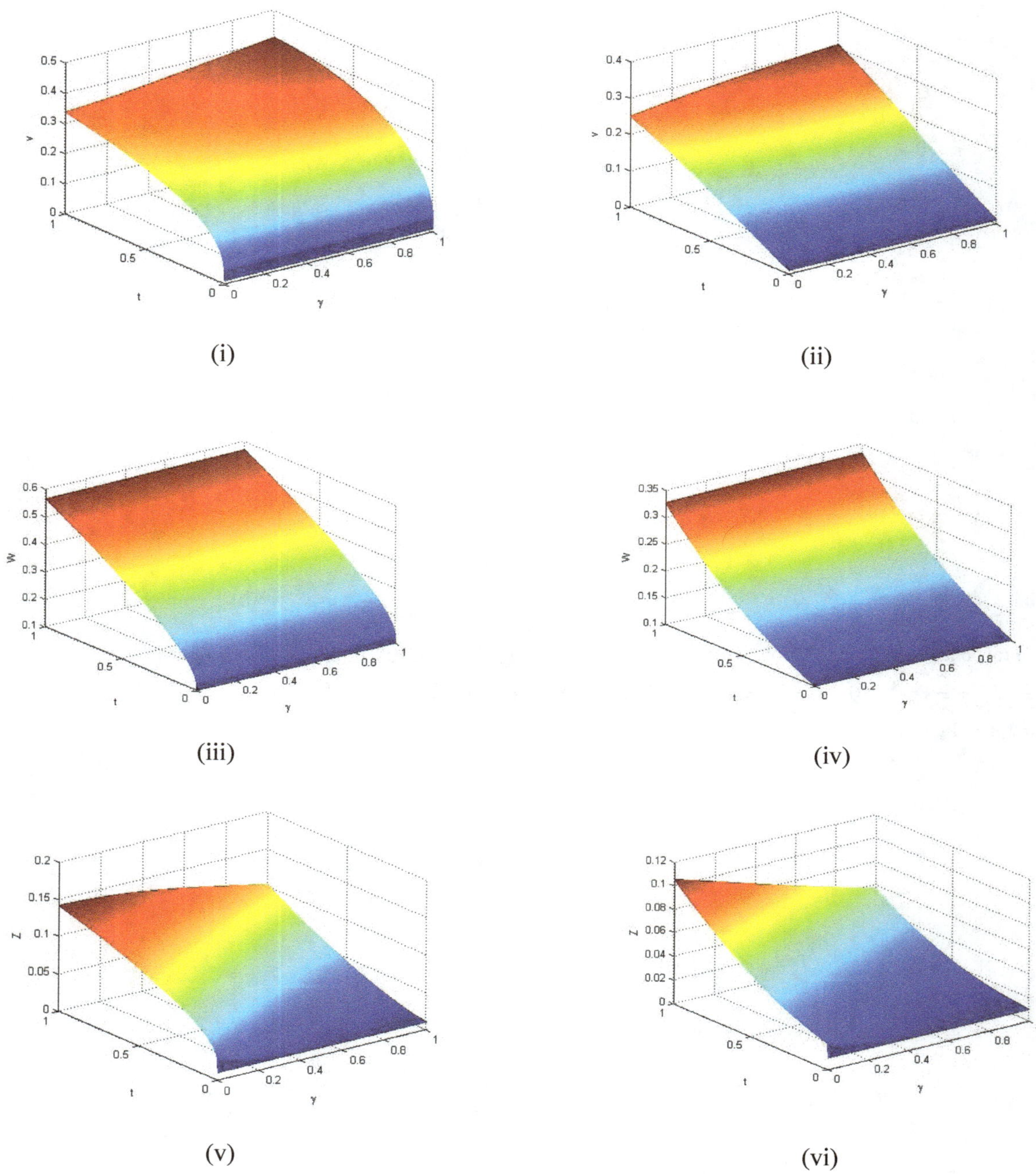

Fig. 4 Fractional order HPM solutions. $V(t)$ for $\gamma \in [0,1]$, $t \in [0,1]$ and (i) $\alpha = \beta = 1/3$, (ii) $\alpha = \beta = 1$. $W(t)$ for $\gamma \in [0,1]$, $t \in [0,1]$ and (iii) $\alpha = \beta = 1/3$, (iv) $\alpha = \beta = 1$. $Z(t)$ for $\gamma \in [0,1]$, $t \in [0,1]$ and (v) $\alpha = \beta = 1/3$, (vi) $\alpha = \beta = 1$.

It is observed from Figs. 1 and 2, that the inhibitory and excitatory neurons are classified on the nature of chemical substances that are released in the synaptic cleft. The mean membrane potential of excitatory and inhibitory neurons increase but the rate of change is different. In Fig. 3, the variations of $V(t), W(t), Z(t)$, with respect to γ (fractional order of (1c)) are plotted. It is observed that $W(t)$ does not change with γ, whereas $V(t)$ increases and $Z(t)$ decreases for $\alpha=\beta=1/3,1/2$. However, for $\alpha=\beta=2/3$ and 1, $V(t)$ and $W(t)$ behave similarly, which remain fixed in the whole range of γ whereas in all the cases, decrease in $Z(t)$ is reported. In Fig. 4, rectangular patterns are observed showing temporal variations of $V(t), W(t), Z(t)$, with respect to γ. In all cases, we observe that $V(t)$ dominates over $W(t)$ and $Z(t)$ and the density of $Z(t)$ is clubbed at left top corner points and it is well below 0.15 mV. However, $W(t)$ is uniformly distributed horizontally and inclined towards the left side in the domain and its value is lying between 0.31 and 0.58. $V(t)$ is concentrated on the upper top corner points and it varies from $0.25mV$ to 0.4 mV.

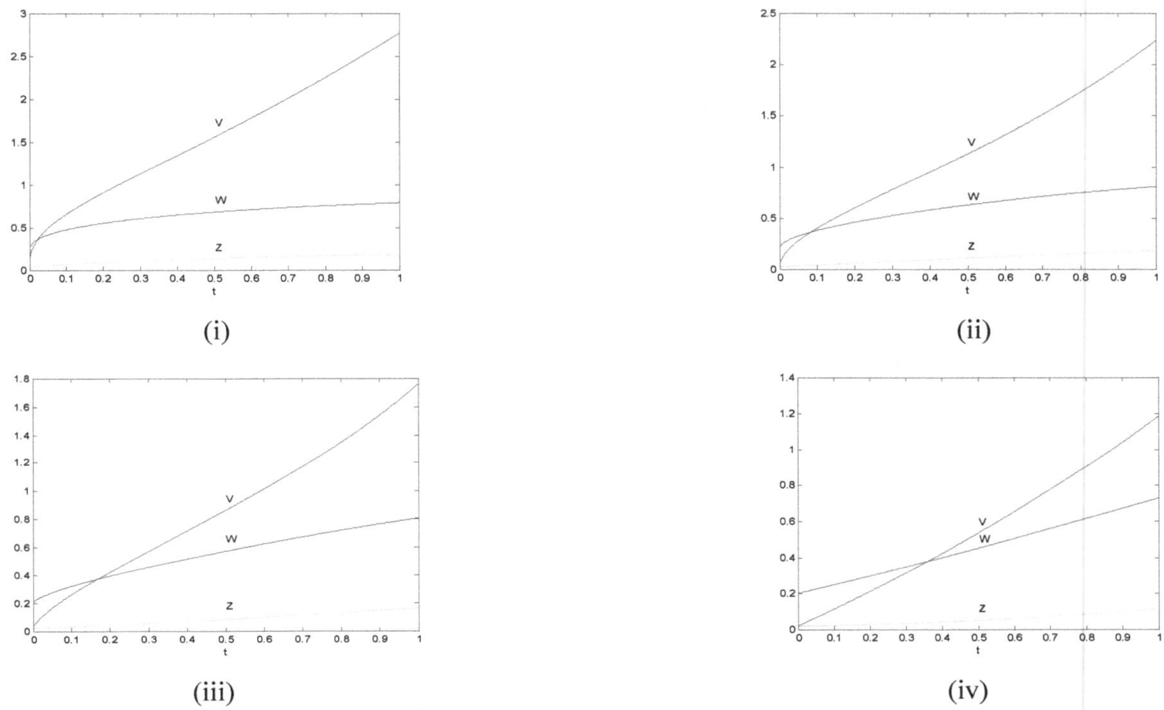

Fig. 5 Fractional order VIM solutions for $V(t), W(t), Z(t)$ for $t\in[0,1]$ and (i) $\alpha=\beta=\gamma=1/3$, (ii) $\alpha=\beta=\gamma=1/2$, (iii) $\alpha=\beta=\gamma=2/3$, (iv) $\alpha=\beta=\gamma=1$.

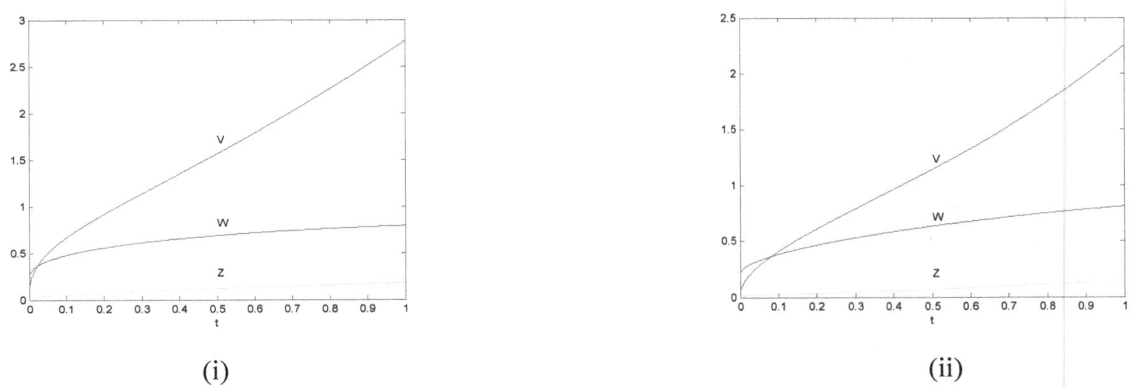

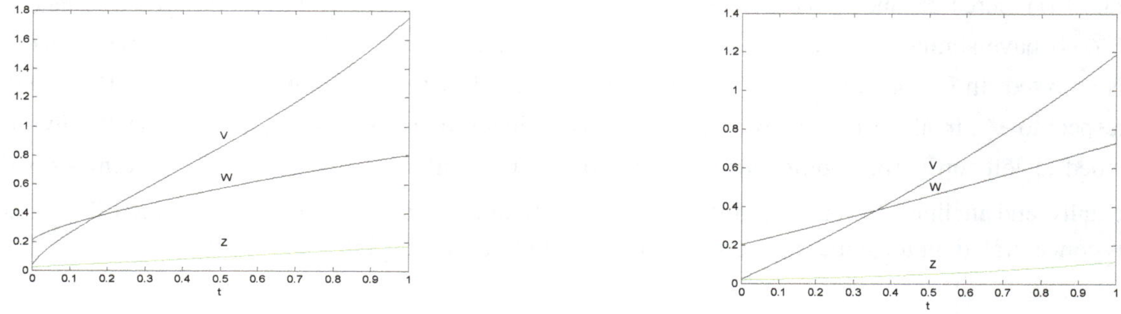

(iii) (iv)

Fig. 6 Fractional order VIM solutions for $V(t), W(t), Z(t)$ for $t \in [0,1]$ and (i) $\alpha = \beta = 1/3, \gamma = 0.5$, (ii) $\alpha = \beta = 1/2, \gamma = 1$, (iii) $\alpha = \beta = 2/3, \gamma = 0.5$, (iv) $\alpha = \beta = 1, \gamma = 1$.

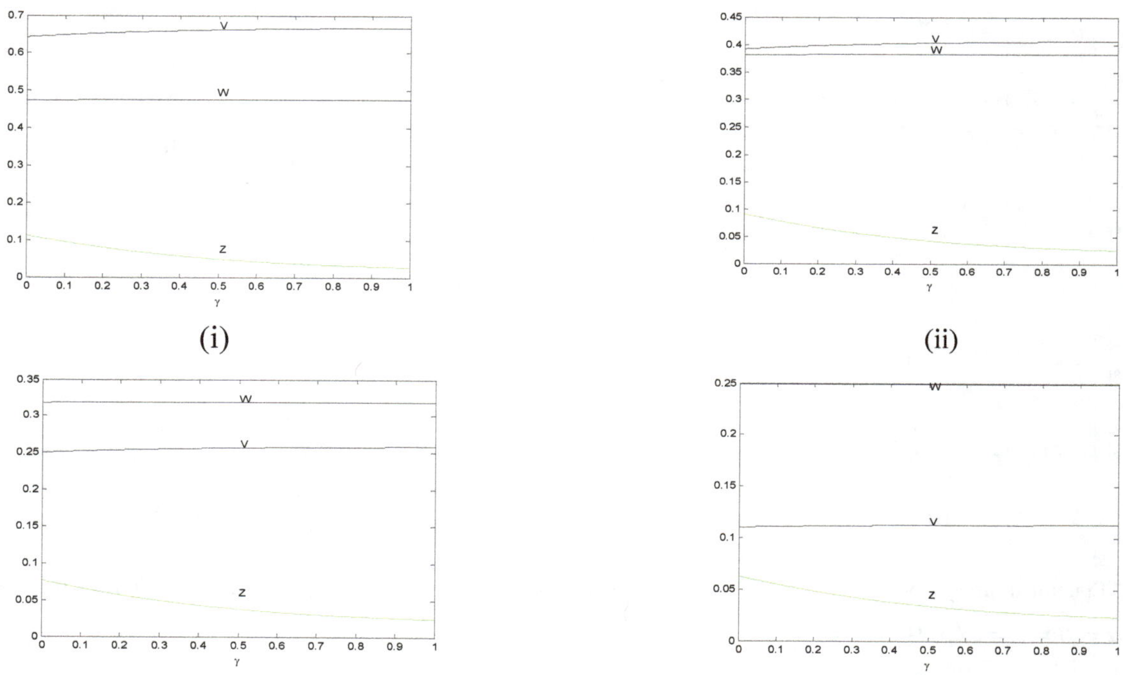

(i) (ii)

(iii) (iv)

Fig. 7 Fractional order VIM solutions for $V(t), W(t), Z(t)$ at $t = 0.1$ and $\gamma \in [0,1]$ (i) $\alpha = \beta = 1/3$, (ii) $\alpha = \beta = 1/2$, (iii) $\alpha = \beta = 2/3$, (iv) $\alpha = \beta = 1$.

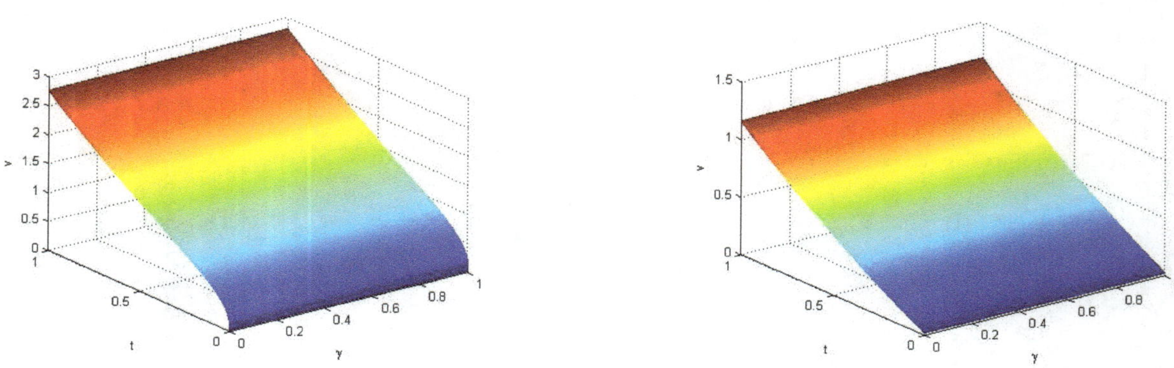

(i) (ii)

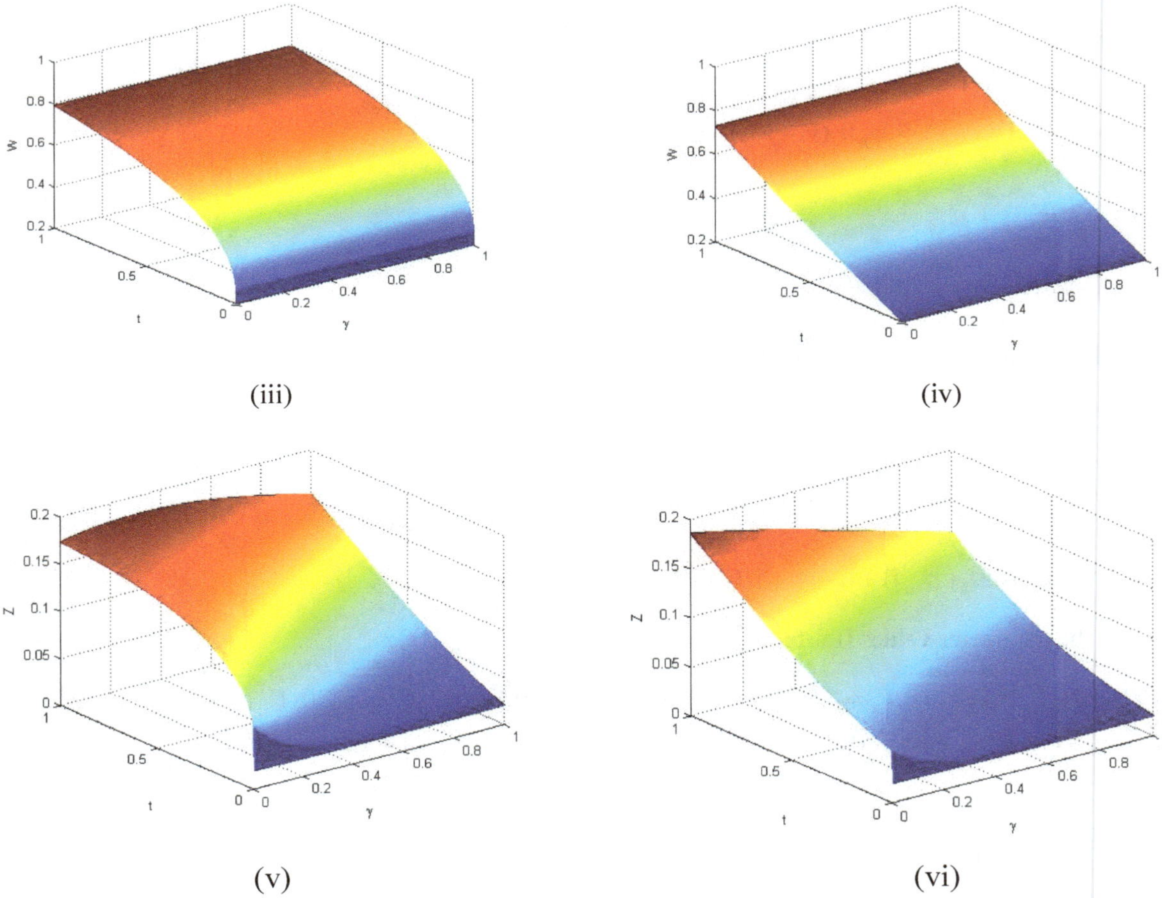

Fig. 8 Fractional order VIM solutions. $V(t)$ for $\gamma \in [0,1]$, $t \in [0,1]$ and (i) $\alpha = \beta = 1/3$, (ii) $\alpha = \beta = 1$. $W(t)$ for $\gamma \in [0,1]$, $t \in [0,1]$ and (iii) $\alpha = \beta = 1/3$, (iv) $\alpha = \beta = 1$. $Z(t)$ for $\gamma \in [0,1]$, $t \in [0,1]$ and (v) $\alpha = \beta = 1/3$, (vi) $\alpha = \beta = 1$.

In VIM solution, we have taken the initial condition as $V(t) = 0.01, W(t) = 0.1$ and $Z(t) = 0.01$ at $t = 0$. It is observed from Figs. 5 and 6 that the mean membrane potential of excitatory and inhibitory neurons, i.e., $V(t)$ and $Z(t)$ increases but the rate of change is different. With the change in mean membrane potential of excitatory cell there is a corresponding change in the opening of potassium channels. The rate of increase of $Z(t)$ and $V(t)$ are different whereas one is very slow and other one is very fast. In Fig. 7, we have plotted the variation of $V(t), W(t)$ and $Z(t)$ with respect to γ. We observe that $W(t)$ and $V(t)$ do not change with γ, whereas $Z(t)$ decreases for all values of α and β. In Fig. 8, rectangular patterns are observed showing the temporal variation of $V(t), W(t)$ and $Z(t)$ with respect to γ. From these figures, we observe that $V(t)$ is uniformly distributed horizontally, $W(t)$ is also uniformly distributed and inclined towards left side and its value is lying between 0.6 to 0.8. The density of $Z(t)$ is clubbed at left top corner points which is well below $0.2mV$.

Now, we present an example for a single neuron as a representative of a subnetwork model for epileptic seizures and apply the homotopy perturbation method.

Example Consider the following homotopy of the model system (7) for a single neuron as a representative of a subnetwork model for epileptic seizures (Larter et al., 1999). We solve the problem using HPM.

$$D_t^{\alpha}V(t) = -g_{Ca}m_{\infty}(V)(V-1) - g_K W(V-V^K) - 0.5(V-V^L) + I - \alpha_{inh}(Z)Z, \tag{17a}$$

$$D_t^{\beta}W(t) = 0.7\,(W_{\infty} - W)/\tau_W\,, \tag{17b}$$

$$D_t^{\beta}Z(t) = 0.1(0.165I + \alpha_{exc}(V)V). \tag{17c}$$

Applying HPM to the problem, we obtain the system as:

$$D_t^{\alpha}V(t) = p\{-0.5g_{Ca}[1+((V-V_1)/V_2)](V-1) - g_K W(V-V^K) - 0.5(V-V^L) + I - \alpha_{inh}[1+((Z-V_7)/V_6)]Z\}, \tag{18a}$$

$$D_t^{\beta}W(t) = p\{0.7(0.5[1+((V-V_3)/V_4)] - W)\}, \tag{18b}$$

$$D_t^{\gamma}Z(t) = p\{0.1(0.165I + \alpha_{exc}[1+((V-V_5)/V_6)]V)\}. \tag{18c}$$

Now, using the parameter values (Larter et al., 1999)

$V_1 = -0.01, V_2 = 0.15, V_3 = 0.0, V_4 = 0.3, V_5 = 0.0, V_6 = 0.6, V_7 = 0.0, g_{Ca} = 1.1, g_K = 2.0, V^K = -0.7, V^L = -0.5,$
$I = 0.3,\ \alpha_{exc} = 1$ and $\alpha_{inh} = 1,$

the above system reduces to

$$D_t^{\alpha}V(t) = p\{-(3.6667V + 0.5867)(V-1) - 2W(V+0.7) - 0.5V + 0.05 - (1+1.6667Z)Z\}, \tag{19a}$$

$$D_t^{\beta}W(t) = p\{0.35 + 1.1655V - 0.7W\}, \tag{19b}$$

$$D_t^{\gamma}Z(t) = p\{0.005 + 0.1V + 0.1667V^2\}. \tag{19c}$$

Assume the initial conditions as $V(0) = 0.01, W(0) = 0.1, Z(0) = 0.01$ (Larter et al., 1999; Nadar and Rai, 2012; Rai et al., 2013) and take $V_0' = a = 0.01, W_0' = b = 0.1, Z_0' = c = 0.01$. Solving the system (19a)-(19c) and equating the coefficients of powers of p, we have

$$p^0 : D_t^{\alpha}V_0' = 0,\ D_t^{\beta}W_0' = 0,\ D_t^{\gamma}Z_0' = 0.$$

$$p^1 : D_t^{\alpha}V_1' = X(a,b,c),\ D_t^{\beta}W_1' = Y(a,b),\ D_t^{\gamma}Z_1' = U(a).$$

$$p^2 : D_t^{\alpha}V_2' = V_1'D(a,b) - 2W_1'(a+0.7) - (3.3334c+1)Z_1',\ D_t^{\beta}W_2' = 1.1655V_1' - 0.7W_1',$$

$$D_t^{\gamma}Z_2' = (0.1 + 0.3334a)V_1'.$$

$$p^3 :\ D_t^{\alpha}V_3' = D(a,b)V_2' - 3.66675V_1'^2 - 2(a+0.7)W_2' - 2V_1'W_1' - Z_2'(3.3334c+1) - 1.6667Z_1'^2,$$

$$D_t^{\beta} W_3' = 1.1655 V_2' - 0.7 W_2',$$

$$D_t^{\gamma} Z_3' = (0.1 + 0.3334a) V_2' + 0.1667 V_1'^2.$$

Solving the above equations, we obtain

$$V_1' = X(a,b,c)\frac{t^{\alpha}}{\Gamma(\alpha+1)},\quad W_1' = Y(a,b)\frac{t^{\beta}}{\Gamma(\beta+1)},\quad Z_1' = U(a)\frac{t^{\gamma}}{\Gamma(\gamma+1)},$$

$$V_2' = D(a,b)X(a,b,c)\frac{t^{2\alpha}}{\Gamma(2\alpha+1)} - 2Y(a,b)(a+0.7)\frac{t^{(\alpha+\beta)}}{\Gamma(\alpha+\beta+1)} - (1+3.3334c)U(a)\frac{t^{(\alpha+\gamma)}}{\Gamma(\alpha+\gamma+1)},$$

$$W_2' = 1.1655X(a,b,c)\frac{t^{(\alpha+\beta)}}{\Gamma(\alpha+\beta+1)} - 0.7Y(a,b)\frac{t^{2\beta}}{\Gamma(2\beta+1)},$$

$$Z_2' = (0.1+0.3334a)X(a,b,c)\frac{t^{(\alpha+\gamma)}}{\Gamma(\alpha+\gamma+1)},$$

$$\begin{aligned}V_3' = {} & D(a,b)(D(a,b)X(a,b,c)\frac{t^{3\alpha}}{\Gamma(3\alpha+1)} - 2Y(a,b)(a+0.7)\frac{t^{(2\alpha+\beta)}}{\Gamma(2\alpha+\beta+1)}) \\ & - D(a,b)(1+3.3334c)U(a)\frac{t^{(2\alpha+\gamma)}}{\Gamma(2\alpha+\gamma+1)} - 3.6667(X(a,b,c))^2\frac{t^{3\alpha}\Gamma(2\alpha+1)}{(\Gamma(\alpha+1))^2\Gamma(3\alpha+1)} \\ & -1.1655(2a+1.4)X(a,b,c)\frac{t^{(2\alpha+\beta)}}{\Gamma(2\alpha+\beta+1)} - 0.7(2a+1.4)Y(a,b)\frac{t^{\alpha+2\beta}}{\Gamma(\alpha+2\beta+1)} \\ & -2X(a,b,c)Y(a,b)\frac{t^{(2\alpha+\beta)}\Gamma(\alpha+\beta+1)}{\Gamma(\alpha+1)\Gamma(\beta+1)\Gamma(2\alpha+\beta+1)} - (1+3.3334c)(0.1+0.3334a)X(a,b,c) \\ & \frac{t^{2\alpha+\gamma}}{\Gamma(2\alpha+\gamma+1)} - 1.6667(U(a))^2\frac{t^{\alpha+2\gamma}\Gamma(2\gamma+1)}{(\Gamma(\gamma+1))^2\Gamma(\alpha+2\gamma+1)},\end{aligned}$$

$$\begin{aligned}W_3' = {} & 1.1655(D(a,b)X(a,b,c)\frac{t^{2\alpha+\beta}}{\Gamma(2\alpha+\beta+1)} - 2Y(a,b)(a+0.7)\frac{t^{(\alpha+2\beta)}}{\Gamma(\alpha+2\beta+1)}) \\ & -1.1655(1+3.3334c)U(a)\frac{t^{(\alpha+\beta+\gamma)}}{\Gamma(\alpha+\beta+\gamma+1)} - 0.7(1.1655X(a,b,c)\frac{t^{(\alpha+2\beta)}}{\Gamma(\alpha+2\beta+1)}) + 0.49Y(a,b)\frac{t^{3\beta}}{\Gamma(3\beta+1)},\end{aligned}$$

$$\begin{aligned}Z_3' = {} & (0.1+0.3334a)(D(a,b)X(a,b,c)\frac{t^{2\alpha+\gamma}}{\Gamma(2\alpha+\gamma+1)} - 2Y(a,b)(a+0.7)\frac{t^{(\alpha+\beta+\gamma)}}{\Gamma(\alpha+\beta+\gamma+1)}) \\ & -(0.1+0.3334a)(1+3.3334c)U(a)\frac{t^{(\alpha+2\gamma)}}{\Gamma(\alpha+2\gamma+1)} + 0.1667(X(a,b,c))^2\frac{t^{2\alpha+\gamma}\Gamma(2\alpha+1)}{(\Gamma(\alpha+1))^2\Gamma(2\alpha+\gamma+1)},\end{aligned}$$

where

$$X(a,b,c) = (-3.6667a^2 + 2.58a - 2ab - 1.4b + 0.6367 - c - 1.6667c^2),\; Y(a,b) = (0.35 + 1.1655a - 0.7b),$$
$$U(a) = (0.005 + 0.1a + 0.1667a^2),\; D(a,b) = (-7.3334a + 2.58 - 2b).$$

Similarly, rest of the components can be obtained. The solutions for $V(t), W(t)$ and $Z(t)$ are then obtained.

We have presented the numerical solution of the Example using HPM in Table 1, for different values of fractional orders α, β, γ for $t \in [0.2, 1]$ with step size $t = 0.2$.

Similarly, we can apply the VIM method and obtain the solutions.

Table 1

$V(t)$				
	$\alpha=\beta=\gamma=1/3$	$\alpha=\beta=\gamma=2/3$	$\alpha=\beta=\gamma=1/2$	$\alpha=\beta=\gamma=1$
$t=0.2$	0.57412	0.263602	0.387733	0.12529
$t=0.4$	0.772912	0.448387	0.589202	0.259881
$t=0.6$	0.918664	0.608484	0.749258	0.402839
$t=0.8$	1.03684	0.746685	0.883042	0.5432225
$t=1$	1.13729	0.863479	0.997026	0.670105
$W(t)$				
	$\alpha=\beta=\gamma=1/3$	$\alpha=\beta=\gamma=2/3$	$\alpha=\beta=\gamma=1/2$	$\alpha=\beta=\gamma=1$
$t=0.2$	0.560412	0.26098	0.366121	0.166946
$t=0.4$	0.818475	0.420615	0.579914	0.254363
$t=0.6$	1.04772	0.605105	0.801581	0.367116
$t=0.8$	1.26203	0.816342	1.03374	0.510066
$t=1$	1.46678	1.05477	1.27662	0.688079
$Z(t)$				
	$\alpha=\beta=\gamma=1/3$	$\alpha=\beta=\gamma=2/3$	$\alpha=\beta=\gamma=1/2$	$\alpha=\beta=\gamma=1$
$t=0.2$	0.0810816	0.0297947	0.0486777	0.014746
$t=0.4$	0.12201	0.0583432	0.0862323	0.0268083
$t=0.6$	0.15638	0.0923855	0.123921	0.0465328
$t=0.8$	0.18711	0.1308	0.161803	0.0742658
$t=1$	0.215403	0.172966	0.199887	0.110353

5 Discussions and Conclusions

In this paper, we have studied a fractional order modified M-L neural model modeling the dynamics of a single neuron that interacts with other neurons through on-off excitatory and inhibitory synapses in a neural system. We have presented the solutions of a modified M-L neural model of fractional-order time derivatives with the help of two different analytical methods HPM, VIM. We observe that the methods are efficient in finding the exact as well as approximate solutions. They provide series solutions that converge very rapidly in real physical problems (Momani and Odibat, 2007). We have tested the analytical results by solving it numerically. It is observed that the inhibitory and excitatory neurons are classified on the nature of chemical substances that are released in the synaptic cleft. With the change in mean membrane potential of excitatory cell there is a corresponding opening of potassium channels. The mean membrane potential of excitatory and inhibitory neurons increases but the rate of change is different (see Figs. 1 and 2) for both HPM and VIM. Fractional differentiation is used to investigate the behavior of neural systems (Magin, 2004). Even processing of external applied stimuli by individual neurons can be expressed by fractional differentiation (Lundstrom et al., 2008). The fractional-order modified M-L neural model may help in understanding the synchronous neural activities occurring during a seizure, stimulus anticipation and in frequency independent phase shifts of oscillatory neuronal firing.

References

Abdulaziz O, Hashim I, Momani S. 2008. Solving systems of fractional differential equations by homotopy perturbation method. Physics Letters A, 372: 451-459

Adomian G. 1998. A review of the decomposition method in applied mathematics. Journal of Mathematical Analysis and Applications, 135: 501–544

Alexander GD, Aurel AL, Jonethan DV. 2011. Information theory in neuroscience. Journal of Computational Neuroscience, 30: 1-5

Anastasio TJ. 1994. The fractional-order dynamics of brainstem vestibulo-oculomotor neurons. Biological Cybernetics, 72: 69-79

Anastasio TJ.1998. Nonuniformity in the linear network model of the oculomotor integrator produces approximately fractional-order dynamics and more realistic neuron behavior. Biological Cybernetics, 79: 377-391

Chakraverty S, Tapaswini S. 2013. Numerical Solution of Fuzzy Arbitrary order Predator Prey Equations. Applications and Applied Mathematics, 8(2): 647-672

Destexhe A. 1994. Oscillations, complex spatiotemporal bahavior, and information transport in nerworks of excitatory and inhibitory neurons. Physical Review E, 50: 1594-1606

Fairhall AL, Lewen GD, Bialek W. de Ruyter van Steveninck RR. 2001. Advances in Neural Information Processing Systems 13. (eds. Leen, T.K., Dietterich, T.G., Tresp, V.)124–130 (MIT Press, Cambridge, Massachusetts)

FitzHugh R. 1955. Mathematical models of threshold phenomena in the nerve membrane. Bulletin of Mathematical Biology, 17: 257–278

Freeman WJ. 1994. Neural networks and chaos. Journal of Theoretical Biology, 171: 13-18

Fusi S, Drew PJ, Abbott LF. 2005. Cascade models of synaptically stored memories. Neuron, 45: 599–611

Gutkin BS, Ermentrout GB. 1998. Dynamics of memberane excitability determine interspike interval variability: a link between spike generation mechanisms and cortical spike train statics. Neural Computation, 10: 1047-1065

He JH. 1999. Homotopy perturbation technique. Computer Methods in Applied Mechanics and Engineering, 178: 257-262

He JH.1999a. Variational iteration method - a kind of non-linear analytical technique: some examples. Nonlinear Mechanics, 34: 699-708

He JH. 2000. A coupling method of a homotopy technique and a perturbation technique for nonlinear problems. International Journal of Nonlinear Mechanics, 35: 37-43

He JH. 2005. A new perturbation technique which is also valid for large parameters. Chaos, Solitons and Fractals, 26: 827-833

He JH. 2005a. Application of homotopy perturbation method to nonlinear wave equations. Chaos, Solitons and Fractals, 26: 695-700

Hindmarsh JL, Rose RM. 1984. A model of neuronal bursting using three coupled first order differential equations. Philosophical Transactions of the Royal Society B. Biological Sciences, 221:87-102

Hodgkin AL, Huxley AF. 1952. A quantitative description of membrane current and its application to conduction and excitation in nerve. Journal of Physiology, 117: 500-544

Hodgkin AL, Rushton WAH. 1946. The electrical constants of a crustacean nerve fiber. Proceedings of the Royal Society B, 133: 444-479

Izhikevich EM. 2000. Neural excitability, spiking and bursting. International Journal of Bifurcation and Chaos, 10(6): 1171-1266

Larter R, Speelman B, Worth RM. 1999. A coupled ordinary differential equation lattice model for the simulation of epileptic seizures. Chaos, 9(3): 795-804

Lundstrom BN, Higgs MH, Spain WJ, Fairhall AL. 2008. Fractional differentiation by neocortical pyramidal neurons. Nature Neuroscience, 11(11): 1335-1342

Magin RL. 2004. Fractional calculus in bioengineering. Critical Reviews in Biomedical Engineering, 32(1): 1-104

Miller KS, Ross B. 1993. An introduction to the fractional calculus and fractional differential equations. John Wiley & Sons, USA

Momani S, Odibat Z. 2007. Homotopy perturbation method for nonlinear partial differential equations of fractional order. Physics Letters A, 365: 345-350

Momani S, Odibat Z. 2007. Comparison between the homotopy perturbation method and the variational iteration method for linear fractional partial differential equations. Computers and Mathematics with Applications, 54: 910–919

Morris C, Lecar H. 1981. Voltage oscillations in the barnacle giant muscle fiber. Journal of Biophys., 35: 193-213

Muhammad A, Keum-Shik H, Myung-Yung J. 2012. Synchronization of coupled chaotic FitzHugh–Nagumo systems. Communications in Nonlinear Science and Numerical Simulation, 17(4): 1615-1627

Nadar SR, Rai V. 2012. Transient Periodicity in a Morris Lecar Neural System. ISRN Biomathematics, 2012: 1-7

Odibat Z, Momani S. 2008. Applications of variational iteration and homotopy perturbation methods to fractional evolution equations. Topol. Meth. Nonlin. Anal., 31: 227-234

Paulin MG, Hoffman LF, Assad C. 2004. Dynamics and the single spike. IEEE Transactions on Neural Networks, 15: 987-994

Podlubny I. 1999. Fractional Differential Equations. Academic Press NY, New York, USA

Rai V, Nadar SR, Khan RA. 2013. Phase-Coupled Oscillations in the Brain: Nonlinear Phenomena in Cellular Signalling. ISRN Biomathematics, 2013: 1-7

Shi M, Wang Z. 2014. Abundant bursting patterns of a fractional order Morris-Lecar neuron model. Communications in Nonlinear Science and Numerical Simulation, 19: 1956–1969

Speelman B. 1997. A dynamical systems approach to the modeling of epileptic seizures. Ph. D thesis, Indiana University Purdue University at Indianapolis, USA

Tateno T, Pakdaman K. 2004. Random dynamics of the Morris-Lecar neural model. Chaos, 14(3): 511-530

Thorson J, Biederman-Thorson M. 1974. Distributed relaxation processes in sensory adaptation. Science, 183: 161-172

Westerlund S, Ekstam L. 1994. Capacitor theory. IEEE Transactions on Dielectrics and Electrical Insulation, 1(8): 26-39

Wilson HR, Cowan JD. 1972. Excitatory and inhibitory interaction in localized populations of model neurons. Journal of Physiology, 12: 1-24

Xie Y, Xu JX, Hu SJ. 2004. A novel dynamical mechanism of neural excitability for integer multiple spiking. Chaos, Solitons and Fractals, 21: 177-184

Appendix A1

Detailed expressions for V_3', W_3' and Z_3' obtained by the VIM are the following:

$$\begin{aligned}
V_3' = {} & 2.0817H\frac{t^{2\alpha}}{\Gamma(2\alpha+1)} + 2.0817K\frac{t^{3\alpha}}{\Gamma(3\alpha+1)} - 7.6367A^2\frac{t^{4\alpha}\Gamma(2\alpha+1)}{(\Gamma(\alpha+1))^2\Gamma(4\alpha+1)} \\
& + 2.0817J\frac{t^{2\alpha+\beta}}{\Gamma(2\alpha+\beta+1)} - 4.1634AB\frac{t^{3\alpha+\beta}\Gamma(\alpha+\beta+1)}{\Gamma(\alpha+1)\Gamma(\beta+1)\Gamma(3\alpha+\beta+1)} + 2.0817P\frac{t^{2\alpha+\gamma}}{\Gamma(2\alpha+\gamma+1)} \\
& -5.2043BG\frac{t^{3\alpha+\gamma}\Gamma(\alpha+\gamma+1)}{\Gamma(\alpha+1)\Gamma(\gamma+1)\Gamma(3\alpha+\gamma+1)} + 0.38685\frac{t^{\alpha}}{\Gamma(\alpha+1)} - 3.6685H^2\frac{t^{3\alpha}\Gamma(2\alpha+1)}{(\Gamma(\alpha+1))^2\Gamma(3\alpha+1)} \\
& -3.6685K^2\frac{t^{5\alpha}\Gamma(4\alpha+1)}{(\Gamma(2\alpha+1))^2\Gamma(5\alpha+1)} - 3.6685^3A^4\frac{t^{7\alpha}(\Gamma(2\alpha+1))^2\Gamma(6\alpha+1)}{(\Gamma(\alpha+1))^4(\Gamma(3\alpha+1))^2\Gamma(7\alpha+1)} \\
& -7.337HK\frac{t^{4\alpha}\Gamma(3\alpha+1)}{\Gamma(\alpha+1)\Gamma(2\alpha+1)\Gamma(4\alpha+1)} + 26.9158A^2K\frac{t^{6\alpha}\Gamma(5\alpha+1)}{(\Gamma(\alpha+1))^2\Gamma(3\alpha+1)\Gamma(6\alpha+1)} \\
& +26.9158A^2H\frac{t^{5\alpha}\Gamma(2\alpha+1)\Gamma(4\alpha+1)}{(\Gamma(\alpha+1))^3\Gamma(3\alpha+1)\Gamma(5\alpha+1)} - 3.66855J^2\frac{t^{3\alpha+2\beta}\Gamma(2\alpha+2\beta+1)}{(\Gamma(\alpha+\beta+1))^2\Gamma(3\alpha+2\beta+1)} \\
& -14.674A^2B^2\frac{t^{5\alpha+2\beta}(\Gamma(\alpha+\beta+1))^2\Gamma(4\alpha+2\beta+1)}{(\Gamma(\alpha+1))^2(\Gamma(\beta+1))^2(\Gamma(2\alpha+\beta+1))^2\Gamma(5\alpha+2\beta+1)} \\
& +14.674ABJ\frac{t^{4\alpha+2\beta}\Gamma(3\alpha+2\beta+1)}{\Gamma(\alpha+1)\Gamma(\beta+1)\Gamma(2\alpha+\beta+1)\Gamma(4\alpha+2\beta+1)} - 3.6685P^2\frac{t^{3\alpha+2\gamma}\Gamma(2\alpha+2\gamma+1)}{(\Gamma(\alpha+\gamma+1))^2\Gamma(3\alpha+2\gamma+1)} \\
& -22.9281B^2G^2\frac{t^{5\alpha+2\gamma}\Gamma(4\alpha+2\gamma+1)(\Gamma(\alpha+\gamma+1))^2}{(\Gamma(\alpha+1))^2(\Gamma(\gamma+1))^2(\Gamma(2\alpha+\gamma+1))^2\Gamma(5\alpha+2\gamma+1)} \\
& +18.3425PBG\frac{t^{4\alpha+2\gamma}\Gamma(3\alpha+2\gamma+1)}{\Gamma(\alpha+1)\Gamma(\gamma+1)\Gamma(2\alpha+\gamma+1)\Gamma(4\alpha+2\gamma+1)} \\
& +36.685JBG\frac{t^{4\alpha+\beta+\gamma}\Gamma(\alpha+\gamma+1)\Gamma(3\alpha+\beta+\gamma+1)}{\Gamma(\alpha+\beta+1)\Gamma(\alpha+1)\Gamma(\gamma+1)\Gamma(2\alpha+\gamma+1)\Gamma(4\alpha+\beta+\gamma+1)} \\
& +14.674ABP\frac{t^{4\alpha+\beta+\gamma}\Gamma(3\alpha+\beta+\gamma+1)\Gamma(\alpha+\beta+1)}{\Gamma(\alpha+\gamma+1)\Gamma(\alpha+1)\Gamma(\beta+1)\Gamma(2\alpha+\beta+1)\Gamma(4\alpha+\beta+\gamma+1)} \\
& -36.685AB^2G\frac{t^{5\alpha+\beta+\gamma}\Gamma(\alpha+\gamma+1)\Gamma(\alpha+\beta+1)\Gamma(4\alpha+\beta+\gamma+1)}{(\Gamma(\alpha+1))^2\Gamma(2\alpha+\beta+1)\Gamma(\beta+1)\Gamma(\gamma+1)\Gamma(2\alpha+\gamma+1)\Gamma(5\alpha+\beta+\gamma+1)} \\
& -7.337HJ\frac{t^{3\alpha+\beta}\Gamma(2\alpha+\beta+1)}{\Gamma(\alpha+1)\Gamma(\alpha+\beta+1)\Gamma(3\alpha+\beta+1)} - 7.337KJ\frac{t^{4\alpha+\beta}\Gamma(3\alpha+\beta+1)}{\Gamma(2\alpha+1)\Gamma(\alpha+\beta+1)\Gamma(4\alpha+\beta+1)} \\
& +26.9158A^2J\frac{t^{5\alpha+\beta}\Gamma(4\alpha+\beta+1)\Gamma(2\alpha+1)}{(\Gamma(\alpha+1))^2\Gamma(\alpha+\beta+1)\Gamma(3\alpha+1)\Gamma(5\alpha+\beta+1)} \\
& +14.674ABH\frac{t^{4\alpha+\beta}\Gamma(\alpha+\beta+1)\Gamma(3\alpha+\beta+1)}{(\Gamma(\alpha+1))^2\Gamma(2\alpha+\beta+1)\Gamma(\beta+1)\Gamma(4\alpha+\beta+1)} \\
& +14.674ABK\frac{t^{5\alpha+\beta}\Gamma(4\alpha+\beta+1)\Gamma(\alpha+\beta+1)}{\Gamma(\alpha+1)\Gamma(\beta+1)\Gamma(2\alpha+1)\Gamma(2\alpha+\beta+1)\Gamma(5\alpha+\beta+1)} \\
& -53.8316A^3B\frac{t^{6\alpha+\beta}\Gamma(5\alpha+\beta+1)\Gamma(2\alpha+1)\Gamma(\alpha+\beta+1)}{(\Gamma(\alpha+1))^3\Gamma(3\alpha+1)\Gamma(\beta+1)\Gamma(2\alpha+\beta+1)\Gamma(6\alpha+\beta+1)}
\end{aligned}$$

$$-7.337HP\frac{t^{3\alpha+\gamma}\Gamma(2\alpha+\gamma+1)}{\Gamma(\alpha+1)\Gamma(\alpha+\gamma+1)\Gamma(3\alpha+\gamma+1)}-7.337KP\frac{t^{4\alpha+\gamma}\Gamma(3\alpha+\gamma+1)}{\Gamma(\alpha+\gamma+1)\Gamma(2\alpha+1)\Gamma(4\alpha+\gamma+1)}$$

$$+26.9158A^2P\frac{t^{5\alpha+\gamma}\Gamma(4\alpha+\gamma+1)\Gamma(2\alpha+1)}{(\Gamma(\alpha+1))^2\Gamma(3\alpha+1)\Gamma(\alpha+\gamma+1)\Gamma(5\alpha+\gamma+1)}$$

$$+18.3425HBG\frac{t^{4\alpha+\gamma}\Gamma(3\alpha+\gamma+1)\Gamma(\alpha+\gamma+1)}{(\Gamma(\alpha+1))^2\Gamma(\gamma+1)\Gamma(2\alpha+\gamma+1)\Gamma(4\alpha+\gamma+1)}$$

$$+18.3425KBG\frac{t^{5\alpha+\gamma}\Gamma(4\alpha+\gamma+1)\Gamma(\alpha+\gamma+1)}{\Gamma(2\alpha+1)\Gamma(\alpha+1)\Gamma(\gamma+1)\Gamma(2\alpha+\gamma+1)\Gamma(5\alpha+\gamma+1)}$$

$$-67.2895A^2BG\frac{t^{6\alpha+\gamma}\Gamma(2\alpha+1)\Gamma(\alpha+\gamma+1)\Gamma(5\alpha+\gamma+1)}{(\Gamma(\alpha+1))^3\Gamma(3\alpha+1)\Gamma(\gamma+1)\Gamma(2\alpha+\gamma+1)\Gamma(6\alpha+\gamma+1)}$$

$$-2LH\frac{t^{2\alpha+\beta}\Gamma(\alpha+\beta+1)}{\Gamma(\alpha+1)\Gamma(\beta+1)\Gamma(2\alpha+\beta+1)}-2KL\frac{t^{3\alpha+\beta}\Gamma(2\alpha+\beta+1)}{\Gamma(2\alpha+1)\Gamma(\beta+1)\Gamma(3\alpha+\beta+1)}$$

$$+7.337A^2L\frac{t^{4\alpha+\beta}\Gamma(3\alpha+\beta+1)\Gamma(2\alpha+1)}{(\Gamma(\alpha+1))^2\Gamma(3\alpha+1)\Gamma(\beta+1)\Gamma(4\alpha+\beta+1)}-2J\frac{t^{2\alpha+2\beta}\Gamma(\alpha+2\beta+1)}{\Gamma(\beta+1)\Gamma(\alpha+\beta+1)\Gamma(2\alpha+2\beta+1)}$$

$$+4ABL\frac{t^{3\alpha+2\beta}\Gamma(2\alpha+2\beta+1)\Gamma(\alpha+\beta+1)}{\Gamma(\alpha+1)(\Gamma(\beta+1))^2\Gamma(2\alpha+\beta+1)\Gamma(3\alpha+2\beta+1)}-2P\frac{t^{2\alpha+\beta+\gamma}\Gamma(\alpha+\beta+\gamma+1)}{\Gamma(\beta+1)\Gamma(\alpha+\gamma+1)\Gamma(2\alpha+\beta+\gamma+1)}$$

$$+0.132AH\frac{t^{3\alpha+\beta}\Gamma(2\alpha+\beta+1)}{\Gamma(\alpha+1)\Gamma(\alpha+\beta+1)\Gamma(3\alpha+\beta+1)}+0.132AK\frac{t^{4\alpha+\beta}\Gamma(3\alpha+\beta+1)}{\Gamma(2\alpha+1)\Gamma(\alpha+\beta+1)\Gamma(4\alpha+\beta+1)}$$

$$-0.4842A^3\frac{t^{5\alpha+\beta}\Gamma(4\alpha+\beta+1)\Gamma(2\alpha+1)}{(\Gamma(\alpha+1))^2\Gamma(3\alpha+1)\Gamma(\alpha+\beta+1)\Gamma(5\alpha+\beta+1)}+0.132AJ\frac{t^{3\alpha+2\beta}\Gamma(2\alpha+2\beta+1)}{(\Gamma(\alpha+\beta+1))^2\Gamma(3\alpha+2\beta+1)}$$

$$-0.264A^2B\frac{t^{4\alpha+2\beta}\Gamma(3\alpha+2\beta+1)}{\Gamma(\alpha+1)\Gamma(\beta+1)\Gamma(2\alpha+\beta+1)\Gamma(4\alpha+2\beta+1)}$$

$$+0.132AP\frac{t^{3\alpha+\beta+\gamma}\Gamma(2\alpha+\beta+\gamma+1)}{\Gamma(\alpha+\beta+1)\Gamma(\alpha+\gamma+1)\Gamma(3\alpha+\beta+\gamma+1)}$$

$$-0.33ABG\frac{t^{4\alpha+\beta+\gamma}\Gamma(3\alpha+\beta+\gamma+1)\Gamma(\alpha+\gamma+1)}{\Gamma(\alpha+1)\Gamma(\gamma+1)\Gamma(2\alpha+\gamma+1)\Gamma(\alpha+\beta+1)\Gamma(4\alpha+\beta+\gamma+1)}$$

$$-0.8BH\frac{t^{2\alpha+2\beta}\Gamma(\alpha+2\beta+1)}{\Gamma(\alpha+1)\Gamma(2\beta+1)\Gamma(2\alpha+2\beta+1)}-0.8BK\frac{t^{3\alpha+2\beta}\Gamma(2\alpha+2\beta+1)}{\Gamma(2\alpha+1)\Gamma(2\beta+1)\Gamma(3\alpha+2\beta+1)}$$

$$+2.9348A^2B\frac{t^{4\alpha+2\beta}\Gamma(3\alpha+2\beta+1)\Gamma(2\alpha+1)}{(\Gamma(\alpha+1))^2\Gamma(3\alpha+1)\Gamma(2\beta+1)\Gamma(4\alpha+2\beta+1)}$$

$$-0.8BJ\frac{t^{2\alpha+3\beta}\Gamma(\alpha+3\beta+1)}{\Gamma(\alpha+\beta+1)\Gamma(2\beta+1)\Gamma(2\alpha+3\beta+1)}$$

$$+1.6AB^2\frac{t^{3\alpha+3\beta}\Gamma(2\alpha+3\beta+1)\Gamma(\alpha+\beta+1)}{\Gamma(\alpha+1)\Gamma(\beta+1)\Gamma(2\beta+1)\Gamma(2\alpha+\beta+1)\Gamma(3\alpha+3\beta+1)}$$

$$-0.8BP\frac{t^{2\alpha+2\beta+\gamma}\Gamma(\alpha+2\beta+\gamma+1)}{\Gamma(\alpha+\gamma+1)\Gamma(2\beta+1)\Gamma(2\alpha+2\beta+\gamma+1)}$$

$$+2B^2G\frac{t^{3\alpha+2\beta+\gamma}\Gamma(\alpha+\gamma+1)\Gamma(2\alpha+2\beta+\gamma+1)}{\Gamma(\alpha+1)\Gamma(\gamma+1)\Gamma(2\beta+1)\Gamma(2\alpha+\gamma+1)\Gamma(3\alpha+2\beta+\gamma+1)}$$

$$-1.4(L\frac{t^{\alpha+\beta}}{\Gamma(\alpha+\beta+1)}-0.066A\frac{t^{2\alpha+\beta}}{\Gamma(2\alpha+\beta+1)}+0.4B\frac{t^{\alpha+2\beta}}{\Gamma(\alpha+2\beta+1)})$$

$$-0.875(M\frac{t^{\alpha+\gamma}}{\Gamma(\alpha+\gamma+1)}+N\frac{t^{2\alpha+\gamma}}{\Gamma(2\alpha+\gamma+1)}+0.375A^2\frac{t^{3\alpha+\gamma}\Gamma(2\alpha+1)}{(\Gamma(\alpha+1))^2\Gamma(3\alpha+\gamma+1)})$$

$$-2.5M^2\frac{t^{\alpha+2\gamma}\Gamma(2\gamma+1)}{(\Gamma(\gamma+1))^2\Gamma(\alpha+2\gamma+1)}-2.5N^2\frac{t^{3\alpha+2\gamma}\Gamma(2\alpha+2\gamma+1)}{(\Gamma(\alpha+\gamma+1))^2\Gamma(3\alpha+2\gamma+1)}$$
$$-0.3516A^4\frac{t^{5\alpha+2\gamma}\Gamma(4\alpha+2\gamma+1)(\Gamma(2\alpha+1))^2}{(\Gamma(\alpha+1))^4(\Gamma(2\alpha+\gamma+1))^2\Gamma(5\alpha+2\gamma+1)}-5MN\frac{t^{2\alpha+2\gamma}\Gamma(\alpha+2\gamma+1)}{\Gamma(\gamma+1)\Gamma(\alpha+\gamma+1)\Gamma(2\alpha+2\gamma+1)}$$
$$-1.875A^2N\frac{t^{4\alpha+2\gamma}\Gamma(3\alpha+2\gamma+1)\Gamma(2\alpha+1)}{\Gamma(\alpha+\gamma+1)(\Gamma(\alpha+1))^2\Gamma(2\alpha+\gamma+1)\Gamma(4\alpha+2\gamma+1)}$$
$$-1.875A^2M\frac{t^{3\alpha+2\gamma}\Gamma(2\alpha+1)\Gamma(2\alpha+2\gamma+1)}{\Gamma(\gamma+1)(\Gamma(\alpha+1))^2\Gamma(2\alpha+\gamma+1)\Gamma(3\alpha+2\gamma+1)},$$

$$W_3'=0.189\frac{t^{\beta}}{\Gamma(\beta+1)}+0.666H\frac{t^{\alpha+\beta}}{\Gamma(\alpha+\beta+1)}+0.666K\frac{t^{2\alpha+\beta}}{\Gamma(2\alpha+\beta+1)}$$
$$-2.4432A^2\frac{t^{3\alpha+\beta}\Gamma(2\alpha+1)}{(\Gamma(\alpha+1))^2\Gamma(3\alpha+\beta+1)}+0.666J\frac{t^{\alpha+2\beta}}{\Gamma(\alpha+2\beta+1)}-1.332AB\frac{t^{2\alpha+2\beta}\Gamma(\alpha+\beta+1)}{\Gamma(\alpha+1)\Gamma(\beta+1)\Gamma(2\alpha+2\beta+1)}$$
$$+0.666P\frac{t^{\alpha+\beta+\gamma}}{\Gamma(\alpha+\beta+\gamma+1)}-1.665BG\frac{t^{2\alpha+\beta+\gamma}\Gamma(\alpha+\gamma+1)}{\Gamma(\alpha+1)\Gamma(\gamma+1)\Gamma(2\alpha+\beta+\gamma+1)}$$
$$-0.4(L\frac{t^{2\beta}}{\Gamma(2\beta+1)}-0.066A\frac{t^{\alpha+2\beta}}{\Gamma(\alpha+2\beta+1)}+0.4B\frac{t^{3\beta}}{\Gamma(3\beta+1)}),$$

$$Z_3'=0.01071\frac{t^{\gamma}}{\Gamma(\gamma+1)}+0.15H\frac{t^{\alpha+\gamma}}{\Gamma(\alpha+\gamma+1)}+0.15K\frac{t^{2\alpha+\gamma}}{\Gamma(2\alpha+\gamma+1)}$$
$$-0.5503A^2\frac{t^{3\alpha+\gamma}\Gamma(2\alpha+1)}{(\Gamma(\alpha+1))^2\Gamma(3\alpha+\gamma+1)}+0.15J\frac{t^{\alpha+\beta+\gamma}}{\Gamma(\alpha+\beta+\gamma+1)}-0.3AB\frac{t^{2\alpha+\beta+\gamma}\Gamma(\alpha+\beta+1)}{\Gamma(\alpha+1)\Gamma(\beta+1)\Gamma(2\alpha+\beta+\gamma+1)}$$
$$+0.15P\frac{t^{\alpha+2\gamma}}{\Gamma(\alpha+2\gamma+1)}-0.375BG\frac{t^{2\alpha+2\gamma}\Gamma(\alpha+\gamma+1)}{\Gamma(\alpha+1)\Gamma(\gamma+1)\Gamma(2\alpha+2\gamma+1)}$$
$$+0.375H^2\frac{t^{2\alpha+\gamma}\Gamma(2\alpha+1)}{(\Gamma(\alpha+1))^2\Gamma(2\alpha+\gamma+1)}+0.375K^2\frac{t^{4\alpha+\gamma}\Gamma(4\alpha+1)}{(\Gamma(2\alpha+1))^2\Gamma(4\alpha+\gamma+1)}$$
$$+5.0467A^4\frac{t^{6\alpha+\gamma}\Gamma(6\alpha+1)(\Gamma(2\alpha+1))^2}{(\Gamma(\alpha+1))^4(\Gamma(3\alpha+1))^2\Gamma(6\alpha+\gamma+1)}+0.75HK\frac{t^{3\alpha+\gamma}\Gamma(3\alpha+1)}{\Gamma(\alpha+1)\Gamma(2\alpha+1)\Gamma(3\alpha+\gamma+1)}$$
$$-2.7514A^2K\frac{t^{5\alpha+\gamma}\Gamma(5\alpha+1)}{(\Gamma(\alpha+1))^2\Gamma(3\alpha+1)\Gamma(5\alpha+\gamma+1)}$$
$$-2.7514A^2H\frac{t^{4\alpha+\gamma}\Gamma(2\alpha+1)\Gamma(4\alpha+1)}{(\Gamma(\alpha+1))^3\Gamma(3\alpha+1)\Gamma(4\alpha+\gamma+1)}+0.375J^2\frac{t^{2\alpha+2\beta+\gamma}\Gamma(2\alpha+2\beta+1)}{(\Gamma(\alpha+\beta+1))^2\Gamma(2\alpha+2\beta+\gamma+1)}$$
$$+1.5A^2B^2\frac{t^{4\alpha+2\beta+\gamma}\Gamma(4\alpha+2\beta+1)(\Gamma(\alpha+\beta+1))^2}{(\Gamma(\alpha+1))^2(\Gamma(\beta+1))^2(\Gamma(2\alpha+\beta+1))^2\Gamma(4\alpha+2\beta+\gamma+1)}$$
$$-1.5ABJ\frac{t^{3\alpha+2\beta+\gamma}\Gamma(3\alpha+2\beta+1)}{\Gamma(\alpha+1)\Gamma(\beta+1)\Gamma(2\alpha+\beta+1)\Gamma(3\alpha+2\beta+\gamma+1)}+0.375P^2\frac{t^{2\alpha+3\gamma}\Gamma(2\alpha+2\gamma+1)}{(\Gamma(\alpha+\gamma+1))^2\Gamma(2\alpha+3\gamma+1)}$$
$$+2.3438B^2G^2\frac{t^{4\alpha+3\gamma}\Gamma(4\alpha+2\gamma+1)(\Gamma(\alpha+\gamma+1))^2}{(\Gamma(\alpha+1))^2(\Gamma(\gamma+1))^2(\Gamma(2\alpha+\gamma+1))^2\Gamma(4\alpha+3\gamma+1)}$$
$$-1.875BGP\frac{t^{3\alpha+3\gamma}\Gamma(3\alpha+2\gamma+1)}{\Gamma(\alpha+1)\Gamma(\gamma+1)\Gamma(2\alpha+\gamma+1)}+0.75JP\frac{t^{2\alpha+\beta+2\gamma}\Gamma(2\alpha+\beta+\gamma+1)}{\Gamma(\alpha+\beta+1)\Gamma(\alpha+\gamma+1)\Gamma(2\alpha+\beta+2\gamma+1)}$$
$$-1.5ABP\frac{t^{3\alpha+\beta+2\gamma}\Gamma(3\alpha+\beta+\gamma+1)\Gamma(\alpha+\beta+1)}{\Gamma(\alpha+\gamma+1)\Gamma(\alpha+1)\Gamma(\beta+1)\Gamma(2\alpha+\beta+1)\Gamma(3\alpha+\beta+2\gamma+1)}$$
$$-1.875BGJ\frac{t^{3\alpha+\beta+2\gamma}\Gamma(3\alpha+\beta+\gamma+1)\Gamma(\alpha+\gamma+1)}{\Gamma(\alpha+\beta+1)\Gamma(\alpha+1)\Gamma(\gamma+1)\Gamma(2\alpha+\gamma+1)\Gamma(3\alpha+\beta+2\gamma+1)}$$

$$+1.875AB^2G\frac{t^{4\alpha+\beta+2\gamma}\Gamma(4\alpha+\beta+\gamma+1)\Gamma(\alpha+\gamma+1)\Gamma(\alpha+\beta+1)}{(\Gamma(\alpha+1))^2\Gamma(\gamma+1)\Gamma(\beta+1)\Gamma(2\alpha+\beta+1)\Gamma(2\alpha+\gamma+1)\Gamma(4\alpha+\beta+2\gamma+1)}$$

$$+0.75HJ\frac{t^{2\alpha+\beta+\gamma}\Gamma(2\alpha+\beta+1)}{\Gamma(\alpha+1)\Gamma(\alpha+\beta+1)\Gamma(2\alpha+\beta+\gamma+1)}+0.75KJ\frac{t^{3\alpha+\beta+\gamma}\Gamma(3\alpha+\beta+1)}{\Gamma(2\alpha+1)\Gamma(\alpha+\beta+1)\Gamma(3\alpha+\beta+\gamma+1)}$$

$$-2.7514A^2J\frac{t^{4\alpha+\beta+\gamma}\Gamma(2\alpha+1)\Gamma(4\alpha+\beta+1)}{(\Gamma(\alpha+1))^2\Gamma(3\alpha+1)\Gamma(\alpha+\beta+1)\Gamma(4\alpha+\beta+\gamma+1)}$$

$$-1.5ABH\frac{t^{3\alpha+\beta+\gamma}\Gamma(3\alpha+\beta+1)\Gamma(\alpha+\beta+1)}{(\Gamma(\alpha+1))^2\Gamma(\beta+1)\Gamma(2\alpha+\beta+1)\Gamma(3\alpha+\beta+\gamma+1)}$$

$$-1.5ABK\frac{t^{4\alpha+\beta+\gamma}\Gamma(4\alpha+\beta+1)\Gamma(\alpha+\beta+1)}{\Gamma(2\alpha+1)\Gamma(\alpha+1)\Gamma(\beta+1)\Gamma(2\alpha+\beta+1)\Gamma(4\alpha+\beta+\gamma+1)}$$

$$+5.5028A^3B\frac{t^{5\alpha+\beta+\gamma}\Gamma(5\alpha+\beta+1)\Gamma(2\alpha+1)\Gamma(\alpha+\beta+1)}{(\Gamma(\alpha+1))^3\Gamma(3\alpha+1)\Gamma(\beta+1)\Gamma(2\alpha+\beta+1)\Gamma(5\alpha+\beta+\gamma+1)}$$

$$+0.75HP\frac{t^{2\alpha+2\gamma}\Gamma(2\alpha+\gamma+1)}{\Gamma(\alpha+1)\Gamma(\alpha+\gamma+1)\Gamma(2\alpha+\gamma+1)}+0.75KP\frac{t^{3\alpha+2\gamma}\Gamma(3\alpha+\gamma+1)}{\Gamma(2\alpha+1)\Gamma(\alpha+\gamma+1)\Gamma(3\alpha+2\gamma+1)}$$

$$-2.7514A^2P\frac{t^{4\alpha+2\gamma}\Gamma(4\alpha+\gamma+1)\Gamma(2\alpha+1)}{(\Gamma(\alpha+1))^2\Gamma(3\alpha+1)\Gamma(\alpha+\gamma+1)\Gamma(4\alpha+2\gamma+1)}$$

$$-1.875BGH\frac{t^{3\alpha+2\gamma}\Gamma(3\alpha+\gamma+1)\Gamma(\alpha+\gamma+1)}{(\Gamma(\alpha+1))^2\Gamma(\gamma+1)\Gamma(2\alpha+\gamma+1)\Gamma(3\alpha+2\gamma+1)}$$

$$-6.8784A^2BG\frac{t^{5\alpha+2\gamma}\Gamma(5\alpha+\gamma+1)\Gamma(2\alpha+1)\Gamma(\alpha+\gamma+1)}{(\Gamma(\alpha+1))^3\Gamma(3\alpha+1)\Gamma(\gamma+1)\Gamma(2\alpha+\gamma+1)\Gamma(5\alpha+2\gamma+1)}$$

$$-1.875BGK\frac{t^{4\alpha+2\gamma}\Gamma(4\alpha+\gamma+1)\Gamma(\alpha+\gamma+1)}{\Gamma(\alpha+1)\Gamma(\gamma+1)\Gamma(2\alpha+1)\Gamma(2\alpha+\gamma+1)\Gamma(4\alpha+2\gamma+1)}.$$

5

EEG-metric based mental stress detection

Gaurav[1], **R. S. Anand**[1], **Vinod Kumar**[2]
[1]Department of Electrical Engineering, Indian Institute of Technology Roorkee, Roorkee 247667, India
[2]Jaypee University of Information Technology, Waknaghat 712100, India
E-mail: gsbmi.dee2014@iitr.ac.in, anandfee@iitr.ac.in, vinodfee@iitr.ac.in

Abstract
Mental stress level is a vital parameter affecting physical well-being, cognition, emotions, and professional efficiency. With growing adversities in modern living standards, causing abnormal mental stress, it is necessary to measure to cure it. Regular personal stress profile generated can be used as neurofeedback for the clinical as well as personal assessment. This paper describes a method to detect mental stress level based on physiological parameters. In this method, an electroencephalogram (EEG)-metric parameters based binary and ternary stress classifier is developed. This is validated through probabilistic stress profiler of differential stress inventory (a questionnaire based evaluation). Nine channel EEG is used to extract physiological signal. EEG-metric based cognitive state and workload outputs are generated for 41 healthy volunteers (37 males and 4 females, age; 24±5 years). All subjects were guided to perform three simple tasks of closed eye, focusing vision on a red dot on center of dark screen and focusing on a white screen. Central tendencies (mean, median and mode) and standard deviation were extracted of EEG-metric (sleep onset, distraction, low engagement, high engagement and cognitive states) as features. Either of the two or three classes of stress are evaluated from probabilistic stress profiler of differential stress inventory and used as training output classes. A supervisory training of multiple layer perceptron based binary support vector machine classifier was used to detect stress class one by one. 40 subject's samples were used for training and interchanging one-by one 41th subjects stress class is determined from the designed classifier. Out of 41 subjects, stress level of 30 subjects is correctly identified by binary classifier and stress level of 26 subjects is correctly identified by ternary classifier, using multi-layer perceptron kernel based SVM.

Keywords EEG-metric; differential stress inventory questionnaire; SVM-MLP.

1 Introduction

Psychological stress is a phenomenon related to thoughts, emotion, physiological changes and everyday social activities. High level of psychological stress can become cause of mental or physiological illness, or other health related problems. It has been observed in recent researches that high level of mental stress adversely affects cardiovascular, endocrine and immune system health, etc. and psychological health (Cohen et al., 2007). Stress is surveyed to be an unavoidable aspect of college student's life. A survey conducted by Nightline Association reported about 65% of university students observe some kind of mental or psychological stress in their daily academic and other activities (Zheng et al., 2016). Cognitive as well as physical performance can also be affected due to daily stress conditions. High risk professions, such as defense services, industrial process plants, vehicle, locomotives and flight operation and control are more prone to get affected due to high stress condition and can be highly dangerous (Driskell et al., 2013). Diagnose and measurement of stress level profiles; quantitatively and qualitatively with abnormality spotting can be achieved by majorly three methods: task oriented, questionnaire based and physiological parameters assessment based evaluations. Psychological questionnaire based methods are widely used to examine the stress profile, but this is mostly only empirical to subject's response and is prone to misrepresentation or manipulation, and it can result to incorrect measurement. Also, evaluation requires an extensive training and expertise.

Physiological and psychological parameters have been proposed to comparatively identify anxiety or stress level (Sharma et al., 2012; Glenn et al., 2014; Hermens et al., 2014). Also, questionnaire based stress evaluation methods have been developed, such as; Cohens Perceived Stress Scale (PSS), Stress Response Inventory (SRI) and Hamilton Depression Rating Scale (HDRS) (Cohen et al., 1983; Koh et al., 2000; Williams, 1988). Few of the methods to induce mental stress in lab settings are as mentioned (Skoluda et al., 2015). Changes in autonomic nervous system responses can be induced by anxiety, this can be observed through changes in physiological factors such as heart rate, blood pressure and respiratory rhythm (Jung et al., 2013). In previous studies, power spectrum density of electroencephalogram (EEG) is observed to changes in specific way with change in emotion or mental states (Alonso et al., 2015; Dogra et al., 2018; Zhang, 2018). Different EEG patterns are seen to be changed with increasing level of stress (Hsieh et al., 2013). Stress evaluation is also possible with examination of electrocardiogram (ECG) based factors, such as heart rate and heart rate variability (Xu et al., 2015).

Cardiac and respiration activity was found to offer better stress assessment biomarkers than speech, galvanic skin response or skin temperature when recorded with wearable biomedical measurement systems (Seoane et al., 2014). For evaluation of anxiety in daily life scenarios, wearable systems which work on physiological parameters have been developed (Wu et al., 2012). A non-invasive EEG sensor for chronic stress evaluation was proposed to be worked in everyday social or professional environments (Hu et al., 2015). Multimodal physiological parameters based mental fatigue prediction methods are developed (Laurentet al., 2013). More studies are needed to validate their efficacy in case of mental stress evaluation.

Present method is an attempt to develop a method to measure overall psychological stress advent through physiological signals and it's parameters, visually EEG. EEG signal is an electrical impression of bioelectric potential from brain, during regular stimulus and triggering of neuronal activity, due to neuronal cell-dendrite current dipole dynamic change. Using EEG as a signature of regular brain neuronal activity, discrete stress levels can be evaluated. In this method, metrics generated from B-Alert X10 based EEG system and questionnaire based; differential stress inventory (DSI), are together used for discrete level stress profile assessment. Features processed and generated through EEG-based metrics and discrete levels of stress profile evaluated from DSI are incorporated on a support vector machine (SVM) based supervisory learning system, which is further used to assess discrete class of stress profile of particular subject.

2 Experiment and Methods

2.1 EEG-based metrics for cognitive states

B-Alert X10® is a wireless hardware system for EEG and ECG acquisition with 9 channel EEG and single channel ECG (B-Alert.com, 2011). B-Alert X-10 system follows 10-20 system of EEG electrodes placement. The nine electrodes are fixed at Fz, F3 and F4 i.e. at center, left and right position of frontal region scalp; Cz, C3 and C4 at center, left and right position of central region of scalp; Poz, P3 and P4 at center of parieto-occipital, and left and right region of parietal region of scalp. ECG electrodes are placed in lead-I region according to Einthoven triangle. Each EEG electrode to scalp impedance is checked before recording and kept below 30KΩ. The sampling frequency for acquisition is kept at 265 samples per second and 16 bit resolution. Before acquisition, an alertness and memory profiler test (AMP) is conducted on each subject which comprises of 3-choice vigilance task (3CVT), visual psychomotor vigilance Task (VPVT) and auditory psychomotor vigilance task (APVT), this takes approximately 20 minutes. After AMP test, the acquisition can be started to get 9 channel raw EEG signal and 1 channel raw ECG signal. Along with EEG and ECG signal, EEG-based metrics including cognitive state and workload outputs is also produced each epoch of second. These EEG-metrics are probability of sleep onset, distraction, low engagement, high engagement, cognitive states (sleep Onset, distraction, low engagement, high engagement), and 2-class model of workload (high workload, low workload) (Berka et al. , 2007). The EEG-metrics are generated as discriminant function analysis of spectral powers of EEG signals in different ranges of frequencies. EEG-metrics is generated from every second as epoch by epoch outputs from ABM model. In this study, probability of sleep onset, distraction, low engagement, high engagement, cognitive states were taken as features to train. For acquisition and analysis, Acknowledge 4.2 software of Biopac™ is used. Acknowledge 4.2 is equipped with filtering, mathematical and analysis tools.

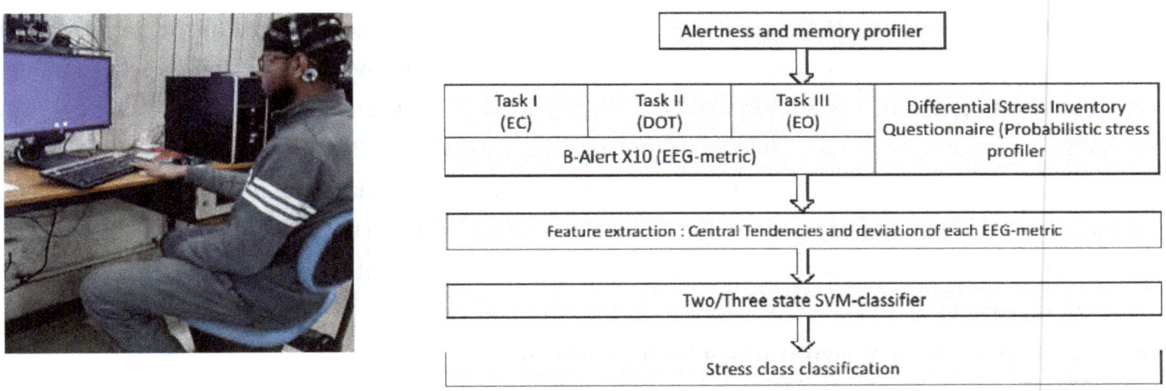

(a) (b)

Fig. 1 (a). A participant fixed with B-Alert X10 system on scalp, perform AMP test. (b) Flowchart of task operation, data acquisition and stress classification.

2.2 Questionnaire based stress profiler evaluation

DSI is a questionnaire based psychological battery test provided by Vienna test system to analyze and differentiate behavioral stress and distribute to the specific category of stress experiences (Rost et al., 1989). The stress questionnaire consists of 52 questions of causes, 21 questions on symptoms, 30 questions on coping and 20 questions on stabilization. The subject has to choose the amount of accordance of themselves, with the condition mentioned in the questionnaire. According to the responses of the subject's, raw score of the evaluation for stress causes, symptoms, coping and stress stability is evaluated. Final evaluation result consists of probabilistic profile classification of the subject in describable classes as Type-I: Normal type, Type-II: Overstressed, Type-III: Stress resistant, Type-IV: Low-stress-successful coping, Type-V: High stress-successful coping, with values between 0 and 1, sum of all profiler is equal to 1. This profiler is used for the training the classes of stress detection system.

2.3 Data acquisition, cognitive tasks and evaluation method

For stress classification study, B-Alert X10 system's EEG and ECG electrodes were attached to the subject's head scalp and chest. Before starting physiological signal acquisition, 15 minutes of baseline, AMP test, is guided to perform by the subject, which generates a learnt, ready to use, definition system for cognitive state and workload outputs. Three tasks are to be performed by all subjects in physically rest condition sitting on a chair. The tasks are, first is 5 minutes of eyes closed (EC), then 5 minutes of eyes open focusing vision on a red color dot on a dark screen (DOT) and 5 minutes of eyes open with focus on a bright screen (EO). First two tasks were performed in a dark environment and third one in a bright environment. During the task, simultaneously the EEG, ECG, cognitive state and workload outputs are acquired and recorded. After the completion of physiological parameters acquisition, a psychological questionnaire of differential stress inventory is conducted on the subjects and stress probabilistic profiling is done. Profiling of 41 subjects (37 males and 4 females) was done with their consent. All the subjects are healthy students from IIT Roorkee with clean medical records, no chronic diseases and age range of 24±5 years. In Fig.1 (a) is shown, a subject is performing AMP task while EEG profile getting recorded.

For training of SVM system, each subject's total 60 features are taken. From the 5 minutes of EC, DOT and EC; five EEG-metrics; sleep onset, distraction, low engagement, high engagement and cognitive state values were used for feature formation. Mean, sum of mean and variance, median and mode of each EEG-based metric for each task was taken as feature. This way total 60 features were accounted for each subjects for training purpose. For output classification, DSI outputs are used. Probabilistic profiler of DSI is combined to form two classes. Profiling is created by adding Type-II: Overstressed and Type-V: High stress to get 'High Stress'. Sum of Type-I: Normal type, Type-III: Stress resistant and Type-IV: Low-stress is to get 'Low Stress'. Similar way a three state stress profile was also created. These 60 features along with stress class of 40 subjects is fed to a multilayer perceptron based SVM classifier. Once the learning is complete, the 60 features of 41st subject are fed to the classifier system to get stress class. Fig. 1 (b) depicts the experimental flow of task performance, DSI evaluation, feature formation and supervised learning based classifier development.

2.4 Linear support vector machine based binary classifier

For classification of the stress profile based on EEG-metrics, support vector machine (SVM) is used. SVM is a supervised learning scheme for classification and regression (Vapnik et al., 1998; Zhao and Hasan, 2013). The principle behind SVM algorithm is theory of structural risk minimization. The hyperplane function of classification is decided based on minimizing of generalization error for decision boundaries. Also SVM is resistant to over-training, and performance increases with generalization. Simplest form of SVM does binary classification, in which few points in the data space is identified to construct a hyperplane which separates two classes of points.

The training data x consists of n data samples each of m dimensions and belonging to class y, is expressed as:

$$(x_1, y_1),\ldots,(x_i, y_i),\ldots,(x_n, y_n),\ \ x \in \Re^m, y \in \{+1,-1\}$$

SVM projects data (x_i, y_i) into an infinite dimensional hyperplane (x_i, y_i) by using dedicated normalized kernel function and defines its decision rule as sign($f(x)$). The discriminant function $f(x)$ creates the optimum hyperplane decision boundary by using weight vector w^* and bias b^*.

$$f(x) = w^* \phi(x) + b^*$$

The optimum values of w^* and b^* are estimated by solving following optimization problem where, C is regularization parameter and ξ_i s are slack variables allowing inseparable data.

$$\min_{w,\xi} \left\{ \frac{1}{2} w^2 + C \sum_{i=1}^{n} \xi_i \right\}; \qquad y_i(w.\phi(x_i) + b) \geq 1 - \xi_i, \qquad \xi_i \geq 0$$

This problem is solved using Lagrange optimization through dual formation, which finally yields optimum value for weight vector w^* and bias b^*. Since SVM estimates an infinite dimensional optimum hyper plane, usually it performs better than the other supervised learning algorithms while solving classification problems even on higher dimensional input features. Other optimal decision boundaries, such as polynomial, radial basis function based and multi-layer perceptron based kernels can also be used in place of linear decision boundary for better classification results.

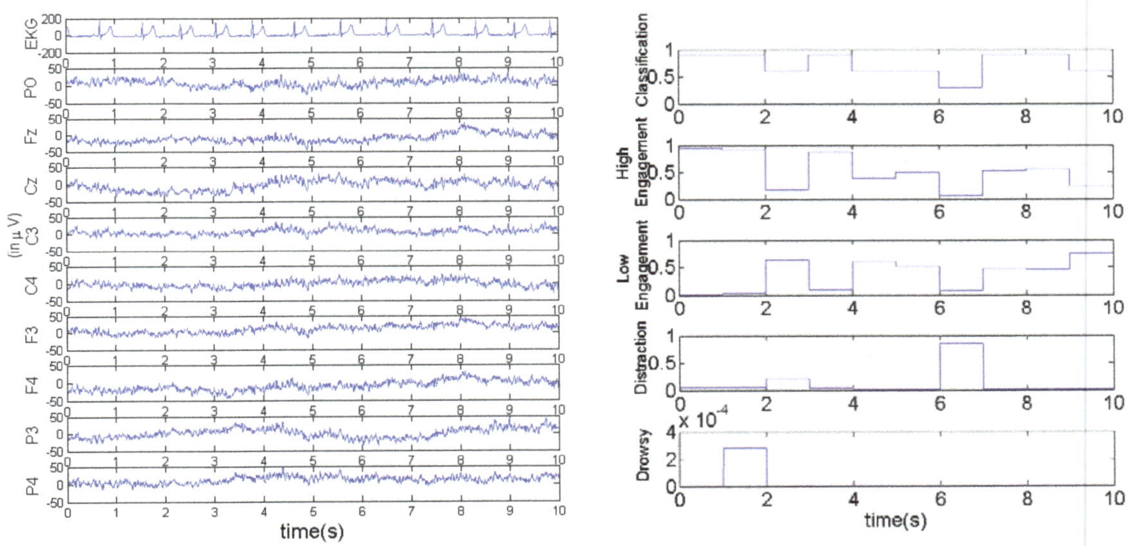

(a) (b)

Fig. 2 (a). Plot of raw ECG and EEG signal sample acquired from B-Alert X10, (b) EEG-metric acquired from B-Alert X10.

3 Results

Fig. 2 (a) shows 10s raw signal sample of one channel ECG and nine channel EEG, acquired through B-Alert X10 system, Fig. 2 (b) shows five of the eight EEG-metric based cognitive states chosen for feature selection, with respect time. As shown in Fig. 2 (b) EEG-metric are generated every second epoch by epoch. The EEG-metrics lies between 0 and 1. Fig. 3 shows boxplot of all EEG-metrics features (mean and median) of all samples of 'High Stress' and 'Low Stress' taken together. Fig. 3 (a) and (b) are box plot of mean and median of metrics: high engagement, low engagement, distraction and drowsy metrics for the task of eyes closed. Similarly, in Fig. 3 (c) and (d), features are for the task of open eyes with vision focused on red dot on dark screen, and Fig. 3 (e) and (d), features are for task of open eyes with vision focused on white screen. Fig. 3 (g) and (h) are Box plot of means and medians of cognitive state classification of less stress and high stress group for three tasks of dot focus, eyes closed and eyes open.

Table 1 is the representation of maximum, minimum, median and mode of all the features, visually, cognitive state classification, high engagement, low engagement, distraction and drowsiness for three task conditions, for two stress classes (Less stress and High stress). Table 1(a) are observation for task of open eyes focusing on a bright dot on dark screen, Table 1(b) are observation for task of closed eyes focusing in a dark room, Table 1(c) are observation for task of open eyes focusing on a bright screen. It is observed that central tendencies (mean and median) for high engagement and low engagement metrics are higher for high stress class than less stress class, whereas distraction and drowsy metrics are lesser for high stress than low stress. Cognitive state classification metric for DOT and EO tasks is seen at 'Low Engagement' for both less and high stress class. Whereas for EC task, less stress class is seen in 'Distraction' state and 'Low Engagement' for high stress class. Binary class SVM with various decision boundaries: linear, quadratic, polynomial, radial-basis function (RBF) and multi-layer perceptron (MLP), were used to develop classifiers using 60 features of each of 40 subjects on two and three class of stresses. Similarly, by interchanging training and prediction matrix of all 41 subjects, the efficacy of the SVM based system is to be checked.

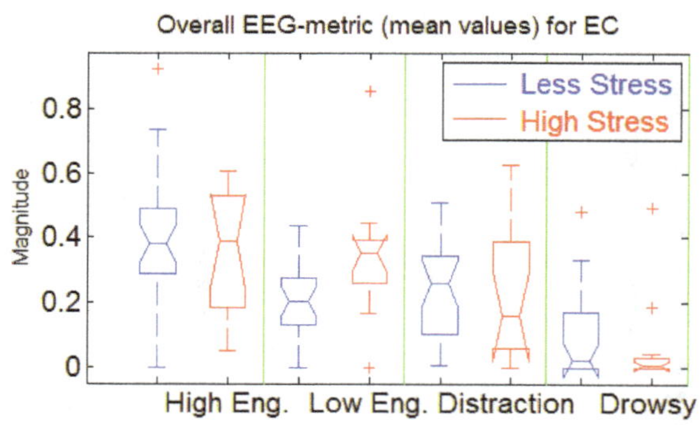

(a)

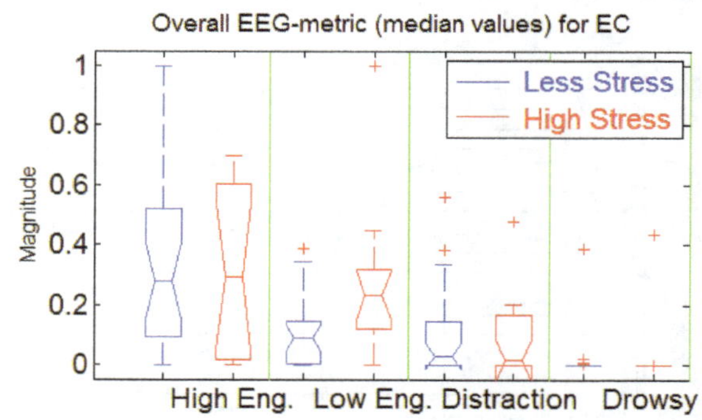

(b)

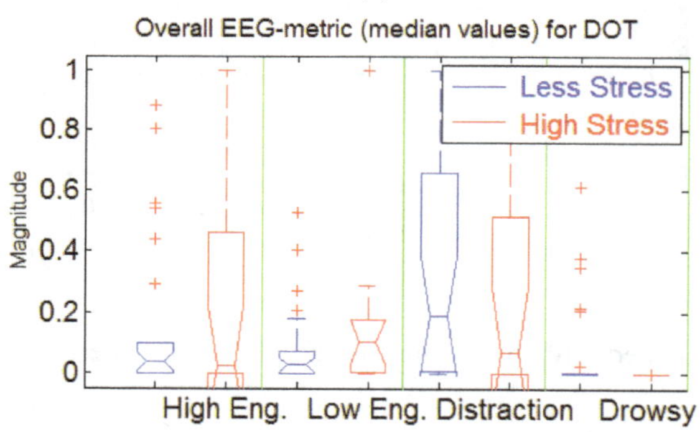

(c)

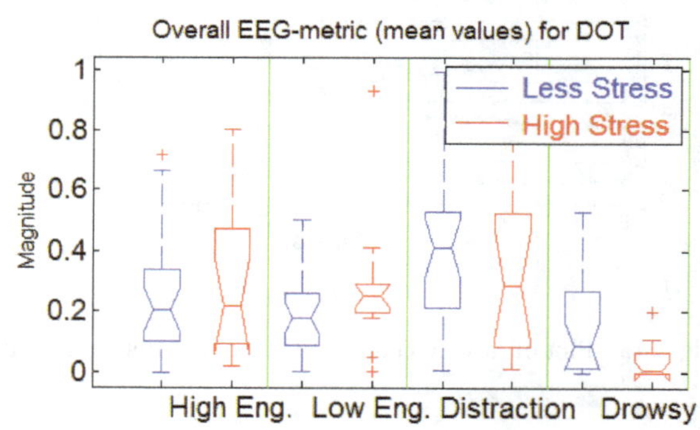

(d)

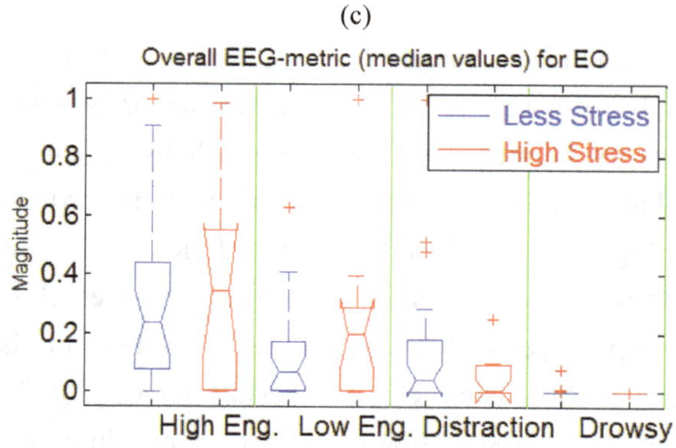

(e)

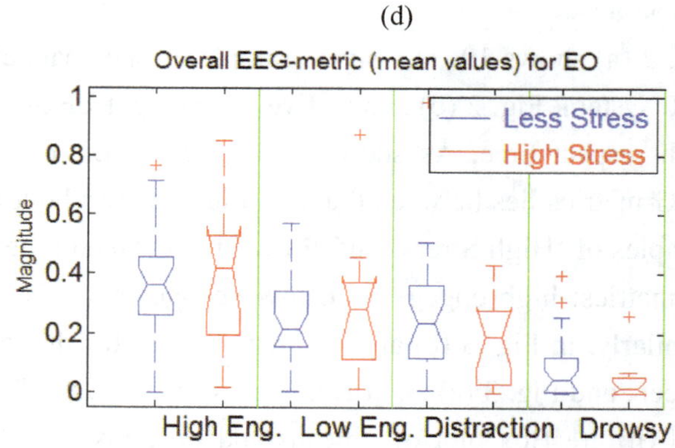

(f)

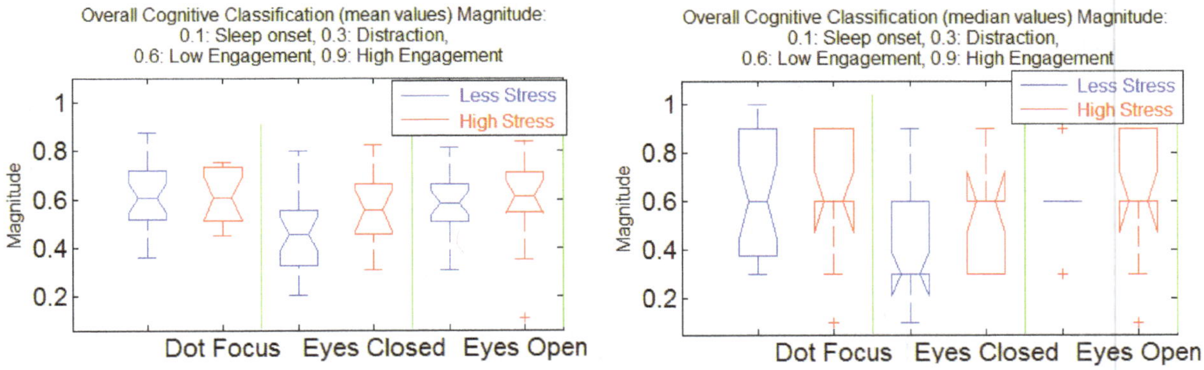

(g) (h)

Fig. 3 Box plot showing contrast out of 41 subjects of either 'Less stress' or 'High Stress' EEG-metrics features of (high engagement, low engagement, Distraction, Drowsy) for the tasks of (a) mean while closed eyes (b) median while closed eyes (c) mean while focusing on a dot on dark screen, (d) median while focusing on a dot on dark screen (e) mean while open eyes (f) median while open eyes, (g) mean of cognitive state classification of less stress and high stress group for three tasks of dot focus, eyes closed and eyes open, (h) medians cognitive state classification.

Altogether, for two states stress classifiers, out of 41 binary outputs, 23 were correct for linear and quadratic, 22 for polynomial, 27 for RBF and 30 for MLP. So accuracy of 56.09% for linear and quadratic, 53.66% for polynomial, 65.85% for RBF and 73.1% for MLP based decision boundary was observed. This evaluates multi-layer perceptron based SVM as the best classifier for two state stress classes.

For three states stress classifiers, out of 41 binary outputs, 16 were correct for linear and quadratic, 17 for polynomial, 23 for RBF and 26 for MLP. So accuracy of 39.02% for linear and quadratic, 41.46% for polynomial, 56.10% for RBF and 63.41% for MLP based decision boundary was observed. This evaluates multi-layer perceptron based SVM as the best classifier for both: two state as well as three stress classes.

Table 1 (a) Comparison of maximum, minimum, median and mode of EEG-metric features for two stress classes (Less stress and High stress) while focusing on a dot on a screen.

Less Stress Class	CC (Mean)*	CC (Median)	CC (Mode)	HE (Mean)	HE (Median)	HE (Mode)	LE (Mean)	LE (Median)	LE (Mode)	Dist (Mean)	Dist (Median)	Dist (Mode)	Dro (Mean)	Dro (Median)	Dro (Mode)
Minimum	0.36	0.3	0.1	0	0	0	0	0	0	0.01	0	0	0	0	0
Maximum	0.88	1	1	0.92	1	1	0.44	0.39	0.62	0.51	0.56	0.19	0.48	0.39	1
Median	0.61	0.6	0.9	0.38	0.28	0	0.2	0.09	0	0.26	0.03	0	0.02	0	0
Mode	0.61	0.6	0.9	0.29	0	0	0.27	0	0	0.01	0	0	0	0	0

High Stress Class	CC (Mean)	CC (Median)	CC (Mode)	HE (Mean)	HE (Median)	HE (Mode)	LE (Mean)	LE (Median)	LE (Mode)	Dist (Mean)	Dist (Median)	Dist (Mode)	Dro (Mean)	Dro (Median)	Dro (Mode)
Minimum	0.45	0.1	0.1	0.05	0	0	0	0	0	0	0	0	0	0	0
Maximum	0.76	0.9	0.9	0.6	0.7	1	0.85	1	1	0.63	0.87	1	0.5	0.43	1
Median	0.605	0.6	0.75	0.395	0.29	0	0.355	0.24	0	0.16	0.015	0	0.005	0	0
Mode	0.58	0.6	0.9	0.55	0.01	0	0.36	0.24	0	0.06	0	0	0	0	0

Table 1 (b) Comparison of maximum, minimum, median and mode of EEG-metric features for two stress classes (Less stress and High stress) while eyes closed.

Less Stress Class	CC (Mean)	CC (Median)	CC (Mode)	HE (Mean)	HE (Median)	HE (Mode)	LE (Mean)	LE (Median)	LE (Mode)	Dist (Mean)	Dist (Median)	Dist (Mode)	Dro (Mean)	Dro (Median)	Dro (Mode)
Minimum	0.21	0.1	0.1	0	0	0	0.01	0	0	0.01	0	0	0	0	0
Maximum	0.8	0.9	0.9	0.72	0.88	0.15	0.5	0.52	0.82	1	1	1	0.53	0.61	1
Median	0.46	0.3	0.3	0.21	0.04	0	0.18	0.03	0	0.41	0.19	0	0.09	0	0
Mode	0.56	0.3	0.3	0.03	0	0	0.02	0	0	0.41	0	0	0	0	0

High Stress Class	CC (Mean)	CC (Median)	CC (Mode)	HE (Mean)	HE (Median)	HE (Mode)	LE (Mean)	LE (Median)	LE (Mode)	Dist (Mean)	Dist (Median)	Dist (Mode)	Dro (Mean)	Dro (Median)	Dro (Mode)
Minimum	0.31	0.3	0	0.02	0	0	0	0	0	0.01	0	0	0	0	0
Maximum	0.83	0.9	0.9	0.8	1	1	0.93	1	1	0.85	1	1	0.2	0	0
Median	0.55	0.6	0.45	0.22	0.025	0	0.255	0.105	0	0.29	0.07	0	0.01	0	0
Mode	0.55	0.3	0.3	0.22	0	0	0.2	0	0	0.01	0	0	0.01	0	0

Table 1 (c) Comparison of maximum, minimum, median and mode of EEG-metric features for two stress classes (Less stress and High stress) while looking at a bright screen.

Less Stress Class	CC (Mean)	CC (Median)	CC (Mode)	HE (Mean)	HE (Median)	HE (Mode)	LE (Mean)	LE (Median)	LE (Mode)	Dist (Mean)	Dist (Median)	Dist (Mode)	Dro (Mean)	Dro (Median)	Dro (Mode)
Minimum	0.31	0.3	0.1	0	0	0	0.01	0	0	0.01	0	0	0	0	0
Maximum	0.82	0.9	0.9	0.77	1	0.71	0.57	0.63	0.25	0.98	1	1	0.4	0.08	1
Median	0.59	0.6	0.6	0.36	0.24	0	0.21	0.07	0	0.24	0.04	0	0.04	0	0
Mode	0.55	0.6	0.9	0.26	0	0	0.13	0	0	0.24	0	0	0	0	0

High Stress Class	CC (Mean)	CC (Median)	CC (Mode)	HE (Mean)	HE (Median)	HE (Mode)	LE (Mean)	LE (Median)	LE (Mode)	Dist (Mean)	Dist (Median)	Dist (Mode)	Dro (Mean)	Dro (Median)	Dro (Mode)
Minimum	0.11	0.1	0.1	0.01	0	0	0.01	0	0	0	0	0	0	0	0
Maximum	0.84	0.9	0.9	0.85	0.99	1	0.87	1	1	0.84	1	1	0.97	1	1
Median	0.62	0.6	0.75	0.42	0.345	0	0.28	0.2	0	0.185	0.005	0	0.015	0	0
Mode	0.62	0.6	0.9	0.53	0	0	0.37	0	0	0.03	0	0	0	0	0

*CC: Cognitive classification state, HE: High engagement metric, LE: Low Engagement metric, Dist: Distraction metric and Dro: Drowsiness metric

4 Discussion

More development in the methods are possible, to detect multiple class and multiple level outputs and to get detailed stress profiler on the basis of physiological signals. Also scopes for better feature choice for classification and feature dimension reduction without affecting the classification mechanism can be further achieved. With mental stress emulator at lab setup, a real time stress or anxiety level detector is also possible. Other than questionnaire based stress profiler, other methods such as task based stress level can be used for training purpose. Further, the number of EEG can be reduced and validity of method with lesser electrodes is to be checked. With other possible EEG-metrics and reduced dimension, the predictor can be developed. The duration of task conduction can be minimized and real-time stress level detection methods can be possible.

5 Conclusion

A simple SVM based classification method for discrete stress level detection was developed using EEG-metrics and cognitive states as features for training and discrete classes generated from evaluation of differential stress inventory questionnaire based stress profiler as training outputs. The experimental study on 41 subjects successfully detected 30 subject's stress level correctly for binary classifier and 26 subject's for ternary classifier, which establishes the efficacy of the method. The method can be used for general purpose stress level detection.

References

Alonso JF, Romero S, Ballester MR, Antonijoan RM, Mañanas MA. 2015. Stress assessment based on EEG univariate features and functional connectivity measures. Physiological Measurement, 36(7): 1351-1365

Berka C, Levendowski DJ, Lumicao MN, Yau A, Davis G, Zivkovic VT, Olmstead RE, Tremoulet PD, Craven PL. 2007. EEG correlates of task engagement and mental workload in vigilance, learning, and memory tasks. Aviation, Space, and Environmental Medicine, 78(5): B231-B244

B-Alert.com. 2011. Advanced Brain Monitoring B-Alert x-10. http://www.b-alert.com/pdf/x10.pdf. Accessed on July 1, 2017

Cohen S, Janicki-Deverts D, Miller GE. 2007. Psychological stress and disease. JAMA, 298(14): 1685-1687

Cohen S, Kamarck T, Mermelstein R. 1983. A global measure of perceived stress. Journal of health and Social Behavior, 24(4): 385-396

Dogra J, Prashar N, Jain S, Sood M. 2018. Improved methods for analyzing MRI brain images. Network Biology, 8(1)

Driskell JE, Salas E. 2013. Stress and Human Performance. Psychology Press, USA

Glenn T, Monteith S. 2014. New measures of mental state and behavior based on data collected from sensors, smartphones, and the Internet. Current psychiatry reports, 16(12): 523

Hermens H, op den Akker H, Tabak M, Wijsman J, Vollenbroek M. 2014. Personalized coaching systems to support healthy behavior in people with chronic conditions. Journal of electromyography and kinesiology, 24(6): 815-826

Hsieh CS, Tai CC. 2013. A fatigue state evaluation system based on the band energy of electroencephalography signals. Sensors and Materials, 25(9): 697-706

Hu B, Peng H, Zhao Q, Hu B, Majoe D, Zheng F, Moore P. 2015. Signal Quality assessment model for wearable EEG sensor on prediction of mental stress. IEEE transactions on nanobioscience, 14(5): 553-561

Jung SJ, Chung WY. 2013. Non-intrusive healthcare system in global machine-to-machine networks. IEEE Sensors Journal, 13(12): 4824-4830

Rost DH, Schermer FJ. 1989. Diagnostik des Leistungsangsterlebens. Diagnostica, 35(4): 287-314

Skoluda N, Strahler J, Schlotz W, Niederberger L, Marques S, Fischer S, Thoma MV, Spoerri C, Ehlert U, Nater UM. 2015. Intra-individual psychological and physiological responses to acute laboratory stressors of different intensity. Psychoneuroendocrinology, 51: 227-236

Sharma N, Gedeon T. 2012. Objective measures, sensors and computational techniques for stress recognition and classification: A survey. Computer methods and programs in biomedicine, 108(3): 1287-1301

Vapnik VN, Vapnik V. 1998. Statistical Learning Theory (Vol. 1). Wiley, New York, USA

Williams JB. 1988. A structured interview guide for the Hamilton Depression Rating Scale. Archives of General Psychiatry, 45(8): 742-747

Wu W, Gil Y, Lee J. 2012. Combination of wearable multi-biosensor platform and resonance frequency training for stress management of the unemployed population. Sensors, 12(10): 13225-13248

Xu Q, Nwe TL, Guan C. 2015. Cluster-based analysis for personalized stress evaluation using physiological signals. IEEE journal of biomedical and health informatics, 19(1): 275-281

Zhang WJ. 2018. Fundamentals of Network Biology. World Scientific, Singapore

Zhao Y, Hasan YA. 2013. Machine learning algorithms for predicting roadside fine particulate matter concentration level in Hong Kong Central. Computational Ecology and Software, 3(3): 61-73

Zheng Y, Wong TC, Leung BH, Poon CC. 2016. Unobtrusive and multimodal wearable sensing to quantify anxiety. IEEE Sensors Journal, 16(10): 3689-3696

6

Evolutionary Network Control also holds for nonlinear networks: Ruling the Lotka-Volterra model

Alessandro Ferrarini
Department of Evolutionary and Functional Biology, University of Parma, Via G. Saragat 4, I-43100 Parma, Italy
E-mail: sgtpm@libero.it, alessandro.ferrarini@unipr.it

Abstract
The proof of our understanding of ecological and biological systems is measured by our skill to rule them, i.e. to channelize them towards a desired state. Control is a cardinal issue in most complex systems, but because a general theory to apply it in a quantitative manner has been absent so far, little was known about how we can rule weighted, directed networks that represent the most common configuration of real systems. To this purpose, Evolutionary Network Control (ENC) has been developed as a theoretical and methodological framework aimed to the control of ecological and biological networks by coupling network dynamics and evolutionary modelling. ENC is a tools to address controllability for arbitrary network topologies and sizes. ENC has proven to cover several topics of network control, e.g. a) the global control from inside and b) from outside, c) the local (step-by-step) control, and the computation of: d) control success, e) feasibility, and f) degree of uncertainty. Taken together, these results indicate that many aspects of controllability can be explored exactly and analytically for arbitrary networks, opening new avenues to deepening our understanding of complex systems. As yet, I have applied ENC only to linear ecological and biological networks. In this work, I show that ENC also holds for any kind of nonlinear networks, and provide an applicative example based on the nonlinear, widely-used, Lotka-Volterra model.

Keywords Evolutionary Network Control; genetic algorithms; global dynamics; nonlinear networks; predator-prey model; sensitivity analysis; stochastic simulations.

1 Introduction

I have recently introduced a methodological framework, named Evolutionary Network Control (ENC; Ferrarini, 2013a; Ferrarini, 2013b), so that ecological and biological networks can be controlled from the outside (Ferrarini, 2013a) but also through the use of endogenous controllers (Ferrarini, 2013b), by coupling network dynamics and evolutionary modelling (Holland, 1975; Goldberg, 1989). The endogenous control requires that the network is optimized at the beginning of its dynamics, by acting upon nodes, edges or both, so that it will inertially go to the desired state. On the contrary, the exogenous control requires that one or more exogenous controllers act upon the network at each time step (Ferrarini, 2011a; Ferrarini, 2011b).

In another paper (Ferrarini, 2013c), I have examined a further issue: how reliable is the achieved solution? This is an important question, because it's not assured that while managing a network-like system we are able to impose to nodes and edges exactly the optimized values we would need in order to achieve the desired control. In order to face this topic, I have coined a 3-parts (network dynamics - genetic optimization - stochastic simulations) solution.

I have further proposed a solution to the choice of the most feasible solution to network control by introducing the concepts of control success and feasibility (Ferrarini, 2013d).

Later, I have faced another pivotal question, i.e. how to locally (step-by-step) drive ecological and biological networks so that also intermediate steps (and not only the final state) are under our control (Ferrarini, 2014). The ratio behind this question is that intermediate dynamics could potentially go below or above critical ecological or biological thresholds, hence invalidating the final global control. To this purpose, I have introduced a solution to the complete (local + global) control of ecological and biological networks by making use of an intermediate control function.

As yet, I have applied ENC only to linear ecological and biological networks. While it is not goal of this paper to discuss the implications of the ENC of ecological and biological networks, I show here that ENC also holds for any kind of nonlinear networks, and provide an applicative example based on the nonlinear, widely-used, Lotka-Volterra model (Lotka, 1925; Volterra, 1926).

2 Subduing the Lotka-Volterra Model via ENC: Mathematical Formulation

It's given a generic ecological (or biological) dynamical system with n interacting actors

$$\frac{d\vec{S}}{dt} = \varphi(\vec{S},t) \tag{1}$$

where S_i is the amount (e.g., number of individuals, total biomass, density, covered surface etc…) of the generic *i-th* actor.

If we also consider inputs (e.g. species reintroductions) and outputs (e.g. hunting) from-to outside, we must write

$$\frac{d\vec{S}}{dt} = \varphi(\vec{S},t) + \vec{I}(t) + \vec{O}(t) \tag{2}$$

with initial values

$$\vec{S}_0 = <S_1(0), S_2(0) \ldots S_n(0)> \tag{3}$$

and co-domain limits

$$\begin{cases} S_{1\min} \le S_1(t) \le S_{1\max} \\ \ldots \\ S_{n\min} \le S_n(t) \le S_{n\max} \end{cases} \quad \forall t \tag{4}$$

The Lotka-Volterra equations (Lotka, 1925; Volterra, 1926), otherwise known as the predator-prey equations, are a combination of two first-order, non-linear, differential equations frequently used to describe the dynamics of biological systems with two interacting species, one as a prey and the other as a predator.

The Lotka-Volterra model makes five assumptions about the environment and the dynamics of the two interacting species: 1) the prey population finds food at any times; 2) the food supply for the predator depends on the size of the prey population; 3) the rate of change of each population is proportional to its size; 4) while interacting, the environment does not change; 5) predators have unbounded appetency. As differential equations are used, the solution is deterministic and continuous so that the generations of both the predator and prey persistently overlap.

The nonlinear Lotka-Volterra model with logistic grow of the prey S_1 is a particular case of eq. (1) and it reads as follows

$$\begin{cases} \dfrac{dS_1}{dt} = \alpha S_1(1-\dfrac{S_1}{\kappa}) - \beta S_1 S_2 \\ \dfrac{dS_2}{dt} = \beta\gamma S_1 S_2 - \delta S_2 \end{cases} \quad (5)$$

with initial values

$$\vec{S}_0 = <S_1(0),\ S_2(0)> \quad (6)$$

and co-domain limits

$$\begin{cases} S_{1\min} \le S_1(t) \le S_{1\max} \\ S_{2\min} \le S_2(t) \le S_{2\max} \end{cases} \quad \forall t \quad (7)$$

In order to get the global control of such model, ENC can act upon the previous Lotka-Volterra model as follows

$$\begin{cases} \dfrac{d\tilde{S}_1}{dt} = \tilde{\alpha} S_1(1-\dfrac{S_1}{\tilde{\kappa}}) - \tilde{\beta} S_1 S_2 \\ \dfrac{d\tilde{S}_2}{dt} = \tilde{\beta}\tilde{\gamma} S_1 S_2 - \tilde{\delta} S_2 \\ \vec{\tilde{S}}_0 = <\tilde{S}_1(0),\ \tilde{S}_2(0)> \end{cases} \quad (8)$$

with steady values at

$$\frac{dS_1}{dt} = \frac{dS_2}{dt} = 0 \quad (9)$$

where the tilde symbol means that the ENC is active over such actors by controlling equation parameters and initial values.

The control equations in (8) are able to globally drive any ecological and biological network, and the nonlinear Lotka-Volterra as a particular case, to the desired final state with an uncertainty degree that can be calculated as proposed in Ferrarini (2013c).

$$\begin{cases} \frac{d\tilde{S}_1}{dt} = \underline{\tilde{\alpha}} S_1 (1 - \frac{S_1}{\underline{\tilde{\kappa}}}) - \underline{\tilde{\beta}} S_1 S_2 \\ \frac{d\tilde{S}_2}{dt} = \underline{\tilde{\beta}\tilde{\gamma}} S_1 S_2 - \underline{\tilde{\delta}} S_2 \\ \tilde{\tilde{S}}_0 = < \underline{\tilde{S}_1}(0), \underline{\tilde{S}_2}(0) > \end{cases} \tag{10}$$

where the underscores represent 1%, 5% or 10% uncertainties about the optimized parameters. Thus for example:

$$\begin{cases} 0.99 * \tilde{\alpha} \le \underline{\tilde{\alpha}} \le 1.01 * \tilde{\alpha} \\ or \\ 0.95 * \tilde{\alpha} \le \underline{\tilde{\alpha}} \le 1.05 * \tilde{\alpha} \\ or \\ 0.90 * \tilde{\alpha} \le \underline{\tilde{\alpha}} \le 1.10 * \tilde{\alpha} \end{cases} \tag{11}$$

If we stochastically vary n times (e.g. 10,000 times) the parameters that have been optimized via ENC, we can compute how many times such uncertainty makes the optimization procedure useless. Hence, uncertainty about network control can be computed as in Ferrarini (2013c)

$$U_{\%} = 100 * \frac{k}{n} \tag{12}$$

where k is the number of stochastic simulations acting upon the optimized parameters that make the optimization procedure useless (i.e. the goal of the optimization procedure is not reached).

In order to apply the ENC framework, the software Control-Lab (Ferrarini, 2013e) has been developed using Visual Basic (Balena, 2001; Pattinson, 1998).

3 An Applicative Example

Let's consider the Lotka-Volterra system of eq. (5) with the following parameters and constants:

$$\begin{cases} S_1(0) = 100 \\ S_2(0) = 10 \\ \alpha = 4 \\ \beta = 0.05 \\ \gamma = 1 \\ \delta = 4 \\ \kappa = 500 \\ dt = 0.01 \end{cases} \tag{13}$$

Fig. 1 shows its dynamical behaviour. Figure 2 depicts its phase plot.

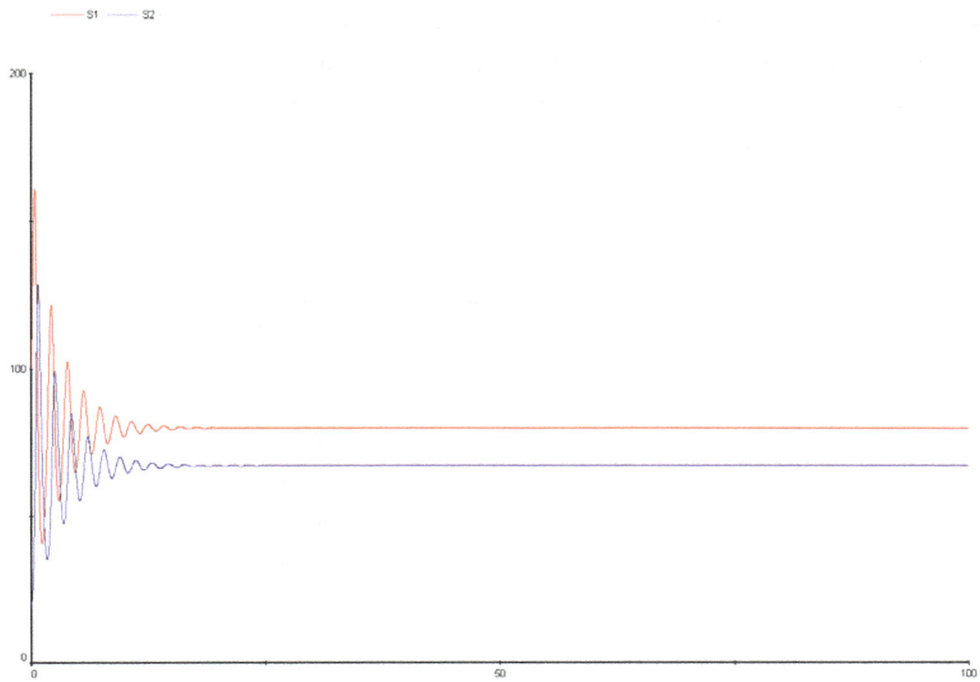

Fig. 1 Time plot of the above-depicted nonlinear Lotka-Volterra dynamical system.

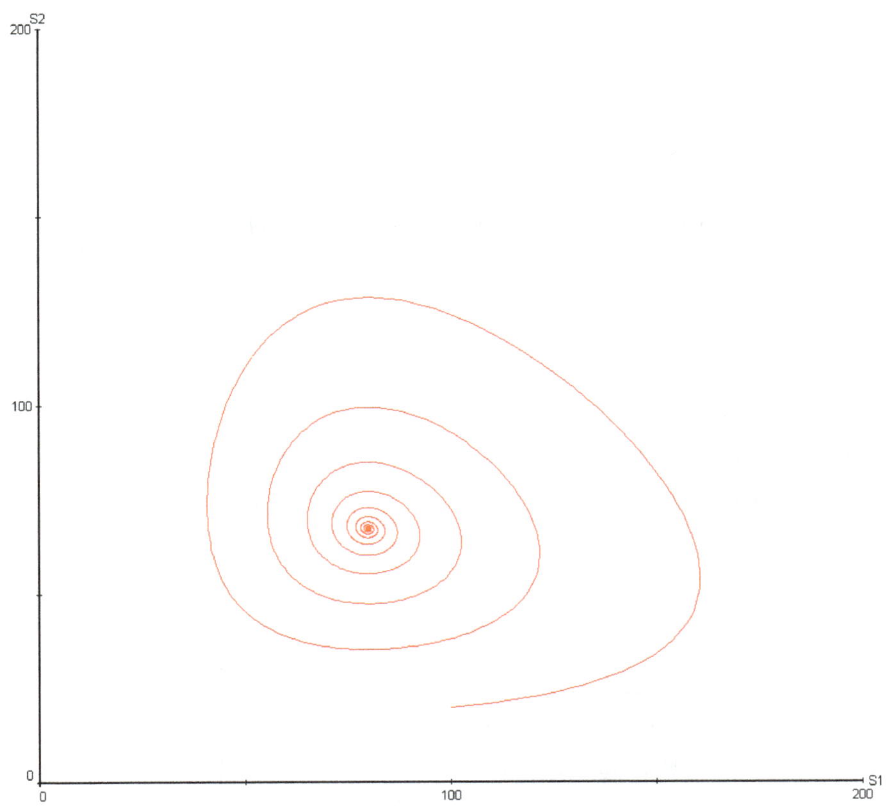

Fig. 2 Phase plot of the above-depicted nonlinear Lotka-Volterra dynamical system.

The previous nonlinear system goes at the steady state with S_1= 80.00 and S_2= 67.20.

Let's suppose we want that both the prey and the predator go to equilibrium with values close to 100. Fig. 3 shows the optimized solution (S_1= 99.45 and S_2= 99.91) detected via ENC by acting upon alpha, beta, gamma and delta. The steady state happens at t= 42.15.

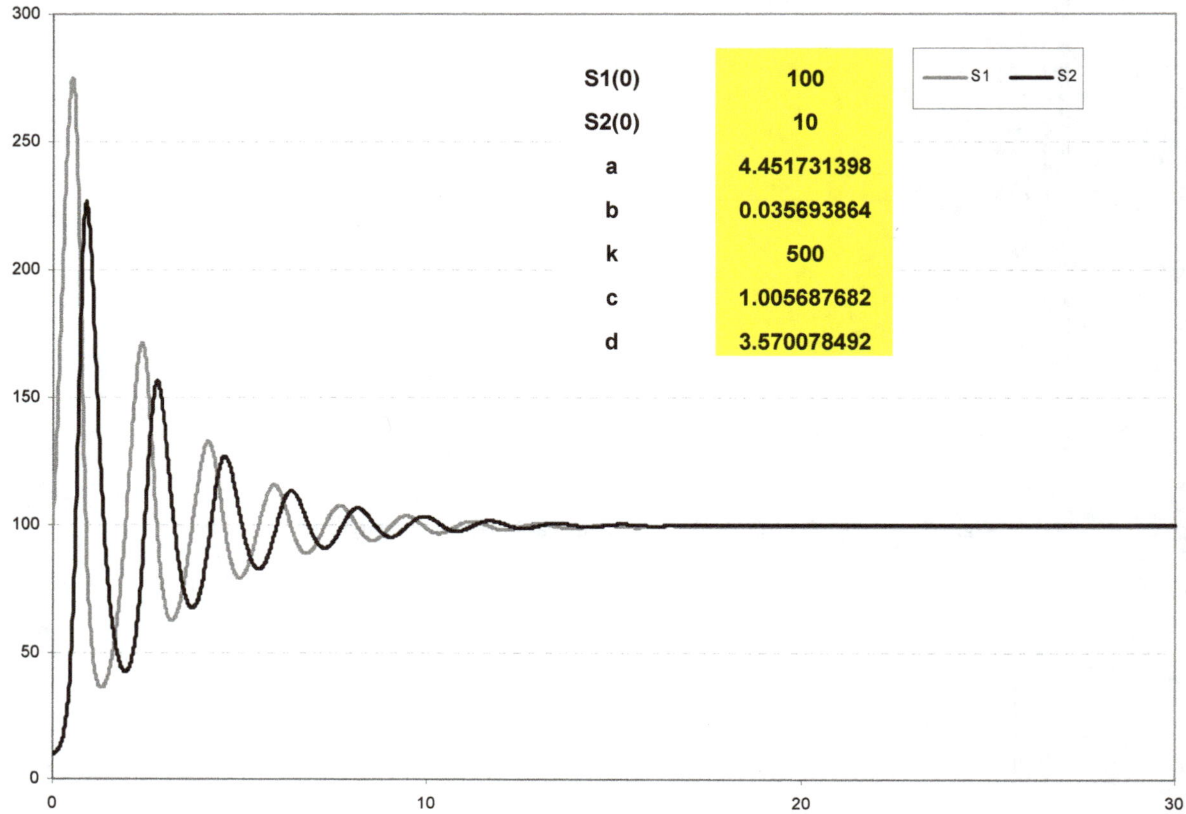

Fig. 3 The ruled Lotka-Volterra system with equilibrium values S_1= 99.45 and S_2= 99.91. The optimized parameters detected via Evolutionary Network Control are indicated above the two curves. The solution has been found through Control-Lab (Ferrarini, 2013e).

After 10,000 simulations (1% uncertainty) on the optimized parameters (alpha, beta, gamma and delta), we achieve the results of Table 1. It is evident that the solution depicted in Fig. 3 is robust with regard to a 1% sensitivity analysis, as mean and median values of the 10,000 simulations are very close to the desired solution S_1= S_2= 100.

Table 1 Sensitivity analysis (1% uncertainty) on the optimized parameters of Fig. 3.

Statistics	S_1	S_2
min	96.73	97.98
max	102.38	101.72
mean	99.52	99.92
median	99.57	99.92
std. dev.	1.02	0.74

Now let's suppose we want that both the prey and the predator go to equilibrium with values close to 50. Fig. 4 shows the optimized solution (S_1= 50.96 and S_2= 50.00) found via ENC by acting upon alpha, beta, gamma and delta. The steady state happens at t= 74.48.

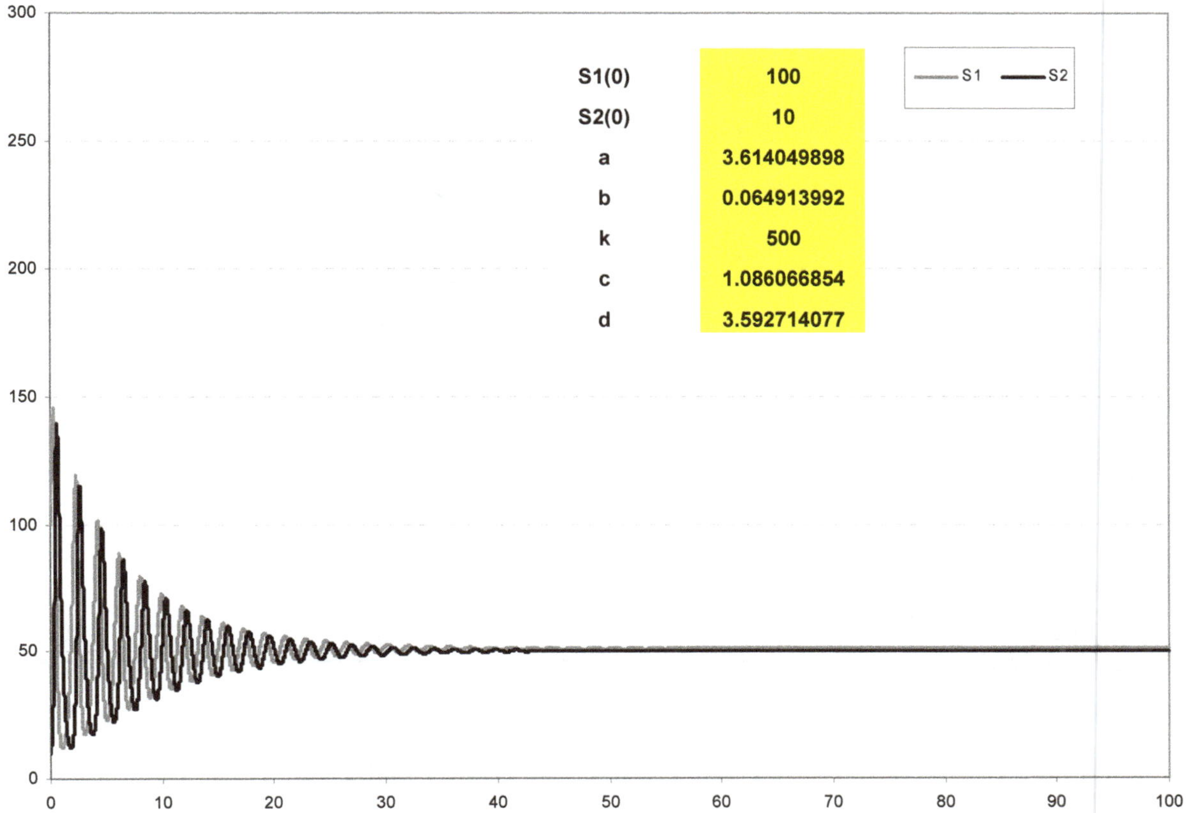

Fig. 4 Ruled Lotka-Volterra system with equilibrium values S_1= 50.96 and S_2= 50.00. The optimized parameters detected via Evolutionary Network Control are indicated above the two curves. The solution has been found through the software Control-Lab (Ferrarini, 2013e).

After 10,000 simulations (5% uncertainty) on the optimized parameters (alpha, beta, gamma and delta), we achieve the results of Table 2. It is clear that the solution depicted in Fig. 4 is robust with regard to a 5% sensitivity analysis, as mean and median values of the 10,000 simulations are very close to the desired solution S_1= S_2= 50.

Table 2 Sensitivity analysis (5% uncertainty) on the optimized parameters of Fig. 4.

Statistics	S_1	S_2
min	44.55	45.50
max	58.45	55.12
mean	50.94	49.95
median	50.85	49.91
std. dev.	2.52	1.95

Last, let's suppose we require that the prey and the predator go to equilibrium with $S_1+S_2= 90$. Fig. 5 shows the optimized solution ($S_1= 54.92$ and $S_2= 35.23$) detected via ENC by acting upon alpha, beta, gamma and delta. The steady state happens at $t= 95.00$.

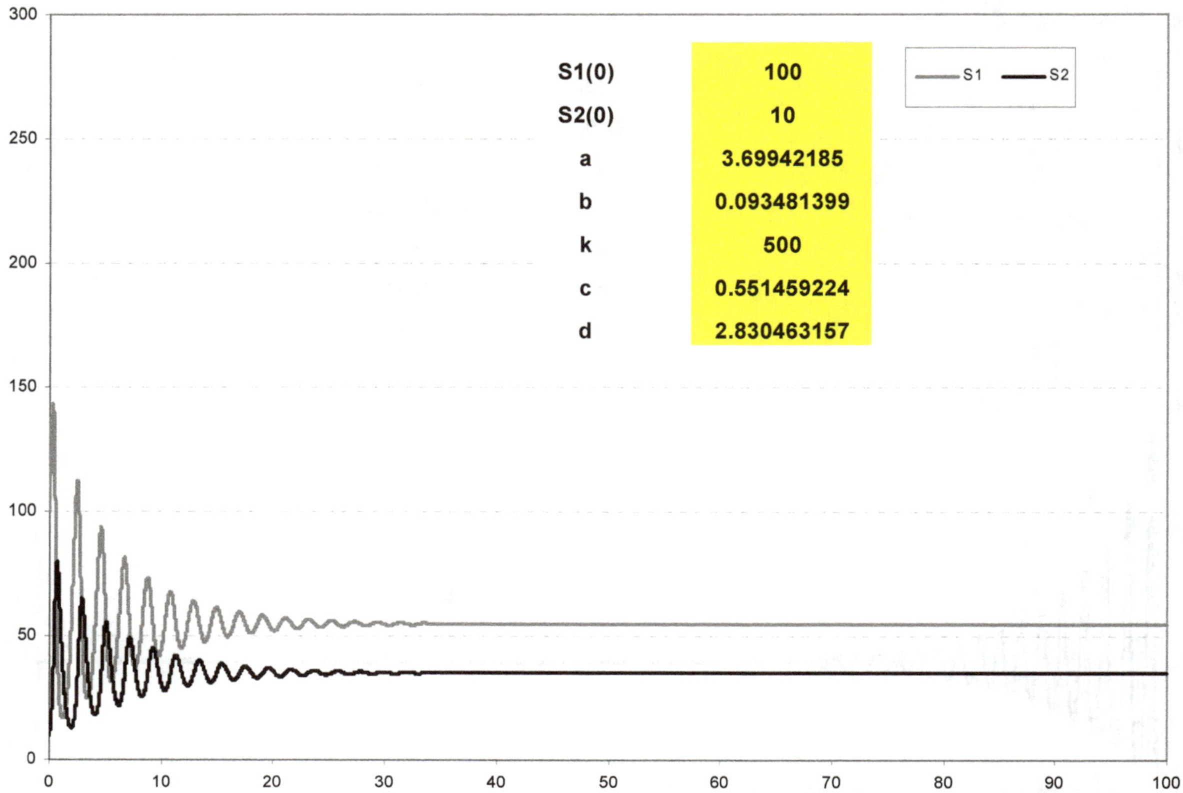

Fig. 5 Ruled Lotka-Volterra system with equilibrium values $S_1= 54.92$ and $S_2= 35.23$ and sum equal to 90.15. The optimized parameters detected via Evolutionary Network Control are indicated above the two curves. The solution has been found through the software Control-Lab (Ferrarini, 2013e).

After 10,000 simulations (10% uncertainty) on the optimized parameters (alpha, beta, gamma and delta), we achieve the results of Table 3. The solution depicted in Fig. 5 is robust with regard to a 10% sensitivity analysis, as mean and median values of the 10,000 simulations are approximately equal to the desired solution $S_1+S_2= 90$.

Table 3 Sensitivity analysis (10% uncertainty) on the optimized parameters of Fig. 5.

Statistics	S_1	S_2
min	42.12	29.52
max	71.29	42.14
mean	55.22	35.27
median	54.91	35.13
std. dev.	5.66	2.65

Any other kind of network control is feasible using ENC, including the control of Lotka-Volterra models with $n>2$ actors. ENC can also be employed to impose early (or late) stability to Lotka-Volterra models in particular, but also to any arbitrary ecological and biological networks more in general (Ferrarini, 2015).

4 Conclusions

The control of ecological and biological networks has unlimited applications: a) neutralize damages to ecological and biological networks, b) safeguard rare and endangered species, c) manage ecological systems at the least possible cost, d) counteract the impacts of climate change.

While in previous papers I have showed how to globally and locally rule linear ecological and biological networks, here I have showed that Evolutionary Network Control is on top of the control of nonlinear networks as well.

References

Balena F. 2001. Programming Microsoft Visual Basic 6.0. Microsoft Press, Redmond, WA, USA

Ferrarini A. 2011a. Some thoughts on the controllability of network systems. Network Biology, 1 (3-4): 186-188

Ferrarini A. 2011b. Some steps forward in semi-quantitative network modelling. Network Biology, 1(1): 72-78

Ferrarini A. 2013a. Exogenous control of biological and ecological systems through evolutionary modelling. Proceedings of the International Academy of Ecology and Environmental Sciences, 3(3): 257-265

Ferrarini A. 2013b. Controlling ecological and biological networks via evolutionary modelling. Network Biology, 3(3): 97-105

Ferrarini A. 2013c. Computing the uncertainty associated with the control of ecological and biological systems. Computational Ecology and Software, 3(3): 74-80

Ferrarini A. 2013d. Networks control: introducing the degree of success and feasibility. Network Biology, 3(4): 115-120

Ferrarini A. 2013e. Control-Lab 5.0: a software for ruling Quantitative Ecological Networks using Ecological Network Control Manual, 137 pages

Ferrarini A. 2014. Local and global control of ecological and biological networks. Network Biology, 4(1): 21-30

Ferrarini A. 2015. Imposing early stability to ecological and biological networks through Evolutionary Network Control. Proceedings of the International Academy of Ecology and Environmental Sciences, 5(1) (in press)

Goldberg DE. 1989. Genetic Algorithms in Search Optimization and Machine Learning. Addison-Wesley, Reading, USA

Holland JH. 1975. Adaptation in natural and artificial systems: an introductory analysis with applications to biology, control and artificial intelligence. University of Michigan Press, Ann Arbor, USA

Lotka AJ. 1925. Elements of Physical Biology. Williams & Wilkins Co., Baltimore, USA

Pattison T. 1998. Programming Distributed Applications with COM and Microsoft Visual Basic 6.0. Microsoft Press, Redmond, WA, USA

Volterra V. 1926. Variazioni e fluttuazioni del numero d'individui in specie animali conviventi. Memoriale Accademia Nazionale dei Lincei di Roma, 2: 31-113

7

Exotic species and the structure of a plant-galling network

Walter Santos de Araújo[1], Julio Miguel Grandez-Rios[2], Leonardo Lima Bergamini[3], Ján Kollár[4]
[1]Departamento de Biologia Geral, Universidade Estadual de Montes Claros, 39401-089, Montes Claros, Brazil
[2]Center for International Forestry Research, 1895, Lima, Peru
[3]Departamento de Ecologia, Universidade Federal de Goiás, 74001-970, Goiânia, Brazil
[4]Department of Planting Design and Maintenance, Slovak University of Agriculture in Nitra, 949-76, Nitra, Slovakia
E-mail: walterbioaraujo@yahoo.com.br, jgrandezrios@yahoo.com.br, llberbamini@gmail.com, jankollar82@gmail.com

Abstract
Gall-inducing insects are highly specialized herbivores and is expected that networks composed by gall-inducing insects and their host plants are also very specialized. However, presence of exotic species might reduce the interaction number for native species, which would lead to changes in the specialization of plant-galling networks. In this study, we use network metrics to describe, for the first time, the structure of a network of gall-inducing insects associated to ornamental host plants. We found that the plant-galling network has a low-connected structure and is more modular than expected by chance. Native insect herbivores were significantly more frequent on native host plant species, while exotic herbivores occurred mostly on exotic host plant species. On the other hand, the number of interactions between insect herbivores and native or exotic plant species did not vary. Our findings show that plant-galling networks are very specialized and structured independently of exotic species presence.

Keywords alien hosts; exotic herbivores; herbivory; network ecology.

1 Introduction

Gall-inducing insects are known as the most specialized insect herbivores of nature (Shorthouse et al., 2005). These are the only arthropod herbivores (together with mites) able to modify the host plant structures at histological and cellular level to induce galls (Stone and Schönrogge, 2003). To induce galls, the insects must deal with the mechanical and chemical defenses of plants, besides being able to control the physiological apparatus of the host (Shorthouse et al., 2005). This high degree of intimacy between galling and plant leads to phylogenetic constraints in the gall induction (Stone and Schönrogge, 2003). Because of this, even the most generalist galling insects are only able to induce galls on host plant species with some degree of phylogenetic relatedness (Price, 2005).

The structure of interactions in plant-galling networks has been largely neglected until very recently in the ecological literature (Barbour et al., 2016). Due to the high specialization degree of insect galls, it is expected that networks composed by gall-inducing insects and their host plants establish very specialized, and often species-specific, interactions. For this reason, is expected for plant-galling networks low connectivity and high compartmentalization in these interactions. The expected high degree of specialization of plant-galling networks, should affect the dynamics of these networks, because each plant species extinction may represent the extinction of a galling species.

A very interesting factor about the structure of plant-galling networks is the occurrence of exotic plant species. For example, Kollár (2011) recorded a network composed by galling insects that feed on ornamental trees and shrubs in the city of Nitra, Slovakia. Since it is located in an urban garden, the flora of this network is composed both by native species from the region as well as exotic plant species. Due their allochthonous origin, exotic plant species have not shared evolutionary history with the native species of galling insects. Many of the species able to induce galls in the exotic plants are also non-indigenous species that were introduced from the same place of origin as these plants (Kollár, 2011). Because of the different evolutionary histories, however, it is expected that native and exotic species both of plants and insects have distinctive ecological interactions in the network.

In the present work, we describe the structure of a network composed by native and exotic gall-inducing insects and ornamental host plants in the city of Nitra, Slovakia. To explore the network structure we calculate connectance, and also modularity and robustness of the interactions in contrast with null models. We also evaluated the structure of a network subset composed exclusively by native species of plants and insects. Additionally, we compare if native and exotic species of plants and insects in the network have differences in their number of interactions. As consequence of the phylogenetic restrictions of plant-galling interactions, we expected that exotic plant and insect species overall interact with a lower number of species than native ones. Finally, we tested whether exotic plant species have more interactions with exotic galling species than with native herbivores.

2 Material and Methods

2.1 Study area

The study was performed from 2004 to 2008 in the Nitra city park, Nitra, Slovakia (48° 19' 7" N, 18° 4' 55" E, 144 m a.s.l.). Nitra city is situated in southwestern Slovakia and its climate is characterized as semi-arid and humid (Kollár, 2011). The average annual total precipitation is about 600 mm and the average annual temperature is about 9.5 °C. The park, which covers 20 ha, has three parts: Sihoť, New Park, and Connecting Park. Samples were collected from all parts. The park is bordered by Nitra River and comprises woody plants in various age stages.

2.2 Sampling of plant-galling interactions

Gall-inducing insects were examined across active search on all native and introduced host woody plants in the park. All galls encountered were recorded and placed individually in labeled plastic bags for transportation to the laboratory. Fragments of each host plant were collected for botanical identification. Insect species were determined in field whenever possible according to the gall morphological characteristics. Some galls were collected to be reared in the laboratory for better determination of the gall-inducers. In these cases the publications of Csóka (1997), Schnaider (1976), Skuhravý and Skuhravá (1998), Blackman and Eastop (1994), Redfern and Shirley (2002) were used for determination. These studies were also used to determine if the insect species status were native or exotic.

We use the Fauna Europaea database (www.faunaeur.org) for taxonomic classification and correct terminology of gall-inducing insects. Plant species were identified using flora catalogs. Host plant status was determined from a checklist of alien plants to Slovakia (Medvecká et al., 2012). We checked plant species nomenclature and synonymy using The Plant List database (www.theplantlist.org). After the species determination, we build the network from the interactions between gall-inducing insect species and their host plant species.

2.3 Data analyses

Network structure analyses were done using the network descriptors connectance, modularity, and robustness. In order to compare changes in network structure caused by the introduction of exotic species, all three indices were computed for the entire network and for a network comprising only native insects and plants. The connectance (C) was calculated as the number of realized interactions (L) divided by number of possible interactions (S_1*S_2, where S_1 is the number of plant species and S_2 is the number of insect species) (Dormann et al., 2009; Zhang, 2011, 2012). Connectance is an inverse measure of overall interaction specialization because, in networks with higher connectance, species tend to have more connections (i.e., to be less specialized); thus, the higher the connectance, the lower the specialization of plant-herbivore networks (Araújo et al., 2015).

In addition to the connectance, we also used modularity to describe the structure of the plant-galling network. The modularity is characterized by presence of species subsets densely connected within the network, also called of modules or compartments (Dormann, 2011; Dormann et al., 2009), indicating the degree of specialization of the interaction arrangement within the network. We computed the bipartite modularity index Q (Barber, 2007) using the DIRTLPAwb+ algorithm to detect network modules (Beckett, 2016). In order to compare modularity values between different networks, we normalized the obtained value by comparing it to the maximum possible value given the network size and marginal totals (Beckett, 2016). The normalized modularity index reaches one when the network has the maximum possible modularity and approaches 0 as the modularity decreases. We then compared the observed normalized value with normalized values computed for 500 null networks generated using the quasiswap algorithm (Dormann, 2009).

Robustness is a measure of the resistance degree of the network to coextinctions (Dormann et al., 2009; Zhang, 2016). To characterize network robustness we used the exponent of the curve generated by the proportion of remaining galling species in function of the proportion of primary extinctions of plant species (Dormann et al., 2009). Robustness values also were compared to null values obtained from the same 500 null networks mentioned above. We performed all network analyses using adaptations of function from the R-package Bipartite (Dormann et al., 2008) and code provided in Beckett (2016).

We also compared if native and exotic species in the network have different average degree (AvD) than is expected from the overall mean number of plant and insect interactions. In this sense we differentiated the AvD between host plants (natives vs. exotics) and galling insects (natives vs. exotics) using t-tests. Additionally we tested whether native insect species interacted more with native plant species than with exotics by using a chi-squared test.

3 Results

The plant-herbivore network contained 90 gall-inducing species belonging to five orders and nine families of insects (Table 1). The orders with most species were Hymenoptera with 32 and Hemiptera with 31 species. The families Cynipidae (Hymenoptera), Aphididae (Hemiptera) and Cecidomyiidae (Diptera) were the most diverse with 27, 25 and 24 species, respectively. Most galling insect species were natives (81) and only nine species were of exotic origin. Seventy-three galling species were monophagous (induced galls on only one plant species) and 16 were oligophagous (10 induced galls on two host species and six on three hosts). Only the species *Rhyacionia buoliana* (Tortricidae: Lepidoptera) can be considered polyphagous, since it induced galls on five host species.

Gall-inducing insects were found on 54 species and 32 genera of host plants. The higher numbers of gall-inducing species were recorded on the host genera *Quercus* (28), *Populus* (8) and *Salix* (7). The native plant *Quercus robur* was the species that hosted the highest diversity with 27 galling species. Seventy-nine species of gall-inducing insects occurred on native plant species and 20 species on exotic host plants. Among all the species, only nine gall-inducing species occurred on both exotic and native host plants. Native insect herbivores were significantly more frequently observed on native host plant species, while exotic herbivores occurred mostly on exotic host plant species ($\chi^2 = 20.956$; p-value < 0.001).

Table 1 Checklist of host plants and gall-inducing insects recorded in Nitra City Park (Nitra, SW, Slovakia). Species marked with * are of exotic origin. Modification of Kollár (2011).

Host plants		Gall-inducing insects			
Family	**Species**	**Order**	**Family**	**Species**	**Occurrence**
Adoxaceae	*Viburnum lantana*	Hemiptera	Aphididae	*Aphis viburni*	Leaf
Adoxaceae	*Viburnum opulus*	Hemiptera	Aphididae	*Aphis viburni*	Leaf
Adoxaceae	*Viburnum rhytidophyllum**	Hemiptera	Aphididae	*Aphis viburni*	Leaf
Betulaceae	*Betula pendula*	Diptera	Cecidomyiidae	*Anisostephus betulinus*	Leaf
Betulaceae	*Carpinus betulus*	Diptera	Cecidomyiidae	*Zygiobia carpini*	Leaf
Buxaceae	*Buxus sempervirens**	Diptera	Cecidomyiidae	*Monarthropalpus flavus**	Leaf
		Hemiptera	Psyllidae	*Psylla buxi**	Leaf
Caprifoliaceae	*Lonicera ligustrina**	Hemiptera	Aphididae	*Hyadaphis tataricae**	Leaf
Caprifoliaceae	*Lonicera xylosteum*	Hemiptera	Aphididae	*Hyadaphis tataricae**	Leaf
		Hemiptera	Aphididae	*Rhopalomyzus lonicerae*	Leaf
Celastraceae	*Euonymus europaeus*	Hemiptera	Aphididae	*Aphis fabae*	Leaf
Cornaceae	*Cornus sanguinea*	Diptera	Cecidomyiidae	*Craneiobia corni*	Leaf
Cupressaceae	*Juniperus communis*	Diptera	Cecidomyiidae	*Oligotrophus juniperinus*	Needle
Fabaceae	*Gleditsia triacanthos**	Diptera	Cecidomyiidae	*Dasineura gleditchiae**	Leaf
Fabaceae	*Robinia pseudoacacia**	Diptera	Cecidomyiidae	*Obolodiplosis robiniae**	Leaf
		Hemiptera	Aphididae	*Aphis craccivora**	Leaf
		Hemiptera	Aphididae	*Aphis fabae*	Leaf
Fagaceae	*Fagus sylvatica*	Diptera	Cecidomyiidae	*Mikiola fagi*	Leaf
Fagaceae	*Quercus cerris*	Hymenoptera	Cynipidae	*Andricus cydoniae*	Leaf
Fagaceae	*Quercus hispanica*	Hymenoptera	Cynipidae	*Andricus anthracina*	Leaf
Fagaceae	*Quercus robur*	Diptera	Cecidomyiidae	*Macrodiplosis pustularis*	Leaf
		Diptera	Cecidomyiidae	*Macrodiplosis roboris*	Leaf
		Hymenoptera	Cynipidae	*Andricus anthracina*	Leaf
		Hymenoptera	Cynipidae	*Andricus conglomeratus*	Bud
		Hymenoptera	Cynipidae	*Andricus coriarius*	Bud
		Hymenoptera	Cynipidae	*Andricus curvator*	Leaf/Bud
		Hymenoptera	Cynipidae	*Andricus fecundator*	Bud
		Hymenoptera	Cynipidae	*Andricus glutinosus*	Bud
		Hymenoptera	Cynipidae	*Andricus grossulariae*	Flower
		Hymenoptera	Cynipidae	*Andricus hungaricus*	Bud

		Hymenoptera	Cynipidae	*Andricus inflator*	Bud
		Hymenoptera	Cynipidae	*Andricus kollari*	Bud
		Hymenoptera	Cynipidae	*Andricus lucidus*	Fruit
		Hymenoptera	Cynipidae	*Andricus mayri*	Fruit
		Hymenoptera	Cynipidae	*Andricus solitarius*	Bud
		Hymenoptera	Cynipidae	*Andricus testaceipes*	Shoot
		Hymenoptera	Cynipidae	*Biorrhiza pallida*	Root/Bud
		Hymenoptera	Cynipidae	*Cynips caputmedusae*	Fruit
		Hymenoptera	Cynipidae	*Cynips disticha*	Leaf
		Hymenoptera	Cynipidae	*Cynips divisa*	Leaf
		Hymenoptera	Cynipidae	*Cynips longiventris*	Leaf
		Hymenoptera	Cynipidae	*Cynips quercuscalicis*	Fruit
		Hymenoptera	Cynipidae	*Cynips quercusfolii*	Leaf
		Hymenoptera	Cynipidae	*Neuroterus laevisculus*	Leaf
		Hymenoptera	Cynipidae	*Neuroterus numismalis*	Leaf
		Hymenoptera	Cynipidae	*Neuroterus quercus-baccarum*	Leaf/Flower
		Hymenoptera	Cynipidae	*Trigonaspis megaptera*	Bud/Shoot
Grossulariaceae	*Ribes aureum**	Hemiptera	Aphididae	*Aphis idaei*	Leaf
		Hemiptera	Aphididae	*Aphis schneideri*	Leaf
Hydrangeaceae	*Philadelphus coronarius**	Hemiptera	Aphididae	*Aphis fabae*	Leaf
Malvaceae	*Hibiscus syriacus**	Hemiptera	Aphididae	*Myzus persicae**	Leaf
Malvaceae	*Tilia cordata*	Diptera	Cecidomyiidae	*Contarinia tiliarum*	Petiole/flower/leaf and shoot
		Diptera	Cecidomyiidae	*Dasineura tiliae*	Leaf
Malvaceae	*Tilia platyphyllos*	Diptera	Cecidomyiidae	*Contarinia tiliarum*	Petiole/flower/leaf and shoot
		Diptera	Cecidomyiidae	*Dasineura tiliae*	Leaf
		Diptera	Cecidomyiidae	*Didymomyia tiliacea*	Leaf
Oleaceae	*Fraxinus excelsior*	Diptera	Cecidomyiidae	*Dasineura acrophila*	Leaf
		Diptera	Cecidomyiidae	*Dasineura fraxini*	Leaf
		Hemiptera	Aphididae	*Prociphilus bumeliae*	Leaf
		Hemiptera	Psyllidae	*Psyllopsis fraxini*	Leaf
Oleaceae	*Fraxinus ornus*	Diptera	Cecidomyiidae	*Dasineura fraxini*	Leaf
Oleaceae	*Ligustrum vulgare*	Hemiptera	Aphididae	*Myzus ligustri*	Leaf
Pinaceae	*Abies alba*	Hemiptera	Adelgidae	*Dreyfusia nordmannianae**	Shoot
Pinaceae	*Abies concolor**	Hemiptera	Adelgidae	*Dreyfusia piceae*	Shoot
Pinaceae	*Larix decidua*	Diptera	Cecidomyiidae	*Dasineura kellneri*	Shoot
		Hemiptera	Aphididae	*Adelges laricis*	Shoot
Pinaceae	*Picea abies*	Hemiptera	Adelgidae	*Sacchiphantes viridis*	Shoot
		Hemiptera	Aphididae	*Adelges laricis*	Shoot

Pinaceae	*Picea glauca**	Hemiptera	Adelgidae	*Sacchiphantes viridis*	Shoot
Pinaceae	*Picea pungens**	Hemiptera	Adelgidae	*Sacchiphantes viridis*	Shoot
Pinaceae	*Pinus contorta**	Lepidoptera	Tortricidae	*Rhyacionia buoliana*	Bud
Pinaceae	*Pinus mugo*	Lepidoptera	Tortricidae	*Rhyacionia buoliana*	Bud
Pinaceae	*Pinus nigra**	Lepidoptera	Tortricidae	*Rhyacionia buoliana*	Bud
Pinaceae	*Pinus ponderosa**	Lepidoptera	Tortricidae	*Rhyacionia buoliana*	Bud
Pinaceae	*Pinus sylvestris*	Diptera	Cecidomyiidae	*Thecodiplosis brachyntera*	Needle
		Lepidoptera	Tortricidae	*Retinia resinella*	Bud
		Lepidoptera	Tortricidae	*Rhyacionia buoliana*	Bud
Pinaceae	*Pseudotsuga menziesii**	Hemiptera	Aphididae	*Gilleteella cooleyi**	Shoot
Rhamnaceae	*Rhamnus cathartica*	Hemiptera	Triozidae	*Trichochermes walkeri*	Leaf
Rosaceae	*Crataegus monogyna*	Hemiptera	Aphididae	*Dysaphis crataegi*	Leaf
Rosaceae	*Prunus avium*	Hemiptera	Aphididae	*Myzus cerasi*	Leaf
Rosaceae	*Rosa canina*	Diptera	Cecidomyiidae	*Dasineura rosae*	Leaf
		Hymenoptera	Cynipidae	*Diplolepis rosae*	Leaf bud
		Hymenoptera	Tenthredinidae	*Blennocampa pusilla*	Leaf
Rosaceae	*Rosa multiflora**	Diptera	Cecidomyiidae	*Dasineura rosae*	Leaf
		Hymenoptera	Cynipidae	*Diplolepis rosae*	Leaf bud
Salicaceae	*Populus x canescens*	Coleoptera	Cerambycidae	*Saperda populnea*	Shoot
Salicaceae	*Populus nigra*	Hemiptera	Aphididae	*Chaitophorus populicola*	Leaf
		Hemiptera	Aphididae	*Pemphigus borealis*	Bud
		Hemiptera	Aphididae	*Pemphigus bursarius*	Petiole
		Hemiptera	Aphididae	*Pemphigus populi*	Leaf
		Hemiptera	Aphididae	*Pemphigus populinigrae*	Leaf
		Hemiptera	Aphididae	*Pemphigus spirothecae*	Petiole
		Hemiptera	Aphididae	*Thecabius affinis*	Leaf
Salicaceae	*Populus simonii**	Hemiptera	Aphididae	*Pemphigus spirothecae*	Petiole
Salicaceae	*Salix alba*	Diptera	Cecidomyiidae	*Rabdophaga salicis*	Leaf
		Diptera	Cecidomyiidae	*Rhabdophaga rosaria*	Leaf
		Hemiptera	Aphididae	*Aphis farinosa*	Leaf
		Hymenoptera	Tenthredinidae	*Euura amerinae*	Leaf
		Hymenoptera	Tenthredinidae	*Pontania proxima*	Leaf
		Hymenoptera	Tenthredinidae	*Pontania vesicator*	Leaf
Salicaceae	*Salix purpurea*	Hymenoptera	Tenthredinidae	*Pontania viminalis*	Leaf
Sapindaceae	*Acer campestre*	Diptera	Cecidomyiidae	*Dasineura rubella*	Leaf
		Diptera	Cecidomyiidae	*Drisina glutinosa*	Leaf
Sapindaceae	*Acer platanoides*	Diptera	Cecidomyiidae	*Acericecis vitrina*	Leaf
		Diptera	Cecidomyiidae	*Drisina glutinosa*	Leaf
Sapindaceae	*Acer pseudoplatanus*	Diptera	Cecidomyiidae	*Acericecis vitrina*	Leaf
		Diptera	Cecidomyiidae	*Drisina glutinosa*	Leaf
Taxaceae	*Taxus baccata*	Diptera	Cecidomyiidae	*Taxomyia taxi*	Bud
Ulmaceae	*Ulmus glabra*	Hemiptera	Aphididae	*Eriosoma ulmi*	Leaf

		Hemiptera	Aphididae	*Tetraneura ulmi*	Leaf
Ulmaceae	*Ulmus laevis*	Hemiptera	Aphididae	*Eriosoma ulmi*	Leaf
		Hemiptera	Aphididae	*Kaltenbachiella pallida*	Leaf
		Hemiptera	Aphididae	*Tetraneura ulmi*	Leaf
Ulmaceae	*Ulmus minor*	Hemiptera	Aphididae	*Eriosoma ulmi*	Leaf
		Hemiptera	Aphididae	*Tetraneura ulmi*	Leaf

Altogether the plant-galling network comprised 116 interactions (Fig. 1), corresponding to only 2.38% of the 4,860 possible interactions (Table 2). Considering only the network of native herbivores and native plants the connectance was 2.72%. The observed modularity to the entire plant-galling network was very high and significantly higher than expected by chance (Table 2). The native-only insect and plant network subset showed a very high and significant modularity. In turn, network robustness considering all herbivores and native-only subset did not differ from the null model values (Table 2).

Native and exotic host plant species showed similar average degrees ($t = -1.56$, $df = 39.05$, p-value = 0.125), with most species interacting with only one species. Similarly, the average degree did not differ between native and exotic insects ($t = -1.44$, $df = 17.91$, p-value = 0.166), with both groups showing a high prevalence of monophagous insects.

Table 2 Network structure analysis results for plant-galling network recorded in Nitra City Park (Nitra, SW, Slovakia).

Analysis	Value observed	Null expectation	ICI null	ICS null	Z value	p
Connectance: all herbivores	0.0238	–	–	–	–	–
Connectance: only natives	0.0272	–	–	–	–	–
Modularity: all herbivores	0.9716	0.7996	0.7726	0.8308	11.3291	0.0020
Modularity: only natives	0.9874	0.8558	0.8420	0.8789	10.9983	0.0020
Robustness: all herbivores	0.5271	0.5393	0.5252	0.5532	-1.6752	0.0539
Robustness: only natives	0.5366	0.5252	0.5095	0.5423	1.3351	0.9022

4 Discussion

Most taxonomic groups of gall-inducing insects have a highly specific and intimate parasitic life-form on plant hosts, and each insect species generally presents an unique gall morphology (Price, 2005). Our results corroborate this high galling specialization since 81% of insect species were recorded on a single host plant species (i.e., monophagous insects). Oligophagous (17.7%) and polyphagous (1.3%) gall-inducing species were recorded on related host plant species, usually in the same plant genus (as is the case for *Rhyacionia buoliana* that induced galls on five species of *Pinus*). Furthermore, most of the recorded galling species induced unique gall morphotypes, even those that induced galls on more than one host species.

The high specificity of plant-galling interactions has led to a highly specialized network structure. The plant-galling network studied here has very little connected (2.38% of potential interactions) and has a low connectance compared to other plant-herbivore networks (Araújo et al., 2015; Cagnolo et al., 2011). Because of high network specialization, the structure of plant-galling network is more modular than expected by chance, but not more robust to random extinctions than randomly structured networks. This means that the removal of plant species causes few changes in the network structure, since each plant species is connected to one or a few species of galling insects. On the other hand, these results also mean that the loss of a plant species will, in most cases, inevitably lead to the extinction of an associated galling species in the network. Because the great dependence of gall-inducing insects on their host plants, the maintenance of galling insect populations necessarily depends on plant species conservation (Araújo, 2013).

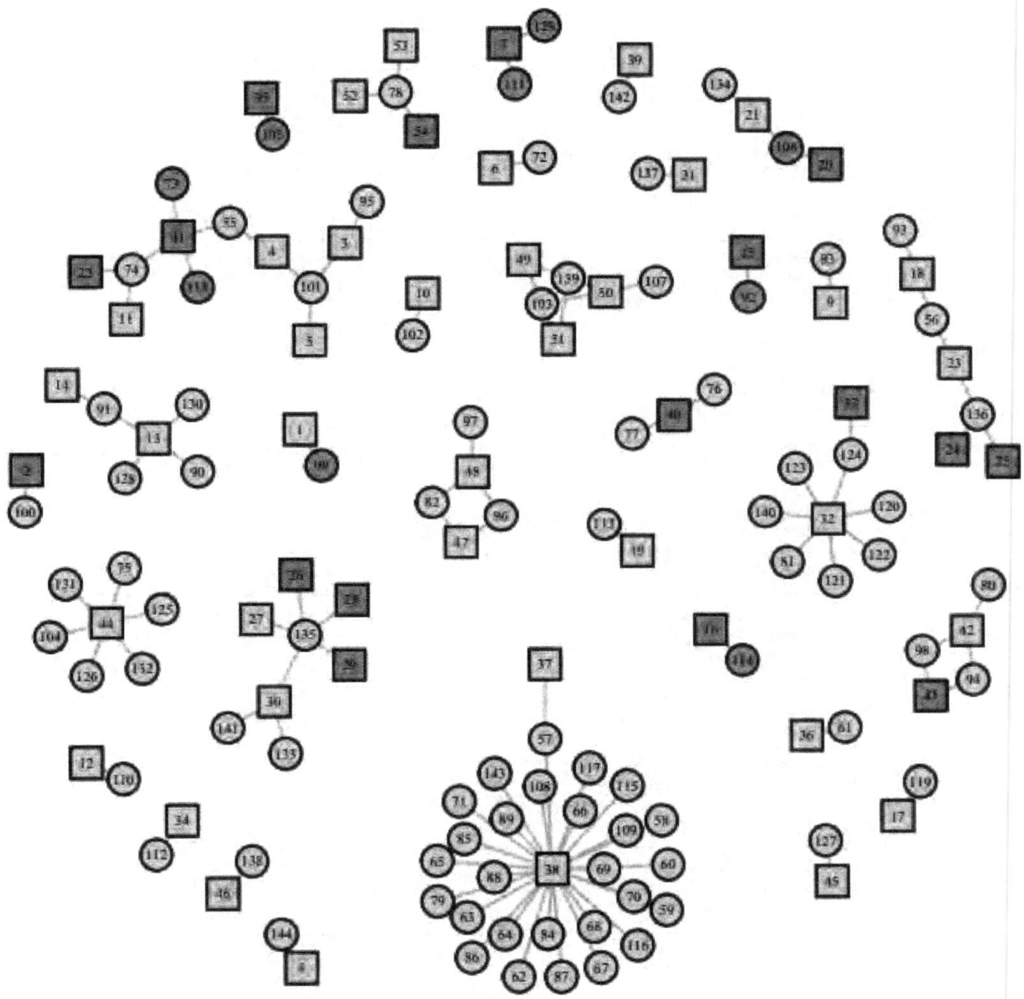

Fig. 1 Representation of the plant-galling network recorded in Nitra City Park (Nitra, SW, Slovakia). Host plant species are represented by squares and gall-inducing insect species are represented by circles. Dark gray indicates exotic species and light gray indicates native species of plants and insects. Codes of plant and insect species names are specified in the Table 3.

Contrary to expectations, native and exotic host plant species did not differ in the number of insect species they interact with. These findings indicate that the galling life-form presents very severe biological restrictions in that only a low number of insect species can induce galls, either on native or exotic plants. We also found that native and exotic insect species had a similar average degree (i.e., number of plant species that interact), counteracting our expectation of lower number of plant species being used by exotic insects than by natives. This absence of difference was observed because most of the insect species are monophagous, which results in average degrees ranging close to one for both native and exotic galling insects. Our findings are in agreement with previous studies that report that most of gall-inducing species are monophagous (e.g., Carneiro et al., 2009; Stone and Schönrogge, 2003).

Our study is the first, to our knowledge, that systematically evaluated the effect of exotic species occurrence on the structure of a plant-galling network. As we expected, native insect herbivores were significantly more frequent on native host plant species, while exotic herbivores occurred mostly on exotic host plant species. This result indicates very specific interactions even for exotic plants and insects, which results in plant-galling networks very specialized and similarly structured independently of exotic species presence. Future studies may investigate whether this pattern is the same for other groups of galling arthropods less specialized than insects (e.g., galling mites) or higher trophic levels (e.g., predators and parasitoids).

Table 3 List of codes for plant and insect species occurring in the plant-galling network recorded in Nitra City Park (Nitra, SW, Slovakia).

Code	Taxon	Species	Code	Taxon	Species
1	Plant	*Abies alba*	73	Insect	*Aphis craccivora*
2	Plant	*Abies concolor*	74	Insect	*Aphis fabae*
3	Plant	*Acer campestre*	75	Insect	*Aphis farinosa*
4	Plant	*Acer platanoides*	76	Insect	*Aphis idaei*
5	Plant	*Acer pseudoplatanus*	77	Insect	*Aphis schneideri*
6	Plant	*Betula pendula*	78	Insect	*Aphis viburni*
7	Plant	*Buxus sempervirens*	79	Insect	*Biorrhiza pallida*
8	Plant	*Carpinus betulus*	80	Insect	*Blennocampa pusilla*
9	Plant	*Cornus sanguinea*	81	Insect	*Chaitophorus populicola*
10	Plant	*Crataegus monogyna*	82	Insect	*Contarinia tiliarum*
11	Plant	*Euonymus europaeus*	83	Insect	*Craneiobia corni*
12	Plant	*Fagus sylvatica*	84	Insect	*Cynips caputmedusae*
13	Plant	*Fraxinus excelsior*	85	Insect	*Cynips disticha*
14	Plant	*Fraxinus ornus*	86	Insect	*Cynips divisa*
15	Plant	*Gleditsia triacanthos*	87	Insect	*Cynips longiventris*
16	Plant	*Hibiscus syriacus*	88	Insect	*Cynips quercuscalicis*
17	Plant	*Juniperus communis*	89	Insect	*Cynips quercusfolii*
18	Plant	*Larix decidua*	90	Insect	*Dasineura acrophila*
19	Plant	*Ligustrum vulgare*	91	Insect	*Dasineura fraxini*
20	Plant	*Lonicera ligustrina*	92	Insect	*Dasineura gleditchiae*
21	Plant	*Lonicera xylosteum*	93	Insect	*Dasineura kellneri*
22	Plant	*Philadelphus coronarius*	94	Insect	*Dasineura rosae*
23	Plant	*Picea abies*	95	Insect	*Dasineura rubella*
24	Plant	*Picea glauca*	96	Insect	*Dasineura tiliae*
25	Plant	*Picea pungens*	97	Insect	*Didymomyia tiliacea*
26	Plant	*Pinus contorta*	98	Insect	*Diplolepis rosae*
27	Plant	*Pinus mugo*	99	Insect	*Dreyfusia nordmannianae*
28	Plant	*Pinus nigra*	100	Insect	*Dreyfusia piceae*
29	Plant	*Pinus ponderosa*	101	Insect	*Drisina glutinosa*
30	Plant	*Pinus sylvestris*	102	Insect	*Dysaphis crataegi*

31	Plant	*Populus canescens*	103	Insect	*Eriosoma ulmi*
32	Plant	*Populus nigra*	104	Insect	*Euura amerinae*
33	Plant	*Populus simonii*	105	Insect	*Gilleteella cooleyi*
34	Plant	*Prunus avium*	106	Insect	*Hyadaphis tataricae*
35	Plant	*Pseudotsuga menziesii*	107	Insect	*Kaltenbachiella pallida*
36	Plant	*Quercus cerris*	108	Insect	*Macrodiplosis pustularis*
37	Plant	*Quercus hispanica*	109	Insect	*Macrodiplosis roboris*
38	Plant	*Quercus robur*	110	Insect	*Mikiola fagi*
39	Plant	*Rhamnus cathartica*	111	Insect	*Monarthropalpus flavus*
40	Plant	*Ribes aureum*	112	Insect	*Myzus cerasi*
41	Plant	*Robinia pseudoacacia*	113	Insect	*Myzus ligustri*
42	Plant	*Rosa canina*	114	Insect	*Myzus persicae*
43	Plant	*Rosa multiflora*	115	Insect	*Neuroterus laevisculus*
44	Plant	*Salix alba*	116	Insect	*Neuroterus numismalis*
45	Plant	*Salix purpurea*	117	Insect	*Neuroterus quercus-baccarum*
46	Plant	*Taxus baccata*	118	Insect	*Obolodiplosis robiniae*
47	Plant	*Tilia cordata*	119	Insect	*Oligotrophus juniperinus*
48	Plant	*Tilia platyphyllos*	120	Insect	*Pemphigus borealis*
49	Plant	*Ulmus glabra*	121	Insect	*Pemphigus bursarius*
50	Plant	*Ulmus laevis*	122	Insect	*Pemphigus populi*
51	Plant	*Ulmus minor*	123	Insect	*Pemphigus populinigrae*
52	Plant	*Viburnum lantana*	124	Insect	*Pemphigus spirothecae*
53	Plant	*Viburnum opulus*	125	Insect	*Pontania proxima*
54	Plant	*Viburnum rhytidophyllum*	126	Insect	*Pontania vesicator*
55	Insect	*Acericecis vitrina*	127	Insect	*Pontania viminalis*
56	Insect	*Adelges laricis*	128	Insect	*Prociphilus bumeliae*
57	Insect	*Andricus anthracina*	129	Insect	*Psylla buxi*
58	Insect	*Andricus conglomeratus*	130	Insect	*Psyllopsis fraxini*
59	Insect	*Andricus coriarius*	131	Insect	*Rabdophaga rosaria*
60	Insect	*Andricus curvator*	132	Insect	*Rabdophaga salicis*
61	Insect	*Andricus cydoniae*	133	Insect	*Retinia resinella*
62	Insect	*Andricus fecundator*	134	Insect	*Rhopalomyzus lonicerae*
63	Insect	*Andricus glutinosus*	135	Insect	*Rhyacionia buoliana*
64	Insect	*Andricus grossulariae*	136	Insect	*Sacchiphantes viridis*
65	Insect	*Andricus hungaricus*	137	Insect	*Saperda populnea*
66	Insect	*Andricus inflator*	138	Insect	*Taxomyia taxi*
67	Insect	*Andricus kollari*	139	Insect	*Tetraneura ulmi*
68	Insect	*Andricus lucidus*	140	Insect	*Thecabius affinis*
69	Insect	*Andricus mayri*	141	Insect	*Thecodiplosis brachyntera*
70	Insect	*Andricus solitarius*	142	Insect	*Trichochermes walkeri*
71	Insect	*Andricus testaceipes*	143	Insect	*Trigonaspis megaptera*
72	Insect	*Anisostephus betulinus*	144	Insect	*Zygiobia carpini*

Acknowledgment

We wish to thank to Jeffrey van Lent (Center for International Forestry Research) for his helpful suggestions on the manuscript and English revision. This research did not receive any specific grant from funding agencies.

References

Araújo WS. 2013. Different relationships between galling and non-galling herbivore richness and plant species richness: a meta-analysis. Arthropod Plant Interaction, 7: 373-377

Araújo WS, Vieira MC, Lewinsohn TM, Almeida-Neto M, 2015. Contrasting effects of land use intensity and exotic host plants on the specialization of interactions in plant-herbivore networks. PLoS One, 10: e0115606

Barber MJ. 2007. Modularity and community detection in bipartite networks. Physical Review E, 76:066102

Barbous MA, Fortuna MA, Bascompte J, Nicholson JR, Julkunen-Tiitto R, Jules ES, Crutsinger GM. 2016. Genetic specificy of a plant-insect food web: implications for linking genetic variation to network complexity. Proceedings of the National Academy of Sciences, 113: 2128-2133

Beckett SJ. 2016. Improved community detection in weighted bipartite networks. Royal Society Open Science, 3: 140536

Blackman RL, Eastop VF. 1994. Aphids on the World's Trees: An Identification and Information Guide. CABI Publishing, Wallingford, UK

Blüthgen N, Menzel F, Hovestadt T, Fiala B, Blüthgen N. 2007. Specialization, constraints & conflicting interests in mutualistic networks. Current Biology, 17: 1-6

Cagnolo L, Salvo A, Valladares G. 2011. Network topology: patterns and mechanisms in plant-herbivore and host-parasitoid food webs. Journal of Animal Ecology, 80: 342-351

Carneiro M, Branco CA, Braga CD, Almada E, Costa MM, Maia V, Fernandes GW. 2009. Are gall midge species (Diptera, Cecidomyiidae) host-plant specialists? Revista Brasileira de Entomologia, 53: 365-378

Csóka G. 1997. Plant Galls. Agroinform kiadó, Budapest, Hungary

Dormann CF. 2011. How to be a specialist? Quantifying specialisation in pollination networks. Network Biology, 1(1): 1-20

Dormann CF, Gruber B, Fründ J. 2008. Introducing the Bipartite Package: Analysing Ecological Networks. R News, 8: 8-11

Dormann CF, Fründ J, Blüthgen N, Gruber B. 2009. Indices, graphs and null models: analyzing bipartite ecological networks. The Open Ecology Journal, 2: 7-24

Dormann CF, Strauss R. 2014. A method for detecting modules in quantitative bipartite networks. Methods in Ecology and Evolution, 5: 90-98

Dunne JA, Williams RJ, Martinez ND. 2002. Network structure and biodiversity loss in food webs: robustness increases with connectance. Ecology Letters, 5: 558-567

Kollár J. 2011. Gall-inducing arthropods associated with ornamental woody plants in a City Park of Nitra (SW Slovakia). Acta Entomologica Serbica, 16: 115-126

Medvecká J, Kliment J, Májeková J, Halada Ľ, Zaliberová M, Gojdičová E, Feráková V, Jarolímek I. 2012. Inventory of the alien flora of Slovakia. Preslia, 84: 257-309

Newman MEJ. 2004. Analysis of weighted networks. Physical Review E, 70: 056131

Price PW. 2005. Adaptive radiation of gall-inducing insects. Basic and Applied Ecology, 6: 413-421

Shorthouse JD, Wool D, Raman A. 2005. Gall-inducing insects – Nature´s most sophisticated herbivores. Basic and Applied Ecology, 6: 407-411

Schnaider Z. 1976. Atlas of tree and shrub damages caused by insects and mites. Państwowe wydawnictwo naukowe, Warszawa

Skuhravý V, Skuhravá M. 1998. Gall midges of forest trees and shrubs. Matice lesnická, Písek

Stone G, Schönrogge K. 2003. The adaptive significance of insect gall morphology. Trends in Ecology and Evolution, 18: 512-522

Zhang WJ. 2011. Constructing ecological interaction networks by correlation analysis: hints from community sampling. Network Biology, 1(2): 81-98

Zhang WJ. 2012. Computational Ecology: Graphs, Networks and Agent-based Modeling. World Scientific, Singapore

Zhang WJ. 2016. Network robustness: Implication, formulization and exploitation. Network Biology, 6(4): 75-85

8

Local and global control of ecological and biological networks

Alessandro Ferrarini
Department of Evolutionary and Functional Biology, University of Parma, Via G. Saragat 4, I-43100 Parma, Italy
E-mail: sgtpm@libero.it, alessandro.ferrarini@unipr.it

Abstract
Recently, I introduced a methodological framework so that ecological and biological networks can be controlled both from inside and outside by coupling network dynamics and evolutionary modelling. The endogenous control requires the network to be optimized at the beginning of its dynamics (by acting upon nodes, edges or both) so that it will then go inertially to the desired state. Instead, the exogenous control requires that exogenous controllers act upon the network at each time step. By the way, all my previous works dealt with the goal of the global optimization of ecological and biological networks, i.e. how to drive them to a desired final state. Here I face another pivotal question: how can we locally (step-by-step) drive ecological and biological networks, so that also intermediate steps (not only the final state) are under our strict control? The ratio behind this question is that intermediate dynamics could potentially go below or above critical ecological-biological thresholds, hence invalidating the final global control. To this purpose, I introduce here a modelling solution to the complete control of ecological and biological networks that couples local and global control.

Keywords edges control; genetic algorithms; global dynamics; local dynamics; network control; nodes control.

1 Introduction

Recently, I introduced a methodological framework so that ecological and biological networks (EBN from now on) can be controlled (Ferrarini, 2011a). EBN can be efficaciously tamed from the outside (Ferrarini, 2013a) but also through the use of endogenous controllers (Ferrarini, 2013b), by coupling network dynamics and evolutionary modelling. The endogenous control requires that the network is optimized at the beginning of its dynamics (by acting upon nodes, edges or both) so that it will then go inertially to the desired state. Instead, the exogenous control requires that exogenous controllers act upon the network at each cycle.

In another paper (Ferrarini, 2013c), I faced a further important question: how reliable is the achieved solution? In other words, which is the degree of uncertainty about getting the desired result if values of edges and nodes were a bit different from optimized ones? This is an important question, because it's not assured that while managing a certain system we are able to impose to nodes and edges exactly the optimized values we would need in order to achieve the desired results. In order to face this topic, I have coined a 3-parts framework (network dynamics - genetic optimization - stochastic simulations).

I have further introduced a solution to the choice of the most feasible solution to network control (Ferrarini, 2013d) by introducing the concepts of control success and feasibility.

By the way, all my previous works dealt with the goal of the global optimization of EBN, i.e. how to drive them to a desired final state. Here I face another pivotal question: how can we locally (step-by-step) drive ecological and biological networks, so that also intermediate steps (not only the final state) are under our control? The ratio behind this question is that intermediate dynamics could potentially go below or above critical ecological-biological thresholds, hence invalidating the final global control. To this purpose, I introduce here a solution to the complete control of EBN by coupling local and global controls.

2 Mathematical Formulation

Given a generic ecological (or biological) dynamic system with n interacting actors

$$\frac{d\vec{S}}{dt} = \varphi(\vec{S},t) \qquad (1)$$

where S_i is the amount (e.g., number of individuals, total biomass, covered surface etc…) of the generic i-th actor, if we also consider inputs (e.g. species reintroductions) and outputs (e.g. hunting) from-to outside, we must write:

$$\frac{d\vec{S}}{dt} = \varphi(\vec{S},t) + \vec{I} + \vec{O} \qquad (2)$$

As noted by numerous authors (Liu et al., 2011; Slotine and Li, 1991) real systems' dynamics of eq. (2) can be modelled and simulated using a system of canonical, linear differential equations

$$\begin{cases} \dfrac{dS_1}{dt} = a_{11}S_1 + \ldots + a_{1n}S_n + I_1 + O_1 \\ \ldots \\ \dfrac{dS_n}{dt} = a_{n1}S_1 + \ldots + a_{nn}S_n + I_n + O_n \end{cases} \qquad (3)$$

that can be written in the compact form

$$\frac{d\vec{S}}{dt} = A * \vec{S} + \vec{I} + \vec{O} \qquad (4)$$

with initial values

$$\vec{S}_0 = \langle S_1(0), S_2(0) \ldots S_n(0) \rangle \qquad (5)$$

and co-domain limits

$$\begin{cases} S_{1\min} \le S_1(t) \le S_{1\max} \\ \ldots \\ S_{n\min} \le S_n(t) \le S_{n\max} \end{cases} \qquad (6)$$

and where

$$A = \begin{pmatrix} a_{11} & \dots & a_{1n} \\ \dots & \dots & \dots \\ a_{n1} & \dots & a_{nn} \end{pmatrix} \quad (7)$$

is the matrix of the per unit-of-time effect on S_i due to unitary S_j.

Such systems can be tamed from outside using the following 1-external-controller model (Ferrarini, 2013a):

$$\begin{cases} \left(\dfrac{dS_1}{dt}\right)_{OPT} = a_{11}S_1 + \dots + a_{1n}S_n + I_1 + O_1 + c_{11*}C_{1*} \\ \dots \\ \left(\dfrac{dS_n}{dt}\right)_{OPT} = a_{n1}S_1 + \dots + a_{nn}S_n + I_n + O_n + c_{n1*}C_{1*} \\ \dfrac{dC_1}{dt} = f_1S_1 + \dots + f_nS_n \end{cases} \quad (8)$$

where asterisks stand for the genetic optimization (Holland, 1975) of exogenous node's edges (i.e., coefficients of interaction with the inner system) and exogenous node's stock, i.e. the modification of such values at the beginning of network dynamics in order to get a certain goal (e.g., maximization of the final value of a certain variable). In case 1 controller is not enough, the model in (8) must be expanded to the following *k*-external-controllers model (Ferrarini, 2013a):

$$\begin{cases} \dfrac{dS_1}{dt} = a_{11}S_1 + \dots + a_{1n}S_n + I_1 + O_1 + c_{11*}C_{1*} + \dots + c_{1k*}C_{k*} \\ \dots \\ \dfrac{dS_n}{dt} = a_{n1}S_1 + \dots + a_{nn}S_n + I_n + O_n + c_{n1*}C_{1*} + \dots + c_{nk*}C_{k*} \\ \dfrac{dC_1}{dt} = f_{11}S_1 + \dots + f_{1n}S_n \\ \dots \\ \dfrac{dC_k}{dt} = f_{k1}S_1 + \dots + f_{kn}S_n \end{cases} \quad (9)$$

Alternatively, an ecological or biological network can be controlled from inside using the following control model (Ferrarini, 2013b):

$$\begin{cases} \left(\dfrac{dS_1}{dt}\right)_{OPT} = a_{11*}S_1^* + \dots + a_{1n*}S_n^* + I_{1*} + O_{1*} \\ \dots \\ \left(\dfrac{dS_n}{dt}\right)_{OPT} = a_{n1*}S_1^* + \dots + a_{nn*}S_n^* + I_{n*} + O_{n*} \end{cases} \quad (10)$$

where asterisks stand for the optimization of edges (i.e., coefficients of interaction among variables) or nodes (i.e., initial stocks), i.e. the modification of their values at the beginning of the network dynamics in order to get a certain goal.

The control models in (8), (9) and (10) are able to globally drive any EBN to the desired final state with an uncertainty degree that can be calculated as proposed in Ferrarini (2013c), and a feasibility degree as in Ferrarini (2013d).

Now, let's assume that we want to tame intermediate dynamics of EBN as well, not only the final state. By introducing an intermediate control function (*ICF*), I conceive several possibilities.

a) we want the generic actor $S_i(t)$ to stay as close as possible to a certain value during its dynamics.
In this case, *ICF* must be as follow

$$ICF = \int_{t=0}^{E} |k - S_i(t)| \, dt \tag{11}$$

where t is time (independent variable), E indicates the time at which the system goes to equilibrium, k is the value at which we want $S_i(t)$ to be tied with, vertical lines indicates the module of the difference.

Now, we can reach our goal by optimizing our network using (8), (9) or (10) in order to achieve the minimization of *ICF*.

As a result we can constrain $S_i(t)$ as close as possible to k along all its dynamics.

b) we want the generic actor $S_i(t)$ to stay as close as possible to a certain value but only during some steps of its dynamics.

In the previous step, I have subdued each dynamical step of $S_i(t)$ from the beginning to the its final state. It could be possible that we want to control $S_i(t)$ just in a certain time interval. In this case, eq. (11) becomes

$$ICF = \int_{t=n}^{m} |k - S_i(t)| \, dt \tag{12}$$

where n and m are generic points along the dynamic timeline. Again, eq. (12) must be minimized using (8), (9) or (10) in order to achieve our goal.

c) we want to tie the generic actor $S_i(t)$ to a certain dynamic $f(t)$ (e.g. exponential). In this case, we need the following equation

$$ICF = \int_{t=0}^{E} |f(t) - S_i(t)| \, dt \tag{13}$$

As a result, by minimizing *ICF* using (8), (9) or (10) we can constrain $S_i(t)$ as close as possible to the desired dynamic function.

d) we want to tie the generic actor $S_i(t)$ to another network's actor $S_j(t)$. In this case, we need the following equation to be minimized

$$ICF = \int_{t=0}^{E} |S_j(t) - S_i(t)| \, dt \tag{14}$$

e) we want to tie the generic actor $S_i(t)$ to a function of another network's actor $S_j(t)$. For instance we could want $S_i(t)$ to be always twice as $Sj(t)$ during network dynamics. In this case, we need the following equation to be minimized using (8), (9) or (10)

$$ICF = \int_{t=0}^{E} \left| f(S_j(t)) - S_i(t) \right| dt \qquad (15)$$

f) we want that the generic actor S_i changes from step to step of a certain value u. To this aim, I use

$$ICF = \int_{t=0}^{E} \left| \frac{dS(t)}{dt} - u \right| dt \qquad (16)$$

By minimizing *ICF* with (8), (9) or (10), we can impose that S_i changes of exactly u at each time step.

3 An Applicative Example

Fig. 1 shows an ecological network borrowed with modification from Ferrarini (2012). Greenish nodes represent positive actors (or events) for the goal of network control, i.e. the conservation of the target species (in green at centre of the network). Reddish nodes represent ecological actors (or events) with negative impact on the target species. Blueish nodes represent resources needed by the target species. The goal is to preserve target species' occurrence in the study area. Stocks stand for the actual amounts of individuals or biomass. Updates (indicated as *upd*) stand for yearly internal dynamics (i.e., intraspecific gains due to births and/or immigration rates minus losses due to deaths and/or emigration rates). Minimum and maximum values stand for lowest and highest values of stock values. For the sake of simplicity, the maximum possible value for each actor (in italic hereafter) has been set to 100. The percent value associated to links represents the percentage of the receiver that is yearly consumed by the transmitter at the beginning of the network simulation. Road mortality and re-introductions accounts for 18 and 10 individuals per year respectively (average values of the last 5 years of field observations).

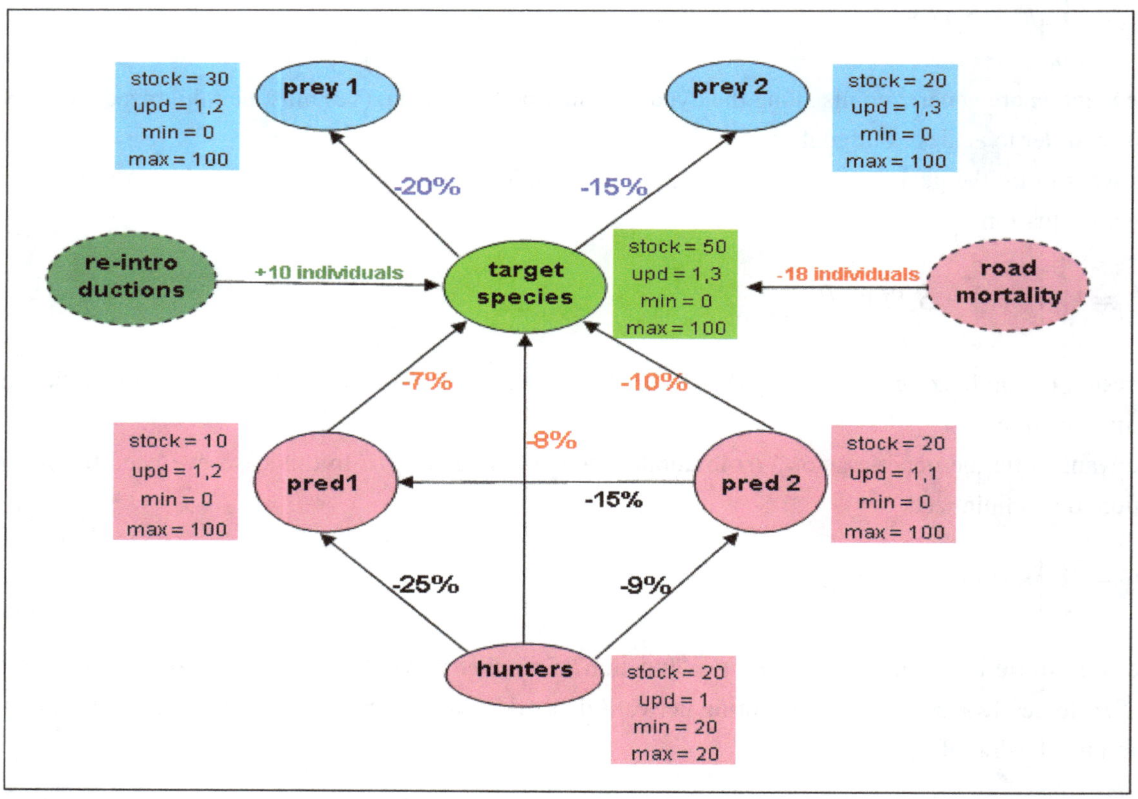

Fig. 1 The ecological network on which evolutionary control has been applied.

Since data are yearly-based, I have calculated inertial dynamics and optimized the above ecological system using a system of difference recurrent equations, instead of differential ones.

Table 1 The interaction matrix relative to the ecological network of Fig. 1.

INTERACTIONS MATRIX (per-cycle changes to receivers per-unit of transmitters)											
		Receivers									
		target	prey1	prey2	pred1	pred2	hunters	var G	var H	var I	var J
Transmitters	target	1.3	-0.12	-0.15	0	0	0	0	0	0	0
	prey1	0	1.2	0	0	0	0	0	0	0	0
	prey2	0	0	1.3	0	0	0	0	0	0	0
	pred1	-0.35	0	0	1.2	0	0	0	0	0	0
	pred2	-0.2	0	0	-0.075	1.1	0	0	0	0	0
	hunters	-0.2	0	0	-0.125	-0.09	1	0	0	0	0
	var G	0	0	0	0	0	0	0	0	0	0
	var H	0	0	0	0	0	0	0	0	0	0
	var I	0	0	0	0	0	0	0	0	0	0
	var J	0	0	0	0	0	0	0	0	0	0

The previous ecological network has the following inertial dynamics (Fig. 2), with the *target species* (green line) going extinction after 7 years.

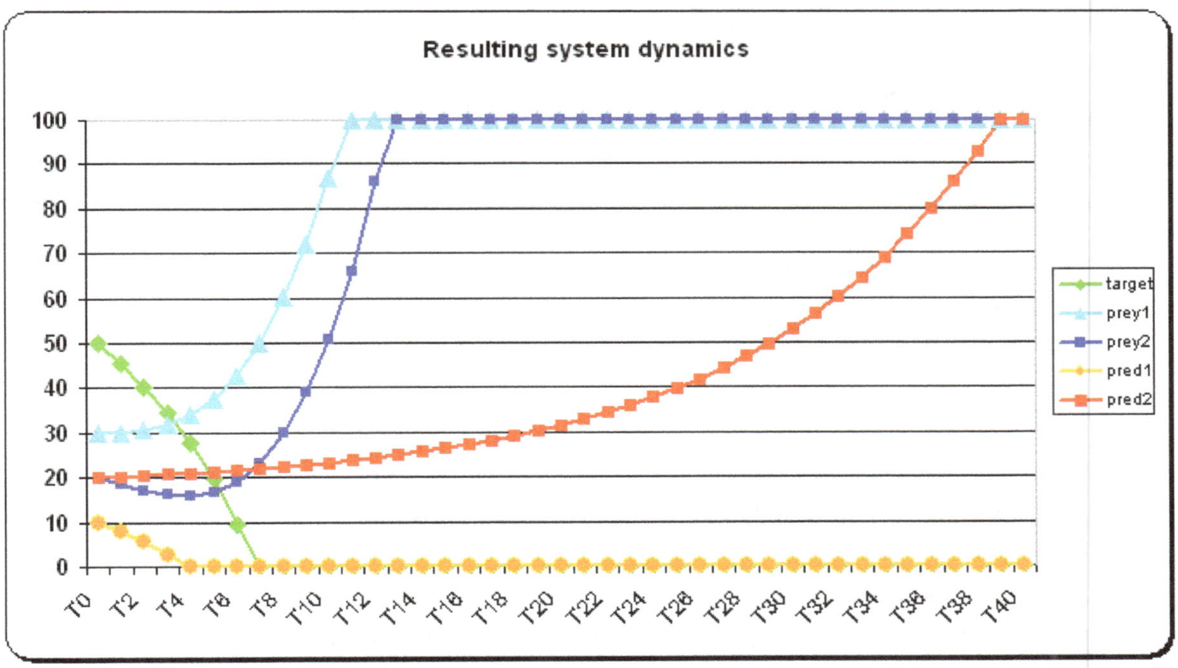

Fig. 2 Resulting dynamics for the network of Fig. 1. X-axis measures time in years.

4 Local Network Control

Previous research revealed that the optimal solution may be to search at a high rate of crossover, a low rate of mutation and proper population size (Kuo et al., 2000). In this study, crossover was set at a probability of 60% while mutations occur with a probability of 5%. This low setting helps to avoid getting trapped local optima during the search (D'heygere et al., 2006). The initial population consisted of 5000 chromosomes that were evolved over minimal 10,000 generations. I applied a steady-state genetic algorithm with a one-point crossover operator (Wall, 1996) to accomplish crossover. In this case the parent genome strings are cut at some random position to produce two "head" and two "tail" segments. The "tail" segments are swapped to produce two new genomes. For parent selection the roulette wheel selection method was used (Goldberg, 1989), where the likelihood of selection is proportionate to the fitness score given by the performance criterion. After crossover and mutation, the individuals with the lowest fitness scores were removed.

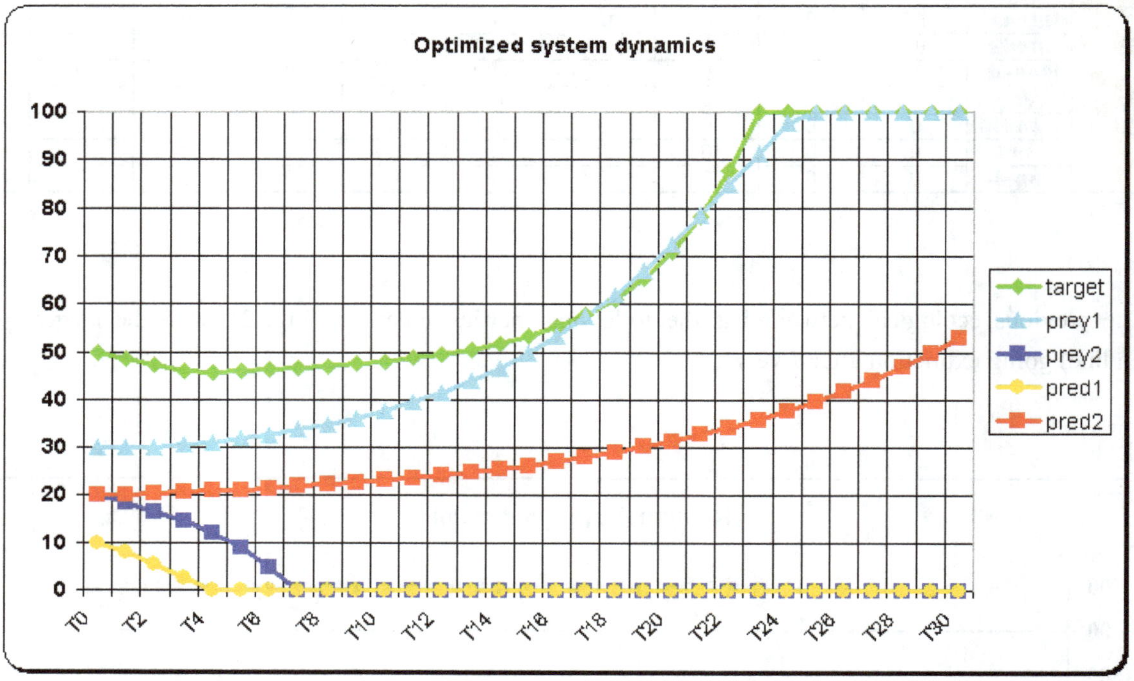

Fig. 3 By using both local and a global optimizations, target species has been constrained close to the stock value 50 for 20 years before going to its maximum.

Now let's suppose we want to stabilize the *target species* to a steady value equal to 50 for at least 20 years before it goes to its maximum. Let's suppose we want to reach it by acting upon *target species*' intraspecific coefficient (actually equal to 1.30). To this aim, I used equations (10) and (12) and, being all other simulation parameters equal, I achieved the results of Fig. 3 where, by setting the intraspecific coefficient equal to 1.36 (solution of the optimization problem), *target species* is constrained to stock values close to 50 for 20 years before going to its maximum. It is not goal of this paper to discuss the ecological and biological meaning of such constrained dynamics. I just want to expose here the methodological aspects of local and global networks control.

Now let's suppose we want to tie *target species*' dynamics to *pred2*'s dynamics so that they are as close as possible as time goes on, and so that they both go to their maximum (i.e. 100) at equilibrium. To this aim, I have used equations (10) and (14) and achieved the optimized matrix of Table 2 and the optimized dynamics of Fig. 4. As it can be seen, I was able to achieve a perfect solution (green and red lines are overlying and they both go to 100 at equilibrium).

Table 2 Optimized matrix in order to tie *target species*' dynamics to *pred2*'s ones, and so that they both go to their maximum. The corresponding dynamics are depicted in Fig. 4.

	target	prey1	prey2	pred1	pred2	hunters
	50	75	46	33	50	8
target	1.91	-0.41	-0.25	0.00	0.00	0.00
prey1	0.00	1.43	0.00	0.00	0.00	0.00
prey2	0.00	0.00	1.95	0.00	0.00	0.00
pred1	-0.14	0.00	0.00	1.00	0.00	0.00
pred2	-0.15	0.00	0.00	-0.60	1.44	0.00
hunters	-0.61	0.00	0.00	-0.32	-0.16	1.00

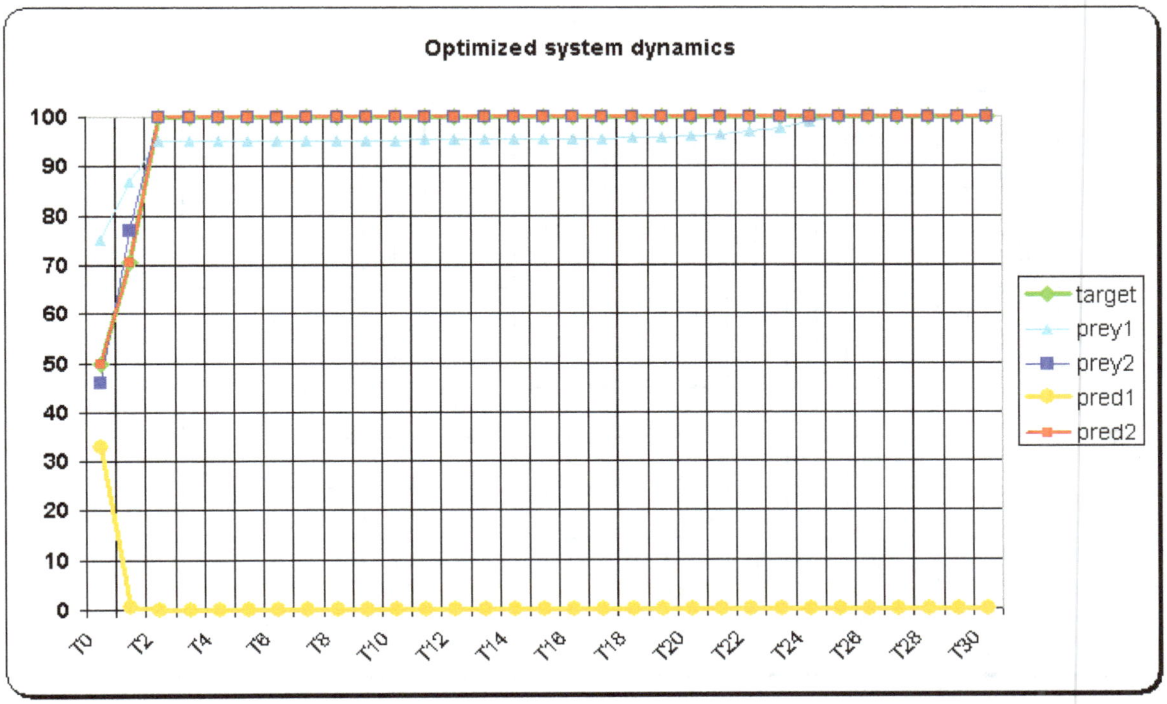

Fig. 4 Optimized dynamics where *target species* and *pred2* are tied together, and they both go to their maximum.

As last example, let's suppose we want the *target species* to go to its maximum with a smooth dynamic (i.e. small increments), let's say close to 2.5 at each time step. To this aim, I used equations (10) and (16), and I achieved the optimized matrix of Table 3 and the optimized dynamics of Fig. 5. As it can be seen, *target species* increases smoothly from its initial stock to the final one as it was purposed. The *target species*' initial value was set to 43 instead of 50 by the optimization model because I stopped GAs simulations after 5 minutes, so the methodological framework was not able to detect a perfect solution, but only something very close to it.

It's clear from these examples that, using the proposed methodological framework, there's no limit to the kind of network control we can search for. It's also clear that the same goal can be achieved via alternative and competing optimized solutions that my framework is able to detect.

Table 3 Optimized matrix that produces the corresponding dynamics depicted in Fig. 5.

	target	prey1	prey2	pred1	pred2	hunters
	43	100	36	16	40	0
target	1.46	-0.78	-0.41	0.00	0.00	0.00
prey1	0.00	1.81	0.00	0.00	0.00	0.00
prey2	0.00	0.00	1.82	0.00	0.00	0.00
pred1	-0.01	0.00	0.00	1.30	0.00	0.00
pred2	-0.24	0.00	0.00	-0.48	1.10	0.00
hunters	-0.35	0.00	0.00	-0.02	-0.14	1.00

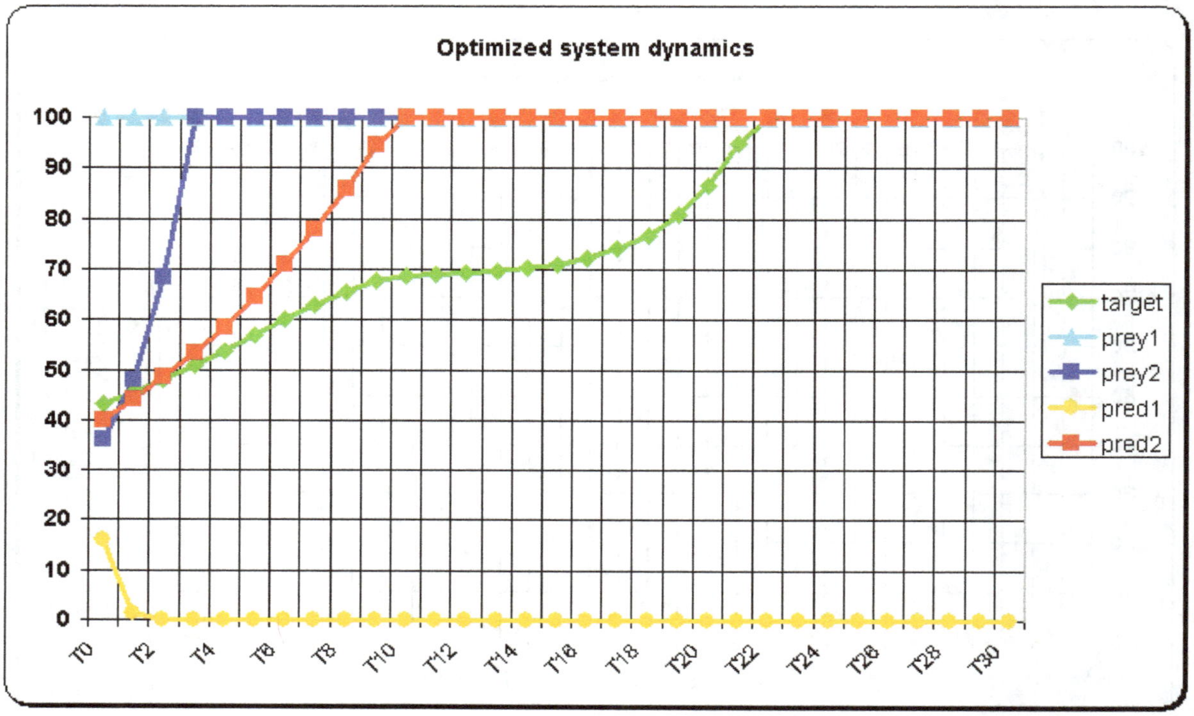

Fig. 5 Optimized dynamics where *target species* goes smoothly to its maximum increasing by about 2.5 at each time step.

Equations from (1) top (16) are powerful enough to allow for a complete control of EBN. Not only, in fact, we can achieve global control by attaining the desired final state, but we can also constrain local network's dynamics, i.e. intermediate states.

As a rule-of-thumb, I suggest that as the local control becomes more and more complex (e.g. we want to locally and globally control more than one variable), one should use a wider control on more switches. For instance one could start by controlling only initial stocks if the optimization goal is simple to achieve, but for tougher goals one might enlarge the control to intra-specific and inter-specific parameters or input-output flows as well. The result is an ecological or biological network with customized local and global dynamics. The framework proposed here might also be applied to semi-quantitative and qualitative networks (Ferrarini, 2011b; Ferrarini, 2011c).

5 Conclusions

The control of EBN is a pivotal topic. Its applications are unlimited, for instance: a) neutralize damages to ecological and biological networks, b) safeguard rare and endangered species, c) manage ecological systems at the least possible cost, and d) counteract the impacts of climate change.

While in previous papers, I showed how to globally control EBN, here I introduced a modelling framework for their total control, by adding local constraints to variables' dynamics. The ratio is that sometimes intermediate dynamics could go below or above critical ecological-biological thresholds, hence invalidating the final global control. By locally controlling EBN, this potential drawback is fixed.

References

D'heygere T, Goethals PLM, Niels De Pauw N. 2006. Genetic algorithms for optimisation of predictive ecosystems models based on decision trees and neural networks. Ecological Modelling, 195: 20-29

Ferrarini A. 2011a. Some thoughts on the controllability of network systems. Network Biology, 1(3-4): 186-188

Ferrarini A. 2011b. Fuzzy cognitive maps outmatch loop analysis in dynamic modeling of ecological systems. Computational Ecology and Software, 1(1): 55-59

Ferrarini A. 2011c. Some steps forward in semi-quantitative network modelling. Network Biology, 1(1): 72-78

Ferrarini A. 2012. Network di biodiversità in alcune aree della provincia di Parma. Technical relation (in Italian)

Ferrarini A. 2013a. Exogenous control of biological and ecological systems through evolutionary modelling. Proceedings of the International Academy of Ecology and Environmental Sciences, 3(3): 257-265

Ferrarini A. 2013b. Controlling ecological and biological networks via evolutionary modelling. Network Biology, 3(3): 97-105

Ferrarini A. 2013c. Computing the uncertainty associated with the control of ecological and biological systems. Computational Ecology and Software, 3(3): 74-80

Ferrarini A. 2013d. Networks control: introducing the degree of success and feasibility. Network Biology, 3(4): 115-120

Goldberg D.E. 1989. Genetic Algorithms in Search Optimization and Machine Learning. Addison-Wesley, Reading, USA

Holland JH. 1975. Adaptation in natural and artificial systems: an introductory analysis with applications to biology, control and artificial intelligence. University of Michigan Press, Ann Arbor, USA

Kuo SF, Merkley GP, Liu CW. 2000. Decision support for irrigation project planning using a genetic algorithm. Agriculture and Water Management, 45: 243–266

Liu YY, Slotine JJ, Barabasi AL. 2011. Controllability of complex networks. Nature, 473:167-173

Slotine JJ, Li W. 1991. Applied Nonlinear Control. Prentice-Hall, USA

Wall M. 1996. GAlib: A C++ Library of Genetic Algorithm Components (Version 2.4). Massachusetts Institute of Technology, USA

9

Network criminology: the criminology based on network science

WenJun Zhang
School of Life Sciences, Sun Yat-sen University, Guangzhou 510275, China; International Academy of Ecology and Environmental Sciences, Hong Kong
E-mail: zhwj@mail.sysu.edu.cn, wjzhang@iaees.org

Abstract
In present study, I proposed the science discipline, network criminology. Network criminology roots in criminology and network science, which focuses on network analysis of criminal networks. It uses the theory and methodology of network science to analyze, predict and control criminal patterns and behavior. The criminal network refers to a criminal group, a terrorism network, etc. Network criminology aims to understand topological structure, organization, function, identification, and control, etc., of criminal networks. Meanwhile, I defined the aims, scope, theory and methodology of network criminology.

Keywords network criminology; criminal networks; scientific discipline.

1 Introduction

In present study, I proposed the science, network criminology. Network criminology is the science on network analysis of criminal networks. The criminal network refers to a criminal group, a terrorism network, etc. In a criminal network, nodes represent individuals (e.g., offenders, victims, potential victims, or areas, etc), and links (i.e., ties) represent relationships between the individuals (Wikipedia, 2016a). Network criminology aims to understand topological structure, organization, function, identification, and control, etc., of criminal networks. Meanwhile, I defined the aims, scope, theory and methodology of network criminology in this study.

2 Definition, Aims and Scope

Network criminology is an interdisciplinary science rooted in criminology, network science, sociology, and other related scientific disciplines. It is a network analysis based (Zhang, 2016c, 2016g, 2016p-q) criminology, and is thus a branch of criminology. It uses the theory and methodology of network science to analyze, predict and control criminal patterns and behavior. Network criminology aims to investigate and understand the structure, properties, organization, function, identification, and control, etc., of criminal networks (e.g., offender networks, offender-victim networks, offender-victim-area networks, etc). The scope of network criminology covers, but not limits to: (1) theories, models, algorithms and software of network criminalogy; (2) network analysis of criminal networks; (3) identification of criminal groups, terrorism networks, offenders, victims, and areas; (4) causes and consequences of criminal behaviors; (5) dynamics and control of criminal networks; (6) methods to prevent the development of criminal groups, terrorism networks, etc.

3 Scientific Foundations

3.1 Criminology

Criminology is the scientific area on the nature, extent, management, causes, control, consequences, and prevention of criminal patterns and behavior, both on the individual and social levels (Wikipedia, 2016a). Criminology is an interdisciplinary field in both the behavioral and social sciences.

Criminology has some branches, e.g., Marxist criminology, conflict criminology, biosocial criminology, and critical criminology, etc. Marxist criminology stated that "defiance is normal - the sense that men are now consciously involved…in assuring their human diversity." (Sparks, 1980). Marxist criminology, conflict criminology and critical criminology argue that most of the relationships between state and citizen are non-consensual and criminal law is not necessarily representative of public beliefs and wishes. Criminal law is exercised in the interests of the ruling or dominant class (Wikipedia, 2016a). Some theories in criminology include (Wikipedia, 2016a)

(1) Sociological positivism. Sociological positivism suggests that societal factors, e.g., poverty, membership of subcultures, or low levels of education, etc., are causes of crimes (Beirne, 1987; Lochner, 2004).

(2) Differential association: It holds that crime is learned through association (Anderson, 1992).

(3) Social structure theories (Hester and Eglin, 1992). There are various viewpoints on social structure theories. Social disorganization theory suggests that neighborhoods with poverty and economic problems tend to have high population turnover (Bursik Jr and Robert, 1988). Further, social ecology argues that crime rates are associated with poverty, disorder, richfull abandoned buildings, and other signs of community deterioration (Bursik Jr and Robert, 1988; Morenoff et al., 2001). Subcultural theory focuses on small cultural groups fragmenting away from the mainstream to generate their own values and meanings about life. Social control theories explain why people do not become criminal (Hirschi, 1969). Finally, social network analysis stresses using network analysis to explain criminal patterns and behavior.

(4) Symbolic interactionism. Symbolic interactionism focuses on the relationship between the powerful state, media, and conservative ruling elite and other less powerful groups.

(5) Individual theories. Traitor theory exploits how a process of brutalization by parents or peers that usually occurs in childhood results in violent crimes in adulthood (Rhodes, 2000). Rational choice theory argues that punishment, if certain, swift, and proportionate to the crime, was a deterrent for crime, with risks outweighing possible benefits to the offender. Routine activity theory draws upon control theories and explains crime in terms of crime opportunities that occur in everyday life (Felson, 1994).

(6) Biosocial criminology is an interdisciplinary field that aims to explain crime and antisocial behavior by exploring both biological and environmental factors (Kevin and Anthony, 2011).

3.2 Network science

The United States National Research Council defined network science as "the study of network representations of physical, biological, and social phenomena leading to predictive models of these phenomena." (Committee on Network Science for Future Army Applications, 2006). More specifically, it is the science dealt with complex networks such as telecommunication networks, computer networks, biological networks, and social networks, considering distinct elements or actors represented by nodes and the connections between the elements or actors as links (Wikipedia, 2016b). Network science draws on theories and methods including graph theory from mathematics, statistical mechanics from physics, data mining and information visualization from computer science, inferential modeling from statistics, and social structure from sociology (Zhang, 2012a; Wikipedia, 2016b).

Some new scientific disciplines, e.g., network biology, network pharmacology, network informatics, network toxicology, etc., are rooted in network science (Barabasi and Otlvai, 2004; Hopkins, 2007, 2008; Zhang, 2011a-b, 2012a-d, 2015a-d, 2016a-r).

4 Methodology

4.1 Data sources

There are two sources of data for research in network criminology, public databases and field investigation/survey. Public databases include unpublished public data and published data. They can be used to construct criminal networks and models. In addition, criminal data can be achieved by investigating the interactions between individuals.

4.2 Big data analytics

Big data is the data sets so large or complex that conventional data processing tecgniques are inadequate. Big data analytics is the process of examining big data to uncover hidden patterns, unknown correlations and other useful information. With big data analytics, e.g., high-performance data mining, predictive analytics, text mining, forecasting and optimization, one can analyze huge volumes of data that conventional analytics can not handle with (Zhang, 2007, 2016g, 2016p-q). Contents of big data analysis include analysis, capture, storage, search, querying, sharing, transfer, visualization, etc (Wikipedia, 2016c; Zhang, 2016g, 2016p-q). Many problems in network criminology may be addressed by using big data analytics.

4.3 Network construction and link prediction

Network construction is the basic procedure for producing a criminal network. There are various methods for constructing a network. Zhang (2011a, 2012a, 2012b) proposed a series of correlation methods to construct networks. Zhang and Li (2015) found that the utilization of Pearson correlation measure may produce false results. For this reason, Zhang (2015c) suggested partial linear correlation and proposed some other partial correlation measures, and used them to jointly predict network links (interactions) (Zhang, 2015b). In addition, there are some other studies and methods about construction and prediction of networks (Goh et al., 2000; Pazos and Valencia, 2001; Guimera and Sales-Pardo, 2009).

In many cases, the criminal networks we achieved are probably incomplete, i.e., some links are missing in the netwok. We may be interesting in finding missing links in the network. Link information in the incomplete network can be used to predict missing links (Clauset et al., 2008; Guimera and Sales-Pardo, 2009; Barzel and Barabási, 2013; Lü et al., 2015; Zhang, 2015d, 2016a, 2016d; Zhang and Li, 2015). Further, network evolution based (Zhang, 2012a, 2015a, 2016b), node similarity based (Zhang, 2015d), and correlation based (Zhang, 2007, 2011a, 2012a, 2012b, 2015b-c, 2016d; Zhang and Li, 2015) methods can be used in link prediction. For example, Zhang (2015d) proposed a node similarity based algorithm to predic missing links in the network. In this algorithm, whether a node v_k can connect to v_i or not, depending on the similarity between v_k and v_i, the similarities between v_i and its adjacent nodes, the similarities between v_k and the adjacent nodes of v_i, and the degree of node v_i, and vice versa. Pearson correlation measure, cosine measure, and (negative) Euclidean distance measure, contingency correlation measure, and Jaccard coefficient measure were used as the between-node similarity.

4.4 Network analysis

Network analysis covers a variety of areas and methods (Fath et al., 2007; Zhang, 2012a). The following methods and theory of network analysis can be used in network criminology.

4.4.1 Attribute analysis

Attribute analysis aims to screen node attributes (e.g., characteristics of offenders or victims) based on their contribution to topological structure of the network (Zhang, 2016e). Zhang (2016e) proposed a method to screen node attributes that significantly influence node centrality in the network. In this method, four node centralities, degree centrality, closeness centrality, betweenness centrality, and circuit centrality, can be used as the dependent variable and attribute-by-node data are used as the data of independent variables. Stepwise linear regression method was applied to screen statistically significant node attributes from candidate attributes.

Brantingham and Brantingham (1994, 1998) argues that four factors – accessibility through high-volume transportation conduits, placement, juxtaposition, and the operation of facilities can account for the criminogenic capacity of specific places (Wikipedia, 2016c).

4.4.2 Topological analysis

Topological analysis of networks mainly includes the following, which are useful in network criminology

Find trees in the network: DFS algorithm, Minty's algorithm, etc (Minty, 1965; Zhang, 2012a, 2016o).

Find the shortest tree: Kruskal algorithm (Zhang, 2012a).

Find the shortest path: Dijkstra algorithm, Floyd algorithm (Dijkstra, 1959; Zhang, 2012a, 2016r).

Find the maximal flow: Ford-Fulkerson algorithm (Ford and Fulkerson , 1956; Zhang, 2012a).

Find circuits (Paton, 1969; Zhang, 2012a).

Calculate network connectedness (connectivity), blocks, cut vertices, and bridges (Zhang, 2012a, 2016i, 2016m), for examples of application, closeness or connectedness of a criminal network, or crucial offenders, victims, potential victims, or areas, etc.

Calculate node centrality and find critical nodes (Zhang, 2012a, 2012c; Zhang and Zhan, 2011; Shams and Khansari, 2014; Khansari et al., 2016). These methods can be used to find critical offenders or victims in criminal networks, etc. Bichler et al. (2010) defined magnetic facilities as the attractiveness of a location for deviant behavior. They looked at the self-nominated hangouts of 5082 delinquent youth living in Southern California. The structure of these networks remained relatively constant over the time during the study period. The centrality statistics used were in-degree and betweenness to identify facilities operating as stable regional convergence locations. Of the locations included in the network, the top twenty with the highest in-degree centrality scores in both the valued and dichotomous networks were investigated for the three characteristics: facility type, place type and school accessibility (Wikipedia, 2016c).

Analyze degree distribution and find critical nodes (Huang and Zhang, 2012; Li and Zhang, 2013; Zhang, 2011a, 2012a, 2012c; Zhang and Zhan, 2011; Rahman et al., 2013; Nuwagaba and Hui, 2015). Critical offenders or victims can be screended by using these methods.

In network criminology, we are interesting in finding criminal groups. The methods for finding modules, mosaics, and sub-networks can be used in this aspect (Kondoh, 2008; Bascompte, 2009; Begum et al., 2014; Nuwagaba and Hui, 2015; Zhang, 2016f, 2016n; Zhang and Li, 2016). Zhang (2016f, 2016n), and Zhang and Li (2016) proposed some methods for identifying hierarchical sub-networks / modules and weighting network links based on their similarity in sub-network / module affiliation. These methods were based on the cluster analysis in which between-node similarity in sets of adjacency nodes is used. Two matrices, linkWeightMat and linkClusterIDs, were achieved by using the algorithm. Two links with both the same weight in linkWeightMat and the same cluster ID in linkClusterIDs belong to the same sub-network / module. Two links with the same weight in linkWeightMat but different cluster IDs in linkClusterIDs belong to two sub-networks / modules at the same hirarchical level. A sub-network / module with the greater weight is the more connected sub-network / modules.

4.4.3 Network stability and robusteness

Network stability and robustness are expected to find some utilization in the network criminology, e.g., analyzing stability and robustness of a criminal network that experienced external destruction or attack. Recently, Zhang (2016h, 2016l) exploited methods and theory on network robustness and network sensitivity. Network stability has been extensively studied in the past (Din, 2014). Studies on ecosystems show that there is a relationship between network connectance and different types of ecosystem stability. Some models suggest that lower connectance involve higher local (May, 1973; Pimm, 1991; Chen and Cohen, 2001) and global (Cohen et al., 1990; Chen and Cohen, 2001) stability, i.e., the system recovers faster after a disturbance. However, another theory suggests that a food web with higher connectance has more numerous reassembly pathways and can thus recover faster from perturbation (Law and Blackford, 1992).

4.4.4 Flow balance analysis

Flow balance analysis analyzes network flows at steady state. Differential equations and other equations are usually used to describe network dynamics (Chen et al., 2010; Schellenberger et al., 2011). As an example, Jain et al. (2011) used mathematical models to decipher balance between cell survival and cell death using insulin. Some standardized indices and matrices can be used in flow balance analysis (Latham, 2006; Fath et al., 2007; Zhang, 2012a), including Average Mutual Information (AMI) (Rutledge et al., 1976) and Ascendency (A) index (Ulanowicz, 1983, 1997). Zorach and Fath and Patten (1999) developed a measure (measures the evenness of flow in a network) for network homogenization. Ulanowicz (2003) presented effective measures (effective connectivity, effective flows, effective nodes, effective rules) for weighted networks. Compartmentalization index is used to measure the degree of well-connected subsystems within a network (Pimm and Lawton, 1980). Constraint efficiency is a measure of a total of constraints that govern flow out of individual compartments (Latham and Scully, 2002). Models and measures above are suggested using in network analysis of criminal networks.

4.5 Network dynamics, evolution and control

Network dynamics can be best described with various mathematical models. Zhang (2015a) proposed a generalized network evolution model and self-organization theory on community assembly, in which the model is a series of differential (difference) equations with different number as the time. In addition, Zhang (2016b) developed a random network based, node attraction facilitated network evolution method. The two dynamic models are useful to study the network evolution, dynamics, and to predict links. Ferrarini (2011, 2011b, 2013a-c, 2014, 2016) have proposed a series of thoughts and methods on the dynamics, controllability and dynamic control of biological networks. A network can be optimized to search for an optimal search plan, and achieve a topological structure so that the network possesses relative stability (Zhang, 2012a, 2016g, 2016p-q).

The dynamic control of network means to change topological structure and key parameters of the network stage by stage so that the goal function of entire network achieves the optimum or suboptimum (Zhang, 2012a, 2016g, 2016p-q).

The control and management of criminal networks are expected to be achieved by using mathematical network models (e.g., differential equations, difference equations) or computer simulation (e.g., addition/deletion of nodes/links following some given rules) of network structure.

Acknowledgment

We are thankful to the support of Discovery and Crucial Node Analysis of Important Biological and Social Networks (2015.6-2020.6), from Yangling Institute of Modern Agricultural Standardization, and High-Quality Textbook *Network Biology* Project for Engineering of Teaching Quality and Teaching Reform of Undergraduate Universities of Guangdong Province (2015.6-2018.6), from Department of Education of Guangdong Province.

References

Anderson F. 1992. Criminological Theory Summaries. Cullen & Agnew. Retrieved November 3, 2011

Barabasi AL, Otlvai ZN. 2004. Network biology: understanding the cell's functional organization. Nature Reviews Genetics, 5(2): 101-113

Barzel B, Barabási AL. 2013. Network link prediction by global silencing of indirect correlations. Nature Biotechnology, 31: 720-725

Begum M, Bagga J, Blakey A, Saha S. 2014. Network motif identification and structure detection with exponential random graph models . Network Biology, 4(4): 155-169

Beirne P. 1987. Adolphe quetelet and the origins of positivist criminology. American Journal of Sociology. 92(5): 1140-1169

Bichler GM, Malm A, Enriquez J. 2010. Magnetic Facilities: Identifying the Convergence Settings of Juvenile Delinquents. Crime and Delinquency, 1-28

Brantingham PJ, Brantingham PL. 1994. The Influence of Street Networks on the Patterning of Property Offenses. In: Crime Prevention Studies (Clarke, ed). Criminal Justice Press, NY, USA

Brantingham PJ, Brantingham PL. 1998. Environmental criminology: From theory to urban planning practice. Studies on Crime and Crime Prevention, 7: 31-60

Bursik Jr, Robert J. 1988. Social disorganization and theories of crime and delinquency: problems and prospects. Criminology, 26(4): 519-539

Chen Q, Wang Z, Wei DQ. 2010. Progress in the applications of flux analysis of metabolic networks. Chinese Science Bulletin, 2010, 55(22): 2315-2322

Chen X, Cohen JE. 2001. Global stability, local stability and permanence in model food webs. Journal of Theoretical Biology, 212: 223-235

Clauset A, Moore C, Newman MEJ. 2008. Hierarchical structure and the prediction of missing links in networks. Nature, 453: 98-101

Committee on Network Science for Future Army Applications. 2006. Network Science. National Research Council

Dijkstra EW. 1959. A note on two problems in connexion with graphs. Numerischemathematik, 1(1): 269-271

Din Q. 2014. Stability analysis of a biological network. Network Biology, 4(3): 123-129

Fath BD, Patten BC. 1999. Quantifying resource homogenization using network flow analysis. Ecological Modelling, 123: 193-205

Fath BD, Scharler UM, Ulanowiczd RE, Hannone B. 2007. Ecological network analysis: network construction. Ecological Modeling, 208: 49-55

Felson M. 1994. Crime and Everyday Life. Pine Forge, Pennsylvania, USA

Ferrarini A. 2011. Some thoughts on the controllability of network systems. Network Biology, 1(3-4): 186-188

Ferrarini A. 2013a. Exogenous control of biological and ecological systems through evolutionary modelling. Proceedings of the International Academy of Ecology and Environmental Sciences, 3(3): 257-265

Ferrarini A. 2013b. Controlling ecological and biological networks via evolutionary modelling. Network Biology, 3(3): 97-105

Ferrarini A. 2013c. Networks control: introducing the degree of success and feasibility. Network Biology, 3(4): 115-120

Ferrarini A. 2014. Local and global control of ecological and biological networks. Network Biology, 4(1): 21-30

Ferrarini A. 2016. Decentralized control of ecological and biological networks through Evolutionary Network Control. Network Biology, 6(3): 65-74

Ford LR Jr, Fulkerson DR. 1956. Maximal flow through a network. Canadian Journal of Mathematics, 8: 399-404

Goh CS, Bogan AA, Joachimiak M, et al. 2000. Co-evolution of proteins with their interactions partners. Journal of Molecular Biology, 299: 283-293

Guimera R, Sales-Pardo M. 2009. Missing and spurious interactions and the reconstruction of complex networks. Proceedings of the National Academy of Sciences of USA, 106: 22073-22078

Hester S, Eglin P. 1992. A Sociology of Crime. Routledge, London, UK

Hirschi T. 1969. Causes of Delinquency. Transaction Publishers, USA

Hopkins AL. 2007. Network pharmacology. Nature Biotechnology, 25(10): 1110-1111

Hopkins AL. 2008. Network pharmacology: the next paradigm in durg discovery. Nature Chemical Biology, 4(11): 682-690

Huang JQ, Zhang WJ. 2012. Analysis on degree distribution of tumor signaling networks. Network Biology, 2(3): 95-109

Jain S, Bhooshan SV, Naik PK. 2011. Mathematical modeling deciphering balance between cell survival and cell death using insulin. Network Biology, 1(1): 46-58

Kevin MB, Anthony W. 2011. Biosocial Criminology. In: The Ashgate Research Companion to Biosocial Theories of Crime (Beaver KM, ed). Routledge, USA

Khansari M, Kaveh A, Heshmati Z, et al. 2016. Centrality measures for immunization of weighted networks. Network Biology, 6(1): 12-27

Latham LG, Scully EP. 2002. Quantifying constraint to assess development in ecological networks. Ecological Modelling, 154: 25-44

Latham LG. 2006. Network flow analysis algorithms. Ecological Modelling, 192: 586-600

Law R, Blackford JC. 1992. Self-assembling food webs. A global view-point of coexistence of species in Lotka-Volterra communities. Ecology, 73: 567–578

Li JR, Zhang WJ. 2013. Identification of crucial metabolites/reactions in tumor signaling networks. Network Biology, 3(4): 121-132

Lochner L. 2004. The effect of Education on crime: Evidence from prison inmates, arrests, and self-reports. The American Economic Review, 94(1): 155-189

Lü LY, Pan LM, Zhou T, et al. 2015. Toward link predictability of complex networks. Proceedings of the National Academy of Sciences of USA, 112: 2325-2330

May RM. 1973. Stability and complexity in model ecosystems. Princeton University Press, USA

Minty GJ. 1965. A simple algorithm for listing all the trees of a graph. IEEE Trans on Circuit Theory, CT-12(1): 120

Morenoff J, Robert S, Stephen R. 2001. Neighborhood inequality, collective efficacy and the spatial dynamics of urban violence. Criminology, 39 (3): 517-560

Nuwagaba S, Hui C. 2015. The architecture of antagonistic networks: Node degree distribution, compartmentalization and nestedness. Computational Ecology and Software, 5(4): 317-327

Paton K. 1969. An algorithm for finding a fundamental set of cycles of a graph. Communications of the ACM, 12(9): 514-518

Pazos F, Valencia A. 2001. Similarity of phylogenetic trees as indicators of protein-protein interaction. Protein Engineering, 14: 609-614

Pimm SL. 1991. The Balance of Nature. University of Chicago Press, Chicago, USA

Rhodes R. 2000. Why They Kill: The Discoveries of a Maverick Criminologist. Vintage Books, New York, USA

Rutledge RW, Basorre BL, Mulholland RJ. 1976. Ecological stability: an information theory viewpoint. Journal of Theoretical Biology, 57: 355-371

Shams B, Khansari M. 2014. Using network properties to evaluate targeted immunization algorithms. Network Biology, 4(3): 74-94

Sparks RF. 1980. A critique of marxist criminology. Crime and Justice, 2: 170-171

Ulanowicz RE. 1983. Identifying the structure of cycling in ecosystems. Mathematical Biosciences, 65: 219-237

Ulanowicz RE. 1997. Ecology, the ascendent perspective. In: Complexity in Ecological Systems Series (Allen TFH, Roberts DW, eds). Columbia University Press, New York, USA

Wikipedia. 2016a. Criminology. https://en.wikipedia.org/wiki/Criminology. Assesed Nov 30, 2016

Wikipedia. 2016b. Network science. https://en.wikipedia.org/wiki/Network_science. Accessed February 3, 2016

Wikipedia. 2016c. Social network analysis (criminology). https://en.wikipedia.org/wiki/Social_network_ analysis_(criminology)?oldid=723893439. Accessed Nov 1, 2016

Zhang WJ. 2007. Computer inference of network of ecological interactions from sampling data. Environmental Monitoring and Assessment, 124: 253-261

Zhang WJ. 2011a. Constructing ecological interaction networks by correlation analysis: hints from community sampling. Network Biology, 1(2): 81-98

Zhang WJ. 2011b. Network Biology: an exciting frontier science. Network Biology, 1(1): 79-80

Zhang WJ. 2012a. Computational Ecology: Graphs, Networks and Agent-based Modeling. World Scientific, Singapore

Zhang WJ. 2012b. How to construct the statistic network? An association network of herbaceous plants constructed from field sampling. Network Biology, 2(2): 57-68

Zhang WJ. 2012c. Several mathematical methods for identifying crucial nodes in networks. Network Biology, 2(4): 121-126

Zhang WJ. 2012d. A Java software for drawing graphs. Network Biology, 2(1): 38-44

Zhang WJ. 2015a. A generalized network evolution model and self-organization theory on community assembly. Selforganizology, 2(3): 55-64

Zhang WJ. 2015b. A hierarchical method for finding interactions: Jointly using linear correlation and rank correlation analysis. Network Biology, 5(4): 137-145

Zhang WJ. 2015c. Calculation and statistic test of partial correlation of general correlation measures. Selforganizology, 2(4): 65-77

Zhang WJ. 2015d. Prediction of missing connections in the network: A node-similarity based algorithm. Selforganizology, 2(4): 91-101

Zhang WJ. 2016a. A node degree dependent random perturbation method for prediction of missing links in the network. Network Biology, 2016, 6(1): 1-11

Zhang WJ. 2016b. A random network based, node attraction facilitated network evolution method. Selforganizology, 3(1): 1-9

Zhang WJ. 2016c. Network chemistry, network toxicology, network informatics, and network behavioristics: A scientific outline. Network Biology, 6(1): 37-39

Zhang WJ. 2016d. Selforganizology: The Science of Self-Organization. World Scientific, Singapore

Zhang WJ. 2016e. Screening node attributes that significantly influence node centrality in the network. Selforganizology, 3(3): 75-86

Zhang WJ. 2016f. A method for identifying hierarchical sub-networks and weighting network links based on their similarity in sub-network affiliation. Selforganizology, 3(2): 43-53

Zhang WJ. 2016g. Network toxicology: A new science. Network Biology, 6(2): 40-49

Zhang WJ. 2016h. Network robustness: Implication, formulization and exploitation. Network Biology, 6(4): 75-85

Zhang WJ. 2016i. Detecting connectedness of network: A Matlab program and application in tumor pathways and a phylogenic network. Selforganizology, 3(4): 117-120

Zhang WJ. 2016j. A mathematical model for dynamics of occurrence probability of missing links in predicted missing link list. Network Pharmacology, 1(4): 86-94

Zhang WJ. 2016k. A node-similarity based algorithm for tree generation and evolution. Network Biology, 6(3): 55-64

Zhang WJ. 2016l. Some methods for sensitivity analysis of systems / networks. Network Pharmacology, 1(3): 74-81

Zhang WJ. 2016m. How to find cut nodes and bridges in the network? A Matlab program and application in tumor pathways. Network Pharmacology, 1(3): 82-85

Zhang WJ. 2016n. A method for identifying hierarchical sub-networks / modules and weighting network links based on their similarity in sub-network / module affiliation. Network Pharmacology, 1(2): 54-65

Zhang WJ. 2016o. Finding trees in the network: Some Matlab programs and application in tumor pathways. Network Pharmacology, 1(2): 66-73

Zhang WJ. 2016p. Network informatics: A new science. Selforganizology, 3(2): 43-50

Zhang WJ. 2016q. Network pharmacology: A further description. Network Pharmacology, 1(1): 1-14

Zhang WJ. 2016r. A Matlab program for finding shortest paths in the network: Application in the tumor pathway. Network Pharmacology, 1(1): 42-53

Zhang WJ, Li X. 2015. Linear correlation analysis in finding interactions: Half of predicted interactions are undeterministic and one-third of candidate direct interactions are missed. Selforganizology, 2(3): 39-45

Zhang WJ, Li X. 2016. A cluster method for finding node sets / sub-networks based on between- node similarity in sets of adjacency nodes: with application in finding sub-networks in tumor pathways. Proceedings of the International Academy of Ecology and Environmental Sciences, 6(1): 13-23

Zhang WJ, Zhan CY. 2011. An algorithm for calculation of degree distribution and detection of network type: with application in food webs. Network Biology, 1(3-4): 159-170

10

Network modelling is strictly required for predicting climate change impacts on biodiversity

Alessandro Ferrarini
Department of Evolutionary and Functional Biology, University of Parma, Via G. Saragat 4, I-43100 Parma, Italy
E-mail: sgtpm@libero.it, alessandro.ferrarini@unipr.it

Abstract
Robust models are pivotal to the prediction of future climate change impacts on biodiversity. A move must be made away from individualistic models of single species toward the implication of synergistically interacting species. The focus should be on the indirect effects due to biotic interactions. Thanks to these kinds of models, counterintuitive results for species could be achieved, emerging from complex biotic feedbacks involving that species-specific expectations are not of necessity consistent with those of their community. In this paper, the proposed approaches can tackle some important limitations of commonly-used individualistic models, as they can: a) deal with an optionally large number of species, b) take into account biotic interactions, c) forecast indirect effects caused by climate change.

Keywords biotic interactions; climate impact models; ecological communities; indirect effects; scenarios simulations.

1 Introduction

Over the next decades, climate change will be one of the major drivers affecting diversity, composition, structure and functioning of ecological communities. Specific changes will include shifts in ecologically crucial factors, such as temperature, rainfall, solar irradiance and wind (Zhang and Chen, 2011). These aspects of climate change are likely to have deep effects on natural communities, with also potential feedbacks from communities to climate (Wu and Zhang, 2012).

Global surface temperature has increased by 0.74 °C on average over the last century with greater warming on land surfaces than on oceans (IPCC, 2001). Future warming is likely to be between 1.1 °C and 6.4 °C by the end of the 21th century, depending on the projection scenario (IPCC, 2001).

An urgent issue is predicting the effects of increased temperature on ecological communities across the globe. Climate change will affect patterns and processes of species and food webs in a variety of ways such as alteration of dispersal rates, spatial uncoupling of interactions and shifted phenology resulting in a rearrangement of species mutual interactions. Studies investigating the consequences of future climate changes on species distributions often apply the assumption that species respond to climate changes in an individualistic way (Baselga et al., 2009; Heegaard and Vandvik, 2004; Tylianakis et al., 2007). Instead, climate-induced scenarios may cause unobvious local-scale alterations to the network of interactions among species, and biotic interactions may complicate the broad-scale control that the environment has on a species' distribution (Suttle et al., 2007). The indirect effects that are potentially sensitive to global climate change and the complex feedbacks that exist among species implicate that species-specific previsions are not necessarily consistent with those of their communities (Ferrier et al., 2006). Accordingly, single-species studies should be expanded to include a more general multi-species assessment based on some kind of synthesis of individualistic models (Ellis et al., 2007).

Changing individual species models to account for complex biotic interactions is challenging. Doing it without data hungry models is even tougher, and inevitably requires some kind of scaling from single species to whole interaction networks. In this paper, several network approaches are proposed that can tackle some important limitations of commonly-used individualistic models, as they can: a) deal with an optionally large number of species, b) take into account biotic interactions among species, c) forecast indirect effects caused by climate change.

2 Four Increasingly Complex Solutions

A full understanding of the effects of climate-induced scenarios on biotic communities necessarily requires the consideration of the whole interaction network among species (Ferrarini, 2012). This requires a methodological approach as follows (Fig. 1): *climate change scenario→ direct impact on each species → indirect impacts on all linked species → feedback impacts at community level.*

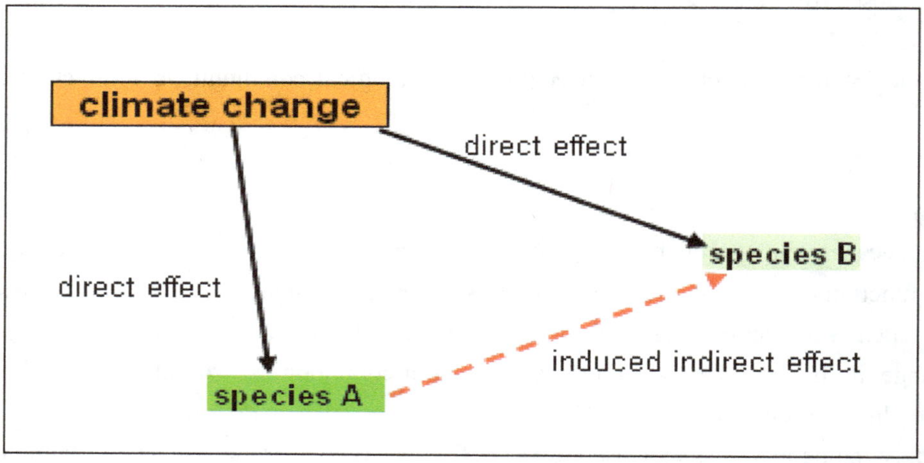

Fig. 1 The proposed conceptual framework for predicting climate change impacts on biodiversity.

This requires a $n \times n$ "climate-induced" interaction matrix whose off-diagonal elements w_{ij} take into account the inter-specific effects of species i upon species j as primed by a climate change scenario, instead elements w_{jj} take into account intra-specific effects (on growth, phenology and reproduction) of climate warming on single species.

In order to do this, I propose here the following methodological framework (Fig. 2) for network modelling of climate change impacts on ecological communities.

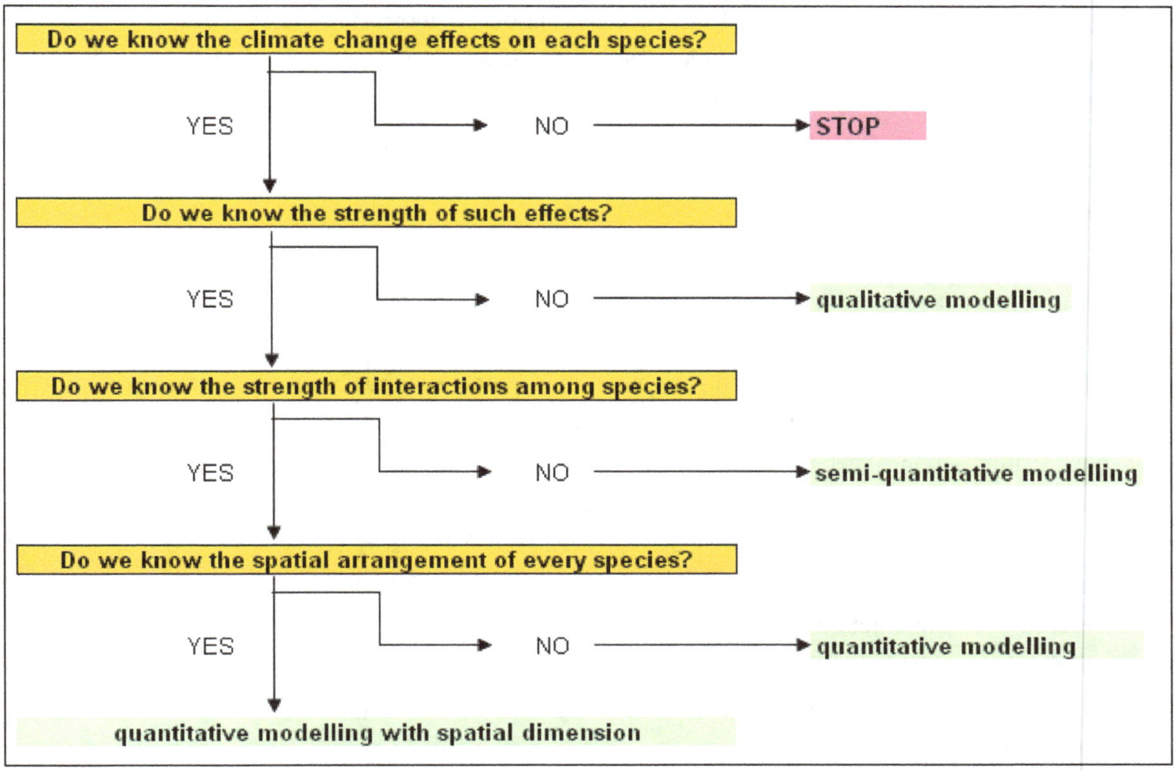

Fig. 2 The proposed framework for the network modeling of climate-induced impacts on biodiversity.

A quantitative network model of interacting species requires a precise estimate of the strength of species interactions induced by climate change, that could be unavailable for most species. On the other hand, a qualitative network model would lose information about strength of interactions. Moreover, a semi-quantitative model could encapsulate "hard" data (e.g., species coverages) based on precise statistical estimates, with "soft" data (e.g., theorized effect of temperature increase on single species), resulting in an overall parsimonious model of the community dynamics under a climate change scenario. The qualitative and semi-quantitative approaches are also worthy where the climate parameters required to feed into the network model are lacking or limited. This often happens for limited-in-size study areas whose extension is smaller or comparable to the spatial resolution of available climatic maps. Instead, most broad-scale studies use coarse resolution (e.g. 50 km × 50 km) data, and can be fed with quantitative climate projections (e.g. Worldclim data).

2.1 Qualitative networks

If precise predictions are not a requirement, I suggest that qualitative models could offer an alternative and cost-effective method for predicting biotic responses to climate change. A qualitative approach, like loop analysis (LA; Puccia and Levins, 1985), would provide predictions on the probable direction of change in species abundances, and would be suitable if only the direction of the effects of climate perturbations is required, not their magnitude (Ferrarini, 2012). The strength of this approach would be in its generality, and its ability to address the complex, nonlinear effects that feedbacks among species determine on single species and on the whole network of species.

LA uses signed digraphs to represent networks of interacting variables. System variables are depicted as nodes in the graph, and each connection between two nodes represents a non-zero coefficient of the community matrix. Press perturbations may act by changing one or more parameters in the growth rate of the variables. Taking the inverse of the community matrix provides an estimate of the direction of change in the equilibrium level of variables in response to these parameter changes. The element *aij* of the matrix represents the effect of variable *j* on the growth variable *i*, when the following equation is solved for a moving equilibrium:

$$\frac{dX_i}{dt} = f_i(X_1, X_2,..., X_n; C_1, C_2,..., C_h) \qquad (1)$$

where $X_1 \ldots X_n$ represent the variables and $C_1 \ldots C_h$ the parameters. Responses of abundances or biomass are arranged in a table of predictions whose signs show the predicted direction of change. The entries in a table denote variations expected in all the column variables when parameter inputs affect each row variable.

For the purposes of this paper, interacting variables (nodes) would represent species, while sign (+, 0, -) of induced connections among species could emerge from the following schema: let's suppose that climate change favors species *i* and disfavor species *j*, this causes species *i* to likely expand its niche hindering species *j*, in case they are in contact. Since each species can undergo 3 kinds of climate-induced effects (positive, negative, null), a couple of species can undergo: (+1, -1) interaction when one species is favorite and the other one disfavored; (-1, -1) interaction when they are both favorite or disfavored; (0, 0) represents no interaction (when the 2 species are both indifferent to climate change). Furthermore, there might be self-damped terms associated to density-dependent control (organisms with spatial limitations) or continuous supply of the species from outside the system.

2.2 Semi-quantitative networks

Fuzzy cognitive maps (FCM; Kosko, 1986) are particularly useful for applications where relationships between concepts cannot be expressed in exact mathematical equations (Ferrarini, 2011a; Ferrarini, 2011b). Biological and environmental quantities and their causal interactions are often described in relative and vague terms. A large proportion of the ecological information is represented in this way, and cannot be used as an input to data-driven mathematical or statistical models. The main advantage of FCM relies in its ability to represent such fuzziness.

FCM are semi-quantitative networks which describe the behaviour of a system in terms of concepts (nodes); each concept represents a variable (e.g., species' abundances) or a characteristic of the system. Values of nodes change over time, and take values in the interval [0, 100]: a value of 0 means that the factor is absent, a value of 100 means that the factor is present to the maximum possible extent, while a value of 50 represents the actual level of system variables. The causal links between nodes are represented by directed weighted edges that illustrate how much one concept influences the interlinked concepts, and the causal weights of the interconnections belong to the [-1, +1] interval. The strength of the weight w_{ij} indicates the degree of influence between concept C_i and concept C_j. The value of each concept at every simulation step is calculated as follows:

$$A_j(t) = A_j(t-1) + \sum_i A_i(t-1) * w_{ij} \qquad (2)$$

where $A_j(t)$ is the value of concept C_j at time *t*, $A_j(t-1)$ is the value of concept C_j at time *t-1*, w_{ij} is the weight of the interconnection from concept C_i to concept C_j.

For the purposes of this paper, while values of the generic *i-th* species C_i at time T_0 (actual surface cover) and diagonal values w_{jj} (direct effects due to climate change) might be assessed through *in situ* monitoring, off-diagonal w_{ij} values (climate-induced indirect effects) could be stochastically simulated. Instead of just using <+1, 0, -1> values as depicted above for quantitative networks, off-diagonal might be simulated using *n* simulations by varying off-diagonal values of a random increase up to 100% for favored species, and a random decrease up to 100% for disfavored species, subsequently achieving both the average expected dynamic and the expected deviations.

2.3 Quantitative networks

If also the strengths of induced interactions among species are known, a dynamical system of n interacting species under climate change impact behaves as follows:

$$\frac{d\vec{S}}{dt} = \varphi(\vec{S}, t) \tag{3}$$

where S_i is the number of individuals (or the total biomass) of the generic *i-th* species. If we also consider inputs and outputs from outer ecosystems, we must write:

$$\frac{d\vec{S}}{dt} = \varphi(\vec{S}, t) + \vec{I} + \vec{O} \tag{4}$$

This can be modelled and simulated using a system of canonical linear equations, as follows:

$$\begin{cases} \dfrac{dS_1}{dt} = a_{11}S_1 + \ldots + a_{1n}S_n + I_1 + O_1 \\ \ldots \\ \dfrac{dS_n}{dt} = a_{n1}S_1 + \ldots + a_{nn}S_n + I_n + O_n \end{cases} \tag{5}$$

that can also be written in a compact form

$$\frac{d\vec{S}}{dt} = A * \vec{S} + \vec{I} + \vec{O} \tag{6}$$

where

$$A = \begin{pmatrix} a_{11} & \ldots & a_{1n} \\ \ldots & \ldots & \ldots \\ a_{n1} & \ldots & a_{nn} \end{pmatrix} \tag{7}$$

is the matrix of the unitary-time effect on S_i due to unitary S_j, with initial values

$$\vec{S}_0 = \langle S_1(0), S_2(0) \ldots S_n(0) \rangle \tag{8}$$

and co-domain limits

$$\begin{cases} S_{1\min} \le S_1(t) \le S_{1\max} \\ \ldots \\ S_{n\min} \le S_n(t) \le S_{n\max} \end{cases} \tag{9}$$

The parameters of such equations could be determined through *in situ* experiments specifically for each study area, or through laboratory experiments where climate variables are accelerated and parameters are extracted specifically for each species and for each kind of interaction.

2.4 Quantitative networks incorporating spatial dimension

If also spatial arrangements of species are known, we need partial differential equations (PDEs). Let $S(x, y, z, t)$ measure species' biomass (or number of individuals) at time t and at the generic location $\langle x, y, z \rangle$, the climate-induced dynamical system of n spatially-explicit interacting species becomes:

$$\begin{cases} \frac{\delta S_1}{\delta t} = a_{11}S_{1(x,y,z)} + \ldots + a_{1n}S_{n(x,y,z)} + I_{1(x,y,z)} + O_{1(x,y,z)} \\ \ldots \\ \frac{\delta S_n}{\delta t} = a_{n1}S_{1(x,y,z)} + \ldots + a_{nn}S_{n(x,y,z)} + I_{n(x,y,z)} + O_{n(x,y,z)} \end{cases} \quad (10)$$

While this kind of network modelling is data hungry, it allows to achieve spatially-explicit predictions of climate change impacts on biotic communities.

3 Conclusions

Animals and plants are embedded in complex networks of interactions with other organisms, and the ways in which climate change works across the whole community is much more complex than the simple direct effects on single species. Local biotic interactions among species complicate the broad-scale control that climate has on species dynamics, and climate change scenarios is likely to cause unobvious alterations to the network of interactions among species. This is particularly true at local scale, where biotic interactions are cardinal to the community response (Trivedi et al., 2008).

Hence, the proposed approaches to the prevision of climate change effects on biotic communities focus on the importance of ecological indirect effects (Krivtsov, 2004; Salas and Borrett, 2011) based on the rationale that climate change determines direct effects on single species, thence inducing complex feedbacks at community-level and leading to indirect effects and unexpected outcomes as well.

References

Baselga A, Araújo MB. 2009. Individualistic vs community modelling of species distributions under climate change. Ecography, 32: 55-65

Ellis CJ, Coppins BJ, Dawson TP. 2007. Predicted response of the lichen epiphyte *Lecanora populicola* to climate change scenarios in a clean-air region of Northern Britain. Biological Conservation, 135: 396-404

Ferrarini A. 2011a. Fuzzy cognitive maps outmatch loop analysis in dynamic modeling of ecological systems. Computational Ecology and Software, 1(1): 55-59

Ferrarini A. 2011b. Some steps forward in semi-quantitative network modelling. Network Biology, 1(1): 72-78

Ferrarini A. 2012. I think different: models of climate change impact on plant species are unrealistic. Environmental Skeptics and Critics, 1(2): 30-33

Ferrier S, Guisan A. 2006. Spatial modelling of biodiversity at the community level. Journal of Applied Ecology, 43: 393-404

Heegaard E, Vandvik V. 2004. Climate change affects the outcome of competitive interactions - an application of principal response curves. Oecologia, 139: 459-466

IPCC 2001. Climate Change 2001: Impacts, Adaptation, and Vulnerability. Summary for Policymakers. Intergovernmental Panel on Climatic Change. Cambridge University Press, Cambridge, UK

Kosko B. 1986. Fuzzy cognitive maps. International Journal of Man-Machine Studies, 24: 65-75

Krivtsov V. 2004. Investigations of indirect relationships in ecology and environmental sciences: a review and the implications for comparative theoretical ecosystem analysis. Ecological Modelling, 174: 37-54

Puccia CJ, Levins R. 1985. Qualitative Modelling of Complex Systems: an Introduction to Loop Analysis and Time Averaging. Harvard University Press, Cambridge, UK

Salas AK, Borrett SR. 2011. Evidence for the dominance of indirect effects in 50 trophic ecosystem networks.

Ecological Modelling, 222: 1192-1204

Suttle KB, Thomsen MA, Power ME. 2007. Species interactions reverse grassland responses to changing climate. Science, 315: 640-642

Trivedi MR, Morecroft MD, Berry PM, et al. 2008. Potential effects of climate change on plant communities in montane nature reserves in Scotland, UK. Biological Conservation, 141: 1665-1675

Tylianakis JM, Tscharntke T, Lewis OT. 2007. Habitat modification alters the structure of tropical host–parasitoid food webs. Nature, 445: 202-205

Wu SH, Zhang WJ. 2012. Current status, crisis and conservation of coral reef ecosystems in China. Proceedings of the International Academy of Ecology and Environmental Sciences, 2(1): 1-11

Zhang WJ, Chen B. 2011. Environment patterns and influential factors of biological invasions: a worldwide survey. Proceedings of the International Academy of Ecology and Environmental Sciences, 1(1): 1-14

11

Using network properties to evaluate targeted immunization algorithms

Bita Shams, Mohammad Khansari

University of Tehran, Faculty of New Science and Technology, Amir Abad, North Kargar Street 14395-1374, Tehran, Iran

E-mail: m.khansari@ut.ac.ir

Abstract

Immunization of complex network with minimal or limited budget is a challenging issue for research community. In spite of much literature in network immunization, no comprehensive research has been conducted for evaluation and comparison of immunization algorithms. In this paper, we propose an evaluation framework for immunization algorithms regarding available amount of vaccination resources, goal of immunization program, and time complexity. The evaluation framework is designed based on network topological metrics which is extensible to all epidemic spreading model. Exploiting evaluation framework on well-known targeted immunization algorithms shows that in general, immunization based on PageRank centrality outperforms other targeting strategies in various types of networks, whereas, closeness and eigenvector centrality exhibit the worst case performance.

Keywords targeted immunization; epidemic spreading; network properties; centrality measures; complex networks.

1 Introduction

The threat of global epidemic spreading has resulted in widespread investigations on how to predict and control epidemics through populations (Gallos et al., 2007; Hadidjojo and Cheong, 2011; Hartvigsen et al., 2007; Salathé and Jones, 2010; Schneider et al., 2012, 2011). Vaccination is one of the most effective solutions which not only protects vaccinated people, but also can prevent transmitting disease among their friends (Cornforth et al., 2011). The traditional vaccination strategy, mass vaccination, might not always be feasible due to its high cost or scarce vaccination resources (Gallos et al., 2007; Hartvigsen et al., 2007; Vidondo et al., 2012). That's the reason why targeted immunization strategies are of the current interest in public health area (Bishop and Shames, 2011; Chen et al., 2008; Cohen et al., 2003; Cornforth et al., 2011; Dezső and Barabási, 2002; Eames et al., 2009; Gallos et al., 2007; Gao et al., 2011, 2010; Hadidjojo and Cheong, 2011; Hartvigsen et al., 2007; Hébert-Dufresne et al., 2013; Masuda, 2009; Miller and Hyman, 2007; Niu et al., 2009; Peng et al., 2010; Restrepo et al., 2006; Salathé and Jones, 2010; Schneider et al., 2012, 2011; Vidondo et al., 2012; Yamada and Yoshida, 2012; Yoshida and Yamada, 2012). Targeted immunization strategies vaccinate intercommunal individuals whose immunization prevent infectious propagation among different communities (Masuda, 2009; Salathé and Jones, 2010).

Impact of close contacts on disease transmission has led to new studies on infectious propagation using social network analysis (Cornforth et al., 2011; Gallos et al., 2007; Hartvigsen et al., 2007; Salathé and Jones, 2010; Weerasinghe, 2013). In these investigations, relations are modeled as an undirected graph whose nodes represent individuals and links represent their relationships (Salathé and Jones, 2010). Node centrality is one of the most studied concepts in social network analysis which determines nodes influence on network flows (Borgatti, 2005; Christley et al., 2005; Freeman, 1978). Accordingly, Applying node centrality measures into targeted immunization have resulted in high performance of vaccination strategies (Chen et al., 2008; Cohen et al., 2003; Dezső and Barabási, 2002; Eames et al., 2009; Gao et al., 2011; Hébert-Dufresne et al., 2013; Masuda, 2009; Miller and Hyman, 2007; Niu et al., 2009; Restrepo et al., 2006; Salathé and Jones, 2010; Schneider et al., 2012, 2011; Vidondo et al., 2012). Several centrality measures have been used in targeted immunizations. Vaccinating highest degree node is the most common strategy (Chen et al., 2008; Cohen et al., 2003; Dezső and Barabási, 2002; Eames et al., 2009; Gallos et al., 2007; Gao et al., 2011, 2010; Miller and Hyman, 2007; Niu et al., 2009; Schneider et al., 2012; Vidondo et al., 2012), whereas, immunizing nodes with highest betweenness centrality is acknowledged as most effective targeted immunization (Chen et al., 2008; Hébert-Dufresne et al., 2013; Salathé and Jones, 2010; Schneider et al., 2012, 2011). Moreover, the relationship between largest eigenvalue of network adjacency matrix and epidemic threshold (Chakrabarti and Faloutsos, 2003; Chakrabarti et al., 2008; Peng et al., 2010) inspired many researchers to use other centrality measures such as eigenvector (Bonacich, 1987; Restrepo et al., 2006) and PageRank (Miller and Hyman, 2007; Page et al., 1999).

Several investigations have been carried out to assess ability of centrality measures in identification of influential spreaders or high risk individuals (see Table 1). In spite of implicit relevance between these problems and network immunization, they are inherently different (Zhang, 2012a, 2012b). Influential spreaders are nodes that lead to faster and wider spreading in complex networks (Chen et al., 2011; Kiss and Bichler, 2008). The high-risk individuals are people with high probability of being infected (Christley et al., 2005). In contrast, targeted node for immunization are nodes whose removal results in the maximum reduction of epidemic spreading in a network (Habiba et al., 2010). As seen in Table 1, the only known research about impact of centrality measures in targeted immunization is conducted by Habiba who compared centrality measures efficiency in identifying spread blockers based on simulation of independent cascade model (Habiba et al., 2010). Regarding this experiments, local centrality measures such as degree are good indicators of blockers. Unfortunately, this research only considered independent cascade model which is not reliable on other epidemic spreading models such as SIR. Additionally, the simulation of epidemic model requires high computational cost (Ventresca and Aleman, 2013).

In spite of many literatures in targeted immunization algorithm, a comprehensive analysis on performance of immunization algorithms has never been reported to our knowledge (See Table 2). Although several experiments have been conducted in order to evaluate immunization algorithms, they have significant shortcomings. First, a small number of efficient immunization algorithms are included (Holme et al., 2002; Schneider et al., 2012, 2011). Second, the experiments are conducted based on a single network (Eames et al., 2009; Miller and Hyman, 2007; Ventresca and Aleman, 2013) or network structure (Hartvigsen et al., 2007). Third, limited number of structural properties are considered (Holme et al., 2002; Masuda, 2009; Schneider et al., 2012; Ventresca and Aleman, 2013). Fourth, the goal of immunization algorithms is not discussed. Fifth, the time complexity of the immunization algorithms is mostly ignored. Finally, above all, only simulation of special epidemic model has been considered which is not extensible to other epidemics.

Table 1 Centrality measures evaluation in the literatures of epidemic spreading.

Purpose	Authors	Centrality measures	Best measure	Approaches
influential spreader	Kiss and Bichler, 2008	In/out Degree, Betweenness, Closeness, PageRank, SenderRank	SenderRank Out Degree	Measuring number of reached customers in marketing message spreading initiated by each measures
	Chen et al., 2011	Degree, Closeness, Betweenness	Closeness	Measuring the number of infected nodes at different time in SIR simulation initiated by each measure
high-risk individuals	Christley et al., 2005	Degree, Closeness, Betweenness, Random Betweenness	Degree	Exploring the relation between node centralities and risk of infection in SIR simulation
Spread blockers	Habiba et al., 2010	Degree, Closeness, Betweenness, PageRank, Clustering Coefficient	Degree	Measuring reduction in number of infected individuals in simulation of independent cascade model spreading

The contributions of this paper are as follows:

- We propose a new evaluation framework regarding amount of available vaccination resources, goal of immunization program and time complexity. Considering available amount of immunization resources, we assess immunization algorithms in case of limited and flexible amount of resources. Largest eigenvalue of network adjacency matrix, size of the largest connected component, and, sum of square partitions. Additionally, immunization threshold and network robustness are measures of evaluating immunization algorithms in case of unlimited or flexible amount of vaccination resources. The measures evaluate immunization algorithms regarding different goal of immunization program which is discussed later.

- We exploit the introduced evaluation framework to study the impact of well-known centrality measures (i.e. degree, betweenness, closeness, eigenvector, and PageRank) on targeted immunization algorithms in various real and artificial networks. Then, we suggest the most preferred immunization algorithms with regard to the network structure.

This paper is organized as follows: In section 2, we introduce our evaluation framework for evaluating the efficiency of immunization algorithms in case of limited amount of vaccination resources. Then, we review network centrality measures and investigate their relationship to epidemic spreading in section 3. Then, in section 4, our simulation setting and numeric result of applying is described. Finally, we conclude the whole paper in section 5.

Table 2 Evaluation of targeted immunization algorithms in literatures.

Literatures	Immunization algorithms	Networks		Efficiency Measures	
		Real	Model	Structural	Epidemic parameter [b]
Holme et al., 2002	Betweenness, Degree	HEP Computer network	Small-world, Scale-free, Erdös-Renyi	LCC [a], Distance	×
Miller and Hyman, 2007	Degree, PageRank	Episims	×	×	SIR $T=1$; $0<\alpha<1$
Hartvigsen,et al., 2007	Degree, Clustering coefficient	×	Small-world	×	SIR Simulation of Influenza $R_0=2$ $T=3$
Chen et al, 2008	Degree, Betweenness	HEP AS Workplace AS Metabolic	Scale-Free, Random-regular, Erdös-Renyi	Immunization threshold , LCC	SIR Simulation $\alpha= 0.2$ $\beta=0.05$
Masuda, 2009	Degree, Eigenvector	HEP PGP WWW email-based	Erdös-Renyi	LCC	×
Eames et al., 2009	Degree, Weight, Secondary case	networks generated based on a conducted survey	×	×	SIR Simulation With different parameters
Schneider et al., 2012	Degree, Betweenness	HEP AS	Scale-free, Erdös-Renyi, Random-regular	LCC, Robustness	×
Ventresca and Aleman, 2013	Degree, PageRank, Total weight	Toronto social networks based on 2006 census data	×	LCC, Degree distribution, Clustering coefficient	×

[a] LCC: Largest connected component of network.

[b] Epidemic parameters (α: infectious rate, β: recovery rate , R_0: reproduction number, T= infectivity Time).

2 Evaluation Framework

To study impact of centrality measures in targeted immunization algorithms, we propose a framework to assess their efficiency in case of limited or flexible amount of vaccination resources regarding different goal of immunization program and time complexity (see Fig. 1). To evaluate immunization algorithms in limited amount of vaccination resources, the framework estimates expected epidemic growth in different situations. To assess their efficiency in case of flexible amount of vaccination resources, it captures their overall efficiency in case of unknown amount of vaccination resource in addition to their ability to minimize total cost of immunization. In the following, we describe this evaluation framework in details.

In case of limited vaccination resources, an immunization algorithm targets and vaccinates appropriate limited set of nodes whose immunization minimize the expected growth of epidemics. In this section, we talk about how to evaluate efficiency of immunized sets by targeted immunization algorithms

Recent investigation showed a strong relationship between epidemic spreading parameters and network structural properties(Ames et al., 2011; Aspnes et al., 2006; Chakrabarti and Faloutsos, 2003; Chakrabarti et al., 2008; Chen et al., 2008; Schneider et al., 2012, 2011; Ventresca and Aleman, 2013). Most important of structural properties are largest eigenvalue of network adjacency matrix, largest connected component of network, and, sum of square partitions. Details of our metrics are elaborated in the following.

2.1 Limited amount of vaccination resources

2.1.1 Increment of network epidemic threshold

Epidemic threshold is a parameter which determines whether an infection dies out over time or becomes an epidemic (Chakrabarti and Faloutsos, 2003; Chakrabarti et al., 2008; Kitchovitch and Liò, 2011; Masuda, 2009; Peng et al., 2010; Restrepo et al., 2006). Epidemic threshold of network is equal to inverse largest eigenvalue of network adjacency matrix (λ) (Chakrabarti and Faloutsos, 2003; Chakrabarti et al., 2008). Thus, an efficient immunization algorithm should reduce λ in order to reduce risk of outbreak contagious (Chakrabarti and Faloutsos, 2003; Chakrabarti et al., 2008; Masuda, 2009; Restrepo et al., 2006). To assess their ability to aim this, we measure the reduction of largest eigenvalue by calculating $E = \frac{\acute{\lambda}}{\lambda}$ where λ is the largest eigenvalue of the initial network and $\acute{\lambda}$ is the largest eigenvalue of the immunized network. Formulation of this problem is shown in Table 3.

Table 3 Formulation of immunization program with goal of increasing epidemic threshold in case of limited amount of vaccination resources.

Input	Social network (G),number of immunization resources (k)	G, k
Output	Subset of k nodes	S ⊆ V(G) such that ‖S‖=k
Goal	Reducing largest eigenvalue of matrix	Minimize λ(G/S)[a]

[a]λ(G/S) is largest eigenvalue of residual network after removal of S.

2.1.2 Reduction of worst expected epidemic size

A network connected component is defined as the set of all nodes which all are reachable from each other. Therefore, an infection that starts in a component cannot propagate to other components. Accordingly, in case of single source of infection, the worst case epidemic size is equal to size of largest connected component of network (LCC)(Chen et al., 2008; Gallos et al., 2007; Hadidjojo and Cheong, 2011; Masuda, 2009; Niu et al., 2009; Restrepo et al., 2006; Schneider et al., 2012, 2011; Ventresca and Aleman, 2013; Yamada and Yoshida, 2012; Yoshida and Yamada, 2012). Hence, the fraction of largest connected component of network is another important factor to assess efficiency of immunization algorithms. The formulation of this problem is shown in Table 4.

Table 4 Formulation of immunization problem with goal of minimizing worst expected epidemic size in case of limited amount of vaccination resources.

Input	Social network (G),number of immunization resources (k)	G, k
Output	Subset of k nodes	S ⊆ V(G) such that ‖S‖=k
Goal	Reducing largest connected component(LCC)	Minimize LCC(G/S)[a]

[a]*LCC*(G/S) is largest connected component of residual network after removal of S.

2.1.3 Reduction of mean expected epidemic size

In addition to worst case epidemic size, we estimate expected mean number of infected persons; if all individuals have equal probability to be initially infected, the probability of outbreak occurring in *ith* component is obtained by $(c_i/N)$where c_i is size of the component and *N* is network size. Therefore, the expected mean number of infected person (i.e. mean epidemic size) is $(\sum |c_i|^2/N)$ called sum of square partition (Aspnes et al., 2006). The formulation of this problem is shown in Table 5.

Table 5 Formulation of immunization problem with goal of minimizing mean expected epidemic size in case of limited amount of vaccination resources

Input	Social network (G),number of immunization resources (k)	G, k
Output	Subset of k nodes	S ⊆ V(G) such that ‖S‖=k
Goal	Reducing sum of square partition	Minimize SSP(G/S)[a]

[a]*SSP*(G/S) is largest connected component of residual network after removal of S.

2.2 Flexible amount of vaccination resources

To compare efficiency of immunization algorithms in case of flexible amount of vaccination resources, two metrics are provided: "immunization threshold" and "robustness". Immunization threshold and Robustness are numeric measures capturing overall efficiency of immunization algorithms where different amount of vaccination resources are available. We consider these measures on artificial networks with different size and degree.

2.2.1 Reduction of whole immunization cost

If sufficient amount of vaccination is available, an efficient immunization algorithm should minimize immunization cost of whole population. In other word, Immunization algorithm should suppress epidemics in very small component by removing minimal number of nodes (i.e. immunization cost). This ability of immunization algorithms is evaluated by their immunization threshold (q_c) which is the fraction of removal nodes when size of largest connected component is zero (Chen et al., 2008). The formulation of this problem is shown in Table 6.

Table 6 formulation of immunization problem with goal of minimizing immunization cost.

Input	Social network (G)	G
Output	A subset of nodes whose removal eradicate epidemic spreading	S $\subseteq$ V(G) such that LCC(G/S)=0
Goal	Minimize number of immunization resource	Minimize $\|\|s\|\|$

2.2.2 Reduction of worst case epidemic size in case of unknown amount of vaccination resources

If the supply of immunization doses is not known, an immunization algorithm is required which prioritize nodes such that average growth of epidemic during the whole immunization process is minimized (Schneider et al., 2012). Formulation of this problem is shown in Table 7.

Table 7 formulation of immunization problem with goal of minimizing worst case epidemic in case of unknown amount of vaccination resources.

Input	Social Network (G)	G
Output	An ordering of node (O)	O={$< o_1, o_2,\ldots,o_n >$}
Goal	Minimizing epidemic spreading in different amount of vaccination resources	Minimize R= $\frac{1}{n}\sum_{i=1}^{n} LCC\ (G/(o_{1:i}))$

To measure efficiency of nodes ranking, Robustness (*R*), is defined by (Schneider et al., 2011) to capture overal efficiency of algorithm in case of unknown amount of vaccination resouces. This metric is calculated by Eq.(1) where *L(Q)* is the fraction of largest component of network when the fraction of immunized nodes is equal to *Q*.

$$R = 1/N\ (\sum_{Q=1}^{N} L(Q)) \quad (1)$$

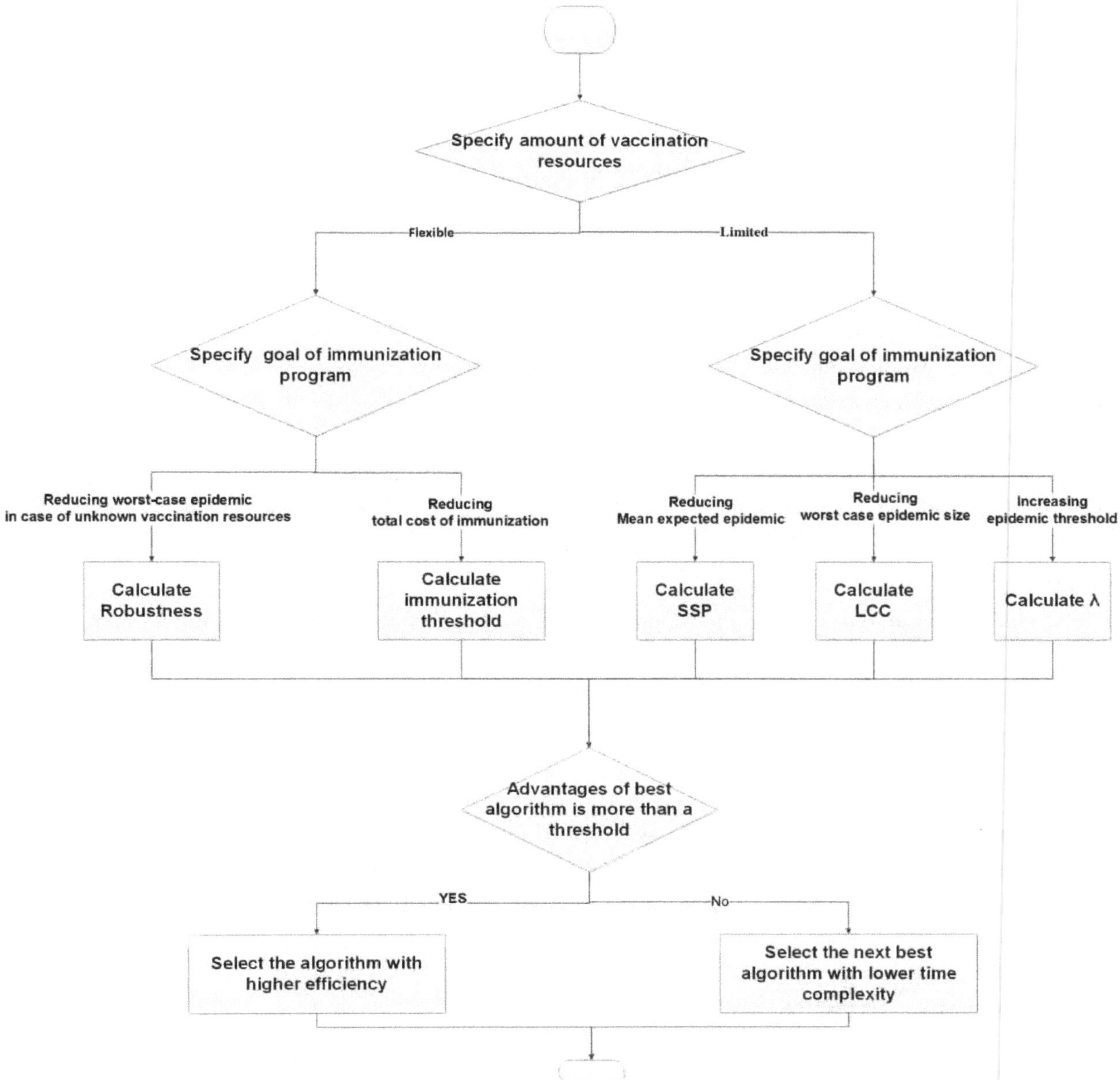

Fig. 1 Evaluation framework for immunization algorithms regarding amount of vaccination resource, goal of immunization program, and, time complexity.

3 Centrality Measures for Targeted Immunization

Regarding various aspects of nodes influence on epidemic spreading, different centrality measures are used in targeted immunization. In this paper, we study various centrality measures including degree, betweenness, closeness, eigenvalue and PageRank which are more applicable in epidemic spreading and targeted immunization (Chakrabarti et al., 2008; Chen et al., 2008; Christley et al., 2005; Cohen et al., 2003; Dezső and Barabási, 2002; Eames et al., 2009; Gao et al., 2011; Habiba et al., 2010; Hébert-Dufresne et al., 2013; Kiss and Bichler, 2008; Masuda, 2009; Miller and Hyman, 2007; Niu et al., 2009; Restrepo et al., 2006; Salathé and

Jones, 2010; Schneider et al., 2012, 2011; Ventresca and Aleman, 2013; Vidondo et al., 2012; Zhang, 2012a, 2012b).

To target and immunize nodes based on their centrality, two kinds of algorithms are generally proposed: initial and adaptive. Initial algorithms calculate node centrality in original networks and immunize *k* most *k* central nodes where *k* is number of available vaccination resource. On the other hand, adaptive algorithms recalculate node centralities in a network of non-immunized nodes after immunizing the most central node. This procedure is iterated until *k* node is immunized (Masuda, 2009; Restrepo et al., 2006; Schneider et al., 2012, 2011). Since the time complexity of initial algorithms is *k* time less than adaptive algorithms, the former one is more applicable for real networks. In this paper, we only consider the initial type algorithms.

A social network is an undirected network represented by adjacency matrix denoted by A; $A_{ij} = 1$ when node i and node j are adjacent, and $A_{ij} = 0$ otherwise. In the following, we review these centrality measures and their relationship with epidemic spreading.

3.1 Degree centrality

Degree centrality which is defined as number of connected links to a node, estimates immediate impact of node infection (Borgatti, 2005). Immunizing highest degree nodes (HD) minimizes network density which is an influential factor in growth rate of epidemic diseases (Hadidjojo and Cheong, 2011). Furthermore, vaccinating high degree nodes is exceedingly effective in scale-free (Dezső and Barabási, 2002) and sparse (Hébert-Dufresne et al., 2013) networks which in high degree nodes play an important role to connect other nodes.

$$d(i) = \sum_{j=1}^{n} A_{ij} \tag{2}$$

3.2 Betweenness centrality

Betweenness centrality of each node is the proportion of times it lies on the geodesic paths between other nodes (Christley et al., 2005; Freeman, 1978).The betweenness centrality of node i is given by

$$BC(i) = \sum_{u \neq v \neq w} (g_{j,k}^{i} / g_{j,k}) \tag{3}$$

where $g_{j,k}$ is the number of geodesic paths from j to k, and $g_{j,k}^{i}$ is the number of these geodesics that pass through node i.

Since nodes with high betweenness centrality bridge network communities (Habiba et al., 2010; Salathé and Jones, 2010), their immunization fragments network to smaller parts. Hence, an epidemic starts in a component cannot infect nodes in other components. Additionally, betweenness centrality is an effective measure to identify high risk individuals as it calculates volume of flow passing through each node (Borgatti, 2005). That is the reason why highest betweenness immunization (HB) is supposed as most effective targeted immunization algorithms.

3.3 Closeness centrality

Closeness centrality of a node is calculated by inverse sum of shortest distances of the node from all other ones.

$$CC(i) = \sum_{j \in V/v} d_G(i,j) \tag{4}$$

where $d_G(v, t)$ is the geodesic distance between v and t (Borgatti, 2005; Freeman, 1978). If all nodes have equal probability to initiate an infection, nodes with highest closeness centrality are more prone to get infected (Borgatti, 2005; Christley et al., 2005). Besides, their infection result in faster epidemic spreading through network (Borgatti, 2005) . Therefore, highest closeness immunization (HC)not only vaccinates high risk individuals, but also postpones epidemic spreading through networks.

3.4 Eigenvector centrality

Eigenvector centrality provides a model of nodal infectious risk according to the risk level of its neighbours (Borgatti, 2005). Let *E(i)* be the eigenvector centrality of node *i* and $\tau(i)$ set of its netighbours, then

$$E(i) = (1/\lambda)\sum_{j\in\tau(i)} E(j) = (1/\lambda)\sum_{j=1}^{N} A_{ij}E(j) \qquad (5)$$

This equation can be rewritten in vector form such as Eq.(6) where $E = \{E(1), E(2), \ldots, E(n)\}$ and λ is the largest eigenvalue of network adjacency matrix.

$$E = (1/\lambda)\ AE \qquad (6)$$
$$\lambda E = AE \qquad (7)$$

Therefore, eigenvector centralityis defined as the principal eigenvector of the adjacency matrix of network (Bonacich, 1987). Moreover, it determines impact of removing nodes on epidemic threshold increment (i.e. reduction of λ)(Masuda, 2009; Restrepo et al., 2006).Let Eq.(8) be the linearized eigenequation after removal of node *k*

$$(\lambda + \Delta\lambda)(E + \Delta E) = (A + \Delta A)\ (E + \Delta E) \qquad (8)$$

It has been provedby (Restrepo et al., 2006) that

$$I_k = -\ (\Delta\lambda/\lambda) \cong E_k^2/\sum_{i=1}^{N} E_k^2 \qquad (9)$$

where I_k is impact of removing node *k*on epidemic threshold and E_k *is* eigenvector centrality of node *k*. Accordingly, immunization of nodes with highest eigenvector (HE) centrality reduces risk of infection through the network.

3.5 PageRank centrality

PageRank centrality, introduced by Google for webpage ranking, determines probability of visiting a node in a random walk (Page et al., 1999). In term of epidemic spreading, nodes with high PageRank centrality are more likely to be infected or infect others along many paths (Miller and Hyman, 2007). Therefore, their immunization eliminates lots of disease transmission routes. Since nodes with high PageRank centrality have many neighbours with low degree, highest PageRank (HP) immunization vaccinate influential nodes whose immunization strongly protects their neighbours (Miller and Hyman, 2007).

4 Experiments and Results

Here, we study the impact of centrality measures in initial targeted immunization using proposed evaluation framework. The proposed framework has been applied to various artificial and real networks to compare efficiency of different targeted immunization algorithms. Table 8 illustrates details of our experiments. We used igraph (Csardi and Nepusz, 2006) package in R.2.15.1 to generate artificial networks, compute nodes centrality, and evaluation metrics.

Table 8 Evaluation framework parameter settings.

<table>
<tr><th>Immunization algorithm</th><th>Vaccination resource</th><th>Efficiency measures</th><th>Datasets</th></tr>
<tr><td rowspan="2">Highest degree immunization (HD)
Highest betweenness immunization (HDB
Highest closeness immunization (HC)
Highest eigenvector immunization (HE
Highest PageRank immunization (HP)</td><td>Limited</td><td>Largest eigenvalue of network adjacency matrix (λ)
Largest connected component of network (LCC)
Sum of square partitions (SSP)</td><td>Real (HEP, FBL, AS)
Artificial(SF, ER, SW)</td></tr>
<tr><td>Flexible</td><td>Immunization threshold (q_c)
Robustness (R)</td><td>Artificial withsize=10^4 and varying degree
Artificial with average degree of 4 and varying size (10^2, 10^3, 10^4, 10^5)</td></tr>
</table>

4.1 Limited amount of vaccination resources

In case of limited vaccination resources, we compare ability of immunization algorithms in reduction of epidemic growth in various real and artificial networks. To aim this, we calculate evaluation criteria including largest eigenvalue of network adjacency matrix, largest connected component of network, and, sum of square partition.In the following, we describe datasets and numeric result.

4.1.1 Datasets

In order to evaluate efficiency of targeted immunization algorithms in case of limited budget, various real and artificial network datasets are used. Our real networks includes HEP (Gehrke et al., 2003; Leskovec et al., 2005) , AS (Shavitt and Shir, 2005), and Facebook-like (FBL) (Opsahl and Panzarasa, 2009), networks which are commonly used in immunization literature(Chen et al., 2008; Gao et al., 2011; Hu and Tang, 2012; Masuda, 2009; Mirzasoleiman et al., 2012; Niu et al., 2009; Schneider et al., 2012, 2011). The artificial network datasets include scale-free (SF) (Cho et al., 2009; Chung and Lu, 2002; Goh et al., 2001), Erdös-Renyi (ER) (Erdos, P. and Renyi, 1959) and small-world (SW) (Watts and Strogatz, 1998) networks. The structural properties of these networks are shown in Table 9.

Real datasets includes HEP, AS, and FBL networks. The HEP network represents citation network of high-energy physics theory derived from e-print Arxiv. It contains 27,770 nodes representing published paper and 352,285 undirected links representing citation between two papers (Gehrke et al., 2003; Leskovec et al., 2005). AS networks containing 25,367 nodes and 75,004 edges captures information of the Internet network at autonomous systems on June 2012 (Shavitt and Shir, 2005). The FBL includes information of 13,838 messaging on an online community of 1,899 students at University of California, Irvine (Opsahl and Panzarasa, 2009).

In addition to the real network, we generate artificial networks of scale-free (SF), small-world (SW) and Erdös-Renyi (ER) models based on parameters mention in(Chen et al., 2008). The SF network is generated using the algorithm presented in (Cho et al., 2009; Chung and Lu, 2002; Goh et al., 2001) by the following settings: network size is 10,000, edge number is 20,000 and γ is 2.5. We construct SW network by

Watts-Strogatz algorithm (Watts and Strogatz, 1998) such that network size is 10,000, mean degree is 4 and switching probability is 0.01. Finally, *"G(n,m)"* implementation of Erdös-Renyi algorithm (Erdos, P. and Renyi, 1959) is used to generate ER network containing 10,000 nodes and 20,000 edges.

Table 9 Dataset properties.

Network	N[a]	M[b]	<d>[c]	Std(d) [d]	CC [e]	λ [f]
HEP	27,770	352,285	14.57	25.371	0.11	111.25
AS	25,367	75,004	5.91	48.03	0.01	103.35
FBL	1899	13,838	12.68	24.46	0.05	48.14
SF	10,000	20,000	4	4.40	0.001	9.90
ER	10,000	20,000	4	1.98	0.000	5.22
SW	10,000	20,000	4	0.27	0.46	4.08

[a] N : Number of vertices [b] M : Number of edges
[c] <d>: Average degree [d] Std(d): Standard deviation of degree
[e] CC: Clustering coefficient [f] λ: Largest eigenvalue of network adjacency matrix

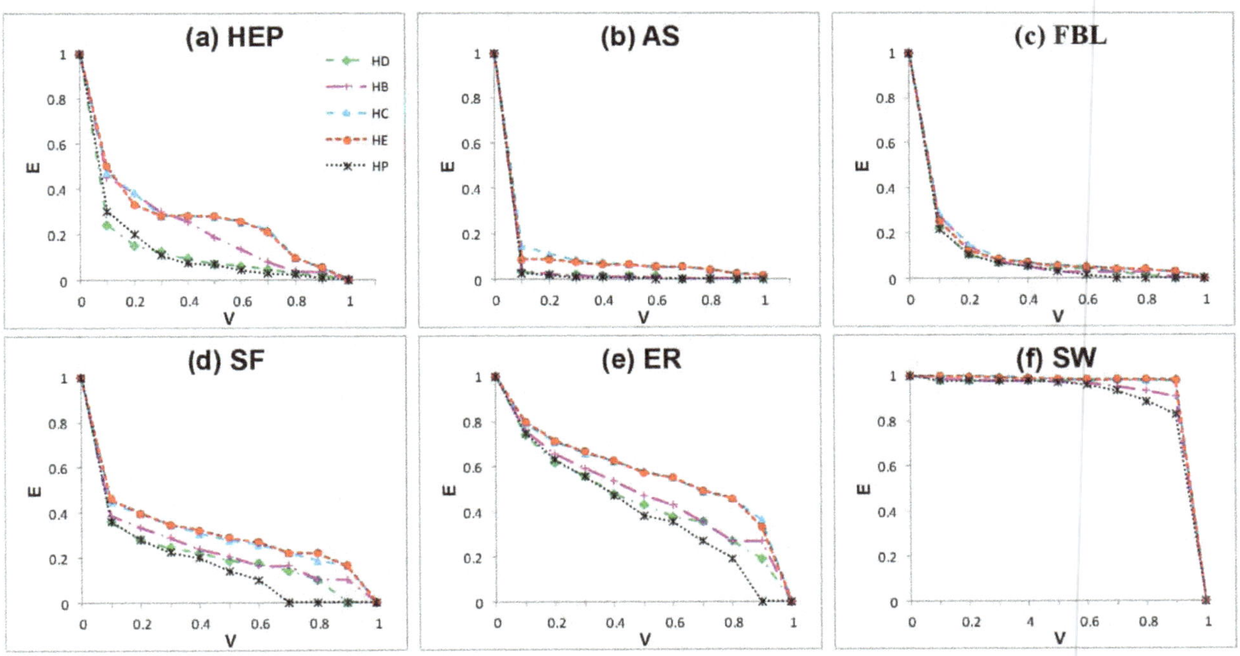

Fig. 2 The fraction of largest eigenvalue ($E=\acute{\lambda} / \lambda$) vs. fraction of immunized nodes (V). (a) HEP network. (b) AS network. (c) FBL network. (d) SF network. (e) ER network. (f) SW network.

4.1.2 Results

To assess their ability to decrease largest eigenvalue of network adjacency matrix, we calculated $E = \frac{\acute{\lambda}}{\lambda}$ where λ is the largest eigenvalue of the initial network and $\acute{\lambda}$ is the largest eigenvalue of the immunized network (Fig. 2). HP outperformed other strategies in all networks. HB and HD exhibited the next best performance in all networks. Their efficiency was very close to each other in all networks except HEP (Fig. 2a) where the initial largest eigenvalue is larger than other network (Table 9). Additionally, our experiments showed that as the largest eigenvalue of network decreases, the slope of *E* curves versus fraction of immunized nodes are also reduced (Fig. 2a, 2b, 2c). For instance, all algorithms required to immunize only 20% of FBL network to reduce *E* below 10%, while, they required to immunize more than 67% in order to reduce *E* to zero (Fig. 2c). This trend is also seen in HEP (Fig. 2a) and AS (Fig. 2b) networks. So we can conclude that there is a threshold below which λ decreases slowly. That could be the reason why targeted immunization did not perform well in artificial networks (Fig. 2d, 2e, 2f). In other words, their λ might be below the threshold.

Moreover, we evaluated efficiency of immunization algorithms in reducing worst-case epidemic size by plotting *L* versus *V* where *L* is fraction of largest connected component of network, and, *V* is fraction of vaccinated nodes (Fig. 3). HP, HD and HB took high advantages of reducing *L* in all networks. The performance of HD and HP were close in all network except SW which is more homogenous than other networks. In SW network, the efficiency of HD is reduced in case of immunization coverage more than 10% since the degree of 90% of nodes have equal degree.On the other hand, HB exhibited high performance in SW networks (Fig. 3f). The high efficiency of HB and HP in this network is explained by high clustering coefficient and low average distance properties of small-world networks (Watts and Strogatz, 1998) (Table 9). In the other word, there are few nodes with high PageRank and betweenness centrality lying on the paths between different clusters of networks. HC and HE performed weaker than other strategies in all network but their weakness is less in HEP (Fig. 3a) and FBL (Fig. 3c) networks which are denser than others (Table 9).

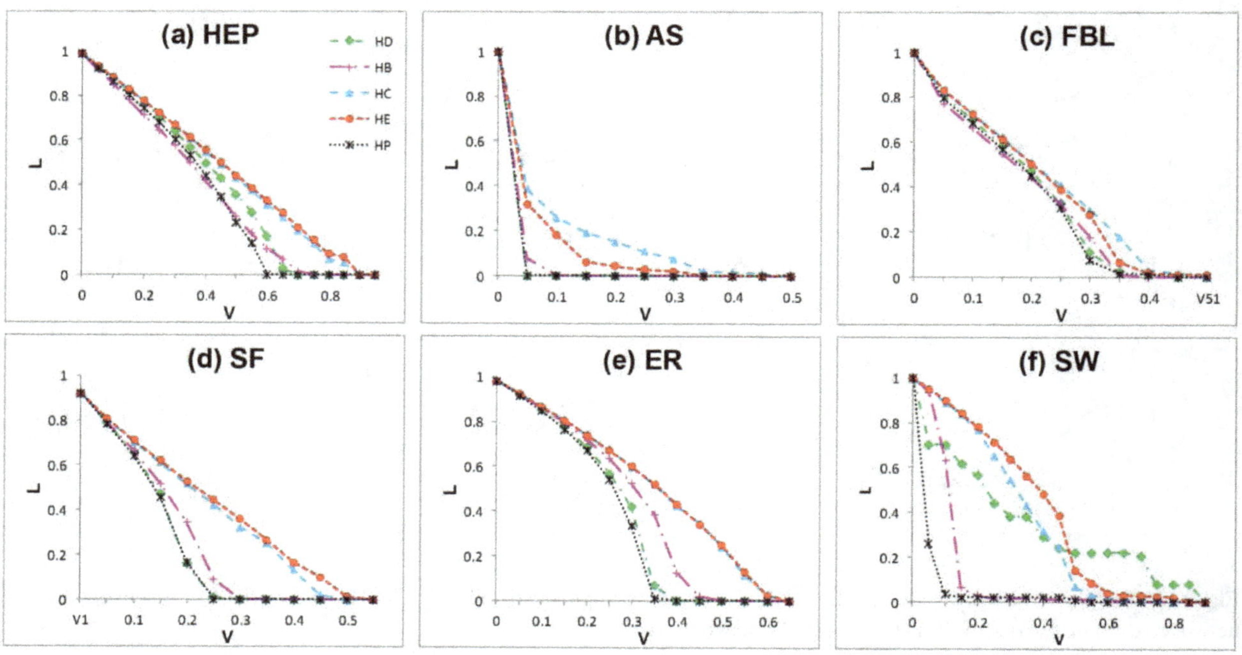

Fig. 3 Fraction of largest connected component (L) vs. fraction of vaccinated nodes (V). (a) HEP network. (b) AS network. (c) FBL network. (d) SF network. (e) ER network. (g) SW network.

In Fig. 4, we plotted *S*, the fraction of expected worst-case epidemic size (i.e. sum of square partitions), versus *V*, the fraction of vaccinated nodes in Fig. . It is obvious that changes of *S* are similar to the changes in worst case epidemic size (Fig. 4) but has a little shift to left. This is because of the fact that expected mean number of infected persons is proportional to sum of square of component size. In the other words, sum of square partitions highlights larger connected components in contrast to smaller ones more clearly. Therefore, the similar trends of *S* and *L* means all centrality measures failed to fragment networks to balanced partitions. Regarding all criteria, HP exhibited the best performance, while, HE and HC exhibited the worst performance in all networks. HD and HB exhibited an acceptable performance in all networks. HD performed better in sparse and heterogeneous networks such as SF network, while, HB performed well in modular and homogenous networks such as SW network.

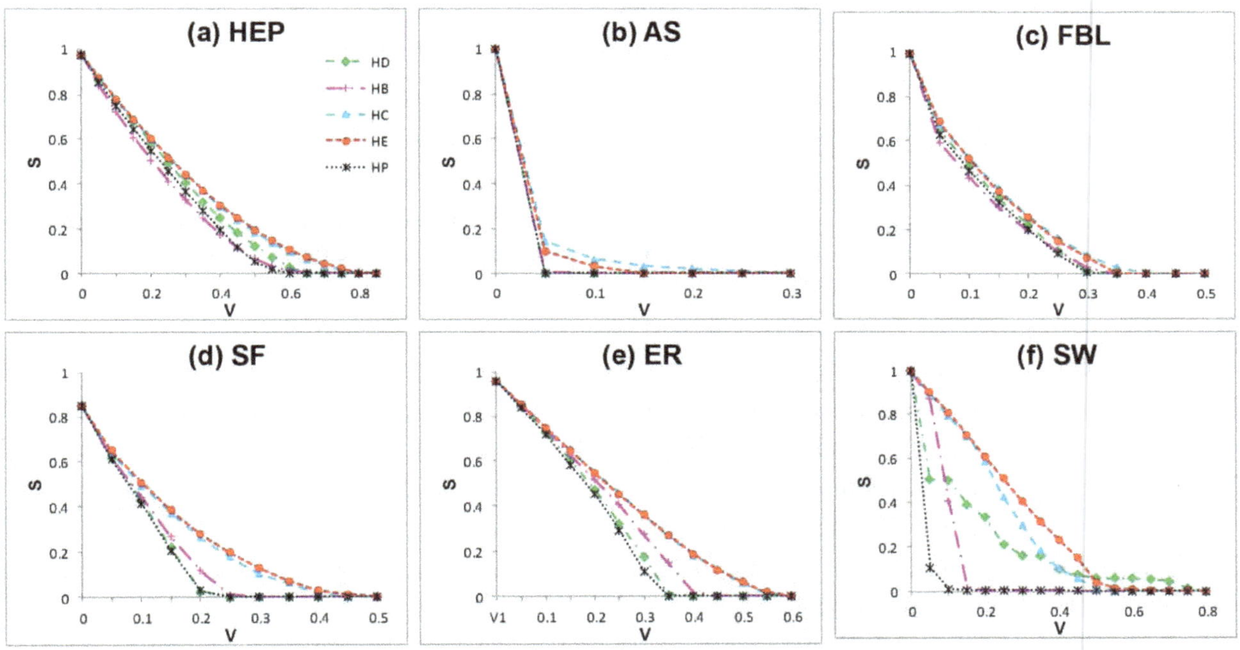

Fig. 4 Fraction of sum of square partitions (S) vs. fraction of vaccinated nodes (V). (a) HEP network. (b) AS network. (c) FBL network. (d) SF network. (e) ER network. (g) SW network.

4.2 Flexible amount of vaccination resources

We evaluate immunization algorithms in case of flexible amount of vaccination resources by computing their "immunization threshold" and "robustness" in artificial networks with different size and degree. The description of datasets and results are explained in the following.

4.2.1 Datasets

In order to assess efficiency of immunization algorithms in case of flexible amount of vaccination resources, artificial networks with different size and degree are generated. We generate scale-free, small-world, and, Erdős-Renyi networks with different size and degree. To compare efficiency of immunization algorithms in networks with different size, each type of artificial networks is generated with 100, 1000, 10,000, and, 100,000 nodes and average degree of 4. Additionally, artificial networks are generated in size of 10,000 and degrees up to 20 to evaluate immunization algorithms in networks with different density.

4.2.2 Results

To compare efficiency of centrality measures in minimizing immunization threshold (q_c), we plotted it as a function of network size (*N*) or average degree (<d>) for all artificial network models. Our experiments showed a directed relationship between network degree and q_c for all algorithms in majority of networks (Fig. 5). That is because of the weak connectivity in sparse networks in contrast to strong relations in dense network. In the other words, sparse networks can be fragmented to small component by removing only a few nodes, while, it is impossible to fragment dense networks.

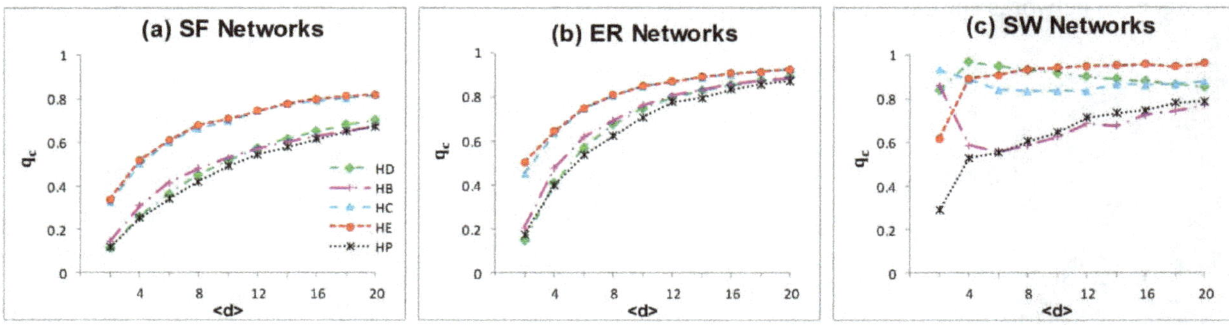

Fig. 5 Immunization threshold (q_c) vs. network average degree (<d>). (a). Scale-free networks (b) Small-world networks. (c) Erdös-Renyi networks. Average degree of all networks is 4. Each point is averaged on 10 different networks.

Considering immunization threshold of networks with different size, HP outperformed other strategies especially in small-world network (Fig.6c). For instance, q_c of HP in largest small-world network is 12% less than HB, 59% less than HC, 71% less than HE, and 85% more than HD. Despite of weak performance of HD in small-world networks, it performed well on scale-free and Erdös-Renyi networks (Fig. 6) such that the maximum difference between its immunization threshold and HP is about 7%. That is because of heterogeneous degree distribution in scale-free and Erdös-Renyi networks compared to homogenous degree distribution in all small-world networks.

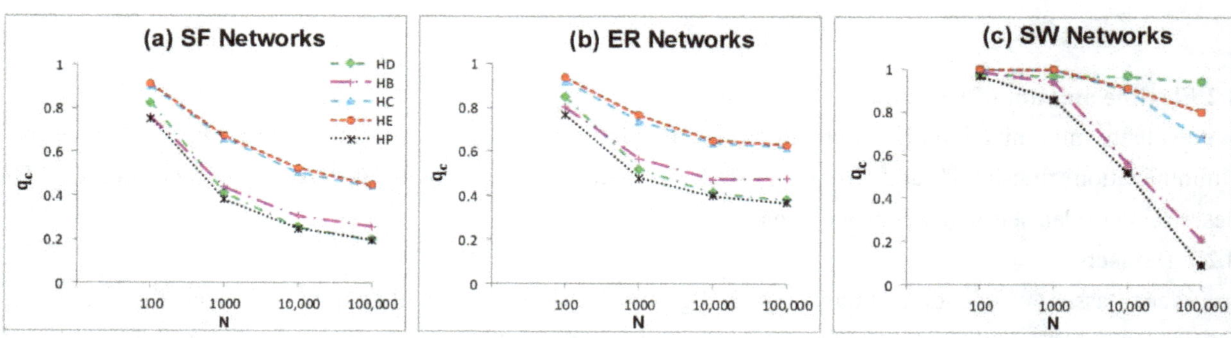

Fig. 6 Immunization threshold (q_c) vs. network size (N). (a). Scale-free networks (b) Small-world networks. (c) Erdös-Renyi networks. Average degree of all networks is 4. Each point is averaged on 10 different networks.

In addition to immunization threshold, we calculate network "robustness", R, of each algorithm. Our experiment showed that impact of degree increment on robustness is similar to its impact on immunization threshold in all networks (Fig. 7). It can be also explained by strong connectivity in dense networks which make it hard to fragment them.

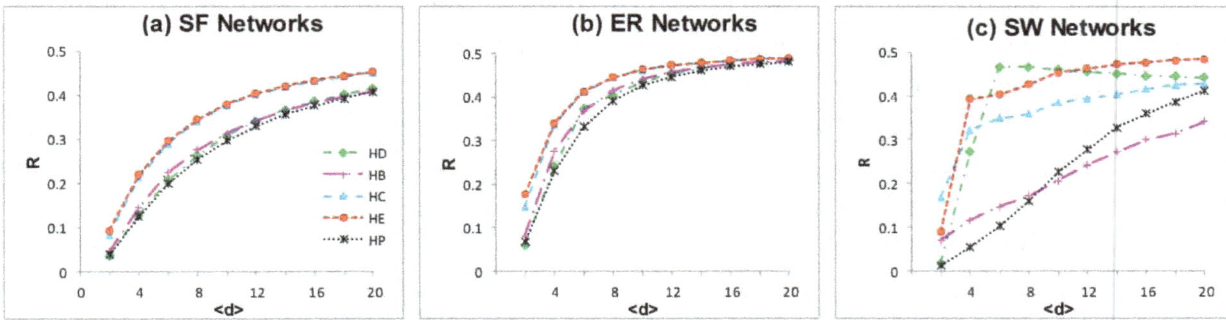

Fig. 7 Network Robustness (R) vs. network average degree (<d>). (a). Scale-free networks (b) Small-world networks. (c) Erdös-Renyi networks. All networks contain 10,000 nodes. Each point is averaged on 10 different networks.

Increasing the network size in scale-free networks reduced *R* about 10% to 14% (Fig. 8a), while, robustness of Erdös-Renyi networks fluctuated about 1% around a fixed value for all algorithms (Fig. 8b). This result is explained by random structure of Erdös-Renyi networks leading to emergence of a giant component which is hardly fragmented. On the other hand, efficiency of algorithms are much better in large scale-free networks compared to small ones with constant degree since central node plays an important role in connecting majority of nodes in large scale-free networks. Increasing small-world networks size had different impacts on robustness of algorithms; the robustness was reduced more than 22% for HP and HB, whereas, increased slightly for HE and HC when network size exceed 1,000 (Fig. 8c).

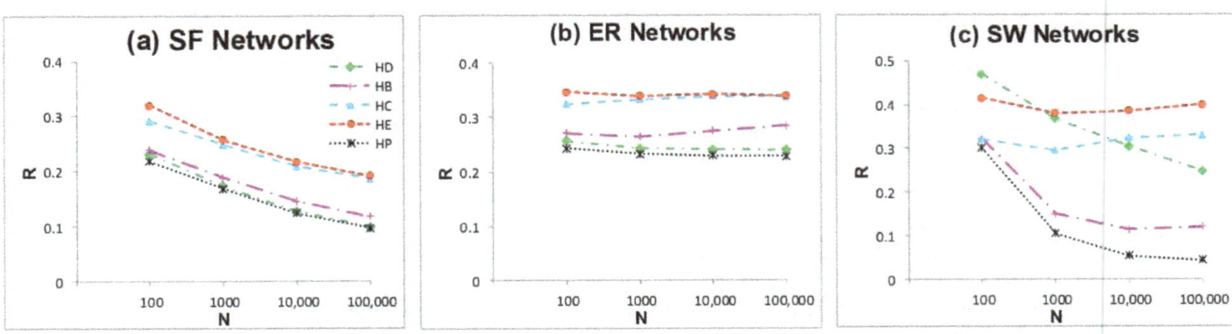

Fig. 8 Network Robustness (R) vs. network size (N). (a). Scale-free networks (b) Small-world networks. (c) Erdös-Renyi networks. Average degree of all networks is 4. Each point is averaged on 10 different networks.

5 Conclusion and Suggested Future Work

Due to high cost and limited resources of vaccination, there is a growing interest in targeted immunization strategies. Targeted immunization strategies vaccinate a subset of individuals whose vaccination minimize epidemic spreading through population. Regarding impact of contact networks on disease transmission, majority of immunization strategies use network centralities to prioritize nodes for vaccination.

In this paper, we first proposed an evaluation framework to study targeted immunization algorithms regarding amount of vaccination resources, goal of immunization program, and time complexity. The framework assesses efficiency of immunization algorithms in case of limited or flexible amount of vaccination resources based on network properties. The evaluation metrics for limited resources consist of the largest eigenvalue (i.e. immunization threshold), the largest connected component of network (i.e. worst case epidemic size), and, sum of square partitions (i.e. mean worst epidemic size. Metrics for flexible resources include immunization threshold (i.e. cost of complete immunization) and robustness (i.e. overall efficiency of immunization algorithms in case of unknown amount of vaccination resource).Next, we studied impact of five centrality measures including degree (HD), betweenness (HB), closeness (HC), eigenvector (HE) and PageRank (HP) in initial targeted immunizations based on the proposed evaluation framework.

In case of limited budget, HP exhibited the best performance, but, HC and HE exhibited the worst performance regarding all criteria. In heterogeneous networks such as AS and scale-free networks, efficiency of degree immunization was very close to PageRank immunization, whereas, it exhibited a weak performance in homogenous networks (e.g. small-world networks). Hence, degree immunization is more preferred in heterogeneous networks (e.g. AS and scale-free networks) due to its lower time complexity in comparison to PageRank immunization (Table 10). Closeness and eigenvector centralities were too weak in comparison to other algorithms in all networks. Additionally, comparing sum of square partitions and largest connected component of immunized network showed that all targeted immunization were unsuccessful to fragment network to balanced component. Furthermore, it should be noted that all immunization algorithms failed to reduce largest eigenvalue of artificial network adjacency matrix. Surprisingly, despite of relationship between eigenvector centrality and largest eigenvalue (Masuda, 2009; Restrepo et al., 2006), it did not show any advantages in comparison to other strategies. Therefore, these algorithms are not given preference in none of networks.

Table 10 Time complexity of node centrality calculation.

Centrality	Complexity [a]	Execution Time [b]					
		FB	AS	HEP	SF	SW	ER
Degree	O(N)	0	0.02	0.05	0	0	0
Betweenness	O(MN)	0.40	174.15	401.46	10.01	9.61	11.93
Closeness	O(MN)	0.77	58.98	170.31	5.65	5.08	6.70
Eigenvector	O(M+N) [c]	0	0.11	0.25	0.06	0.36	0.07
PageRank	O(M) [d]	0.02	0.27	0.74	0.07	0.06	0.06

[a] The Complexity times are based on implementation of igraph package where N is network size and M is number of edges.

[b] The execution times are measured in millisecond on a computer running an Intel core i7- i2670QM processor at 2.20Hz with 8GB of RAM.

[c] Depends on the input graph, usually it is O(|M|+|N|).

[d] Depends on the input graph, usually it is O(M).

In case of flexible amount of vaccination resources, we evaluated immunization threshold and robustness of algorithms in artificial networks with different degree and size. Considering both metrics, our experiments showed that PageRank outperforms other strategies in all networks except dense small-world network which in betweenness overtook it; nevertheless, betweenness immunization is still not an appropriate strategy for these networks due to its high time complexity. The best implementation of betweenness calculation of networks with *N* nodes and *M* edges goes as complexity of *O (MN)* (Brandes, 2001) that converges to $O(N^3)$ for dense network which is far more than PageRank time complexity (Table 10). Therefore, PageRank immunization is strongly recommended for all networks especially in dense, homogenous, and clustered networks. Our recommendation for each model network considering time complexity and goal of immunization is given in Table 11.

Table 11 Recommended algorithm for artificial networks considering evaluation framework.

	Limited amount of vaccination resources			**Flexible amount of vaccination resources**	
Goal	Maximizing epidemic threshold (λ)	Minimizing worst-case epidemic (LCC)	Minimizing expected worst epidemic size (SSP)	Minimizing immunization cost (q_c)	Minimizing average growth of epidemic in case of unknown amount of resources (R)
Scale-Free	HP	HD	HD	HD	HD
Erdös-Renyi	HP	HP	HD	HP	HD
Small-World	HP	HP	HP	HP	HP [a]

[a] HB is more preferred than HP in small dense small-world network.

Although, these targeted immunization algorithms show high advantages of lowering cost in comparison to mass and random vaccination, their performance is still so far from an optimal solution especially in dense networks where strong connectivity makes it hard to fragment. Therefore, it is a new line of research to present immunization algorithms to optimize evaluation criteria. Additionally, it is necessary to study about the reasons why immunization algorithms failed to reduce largest eigenvalue of artificial network adjacency matrix. Furthermore, it is valuable for future research to evaluate targeted immunization in improving epidemic parameters (e.g. epidemic period and size) by simulating epidemics model (e.g. SIR, SIS, and SI) models. Finally, it is helpful to suggest the most preferred immunization algorithms with regard to network structure and epidemic spreading model

References

Ames GM, George DB, Hampson CP, et al. 2011. Using network properties to predict disease dynamics on human contact networks. Proceedings of the Royal Society Biological Sciences, 278: 3544-3550. doi:10.1098/rspb.2011.0290

Aspnes J, Chang K, Yampolskiy A. 2006. Inoculation strategies for victims of viruses and the sum-of-squares partition problem. Journal of Computer and System Sciences 72, 1077-1093. doi:10.1016/j.jcss.2006.02.003

Bishop AN, Shames I. 2011. Link operations for slowing the spread of disease in complex networks. Europhysics Letters, 95: 18005. doi:10.1209/0295-5075/95/18005

Bonacich P. 1987. Power and centrality: A family of measures. American Journal of Sociology, 92: 1170-1182

Borgatti SP. 2005. Centrality and network flow. Social Networks, 27: 55-71. doi:10.1016/j.socnet.2004.11.008

Brandes U. 2001. A faster algorithm for betweenness centrality. The Journal of Mathematical Sociology, 25: 163-177. doi:10.1080/0022250X.2001.9990249

Chakrabarti D, Faloutsos C. 2003. Epidemic spreading in real networks: an eigenvalue viewpoint. In: 22nd International Symposium on Reliable Distributed Systems, 2003. Proceedings of the IEEE Computer Soc, 25-34. doi:10.1109/RELDIS.2003.1238052

Chakrabarti D, Wang Y, Wang C, et al. 2008. Epidemic thresholds in real networks. ACM Transactions on Information and System Security, 10: 1-26. doi:10.1145/1284680.1284681

Chen D, Lü L, Shang MS, et al. 2011. Identifying influential nodes in complex networks. Physica A: Statistical Mechanics and its Applications, 391: 1777-1787. doi:10.1016/j.physa.2011.09.017

Chen Y, Paul G, Havlin S, et al. 2008. Finding a Better Immunization Strategy. Physical Review Letters, 101: 2-5. doi:10.1103/PhysRevLett.101.058701

Cho YS, Kim JS, Park J, et al. 2009. Percolation Transitions in Scale-Free Networks under the Achlioptas Process. Physical Review Letters, 103: 135702. doi:10.1103/PhysRevLett.103.135702

Christley RM, Pinchbeck GL, Bowers RG, et al. 2005. Infection in social networks: using network analysis to identify high-risk individuals. American journal of Epidemiology, 162: 1024-1031. doi:10.1093/aje/kwi308

Chung F, Lu L. 2002. Connected Components in Random Graphs with Given Expected Degree Sequences. Annals of Combinatorics, 6: 125-145. doi:10.1007/PL00012580

Cohen R, Havlin S, Ben-Avraham D. 2003. Efficient Immunization Strategies for Computer Networks and Populations. Physical Review Letters, 91: 2-5. doi:10.1103/PhysRevLett.91.247901

Cornforth DM, Reluga TC, Shim E, et al. 2011. Erratic flu vaccination emerges from short-sighted behavior in contact networks. PLoS Computational Biology, 7: e1001062. doi:10.1371/journal.pcbi.1001062

Csardi G, Nepusz T. 2006. The igraph software package for complex network research. International Journal of Complex Systems, 1695

Dezső Z, Barabási AL. 2002. Halting viruses in scale-free networks. Physical Review E, 65: 55-103. doi:10.1103/PhysRevE.65.055103

Eames KTD, Read JM, Edmunds WJ. 2009. Epidemic prediction and control in weighted networks. Epidemics, 1: 70-76. doi:10.1016/j.epidem.2008.12.001

Erdos P, Renyi A. 1959. On random graphs. Publicationes Mathematica, 6: 290-297

Freeman LC. 1978. Centrality in social networks conceptual clarification. Social Networks, 1: 215-239. doi:10.1016/0378-8733(78)90021-7

Gallos L, Liljeros F, Argyrakis P, et al. 2007. Improving immunization strategies. Physical Review E, 75: 1-4. doi:10.1103/PhysRevE.75.045104

Gao C, Liu J, Zhong N. 2011. Network Immunization with Distributed Autonomy-Oriented Entities. IEEE Transactions on Parallel and Distributed Systems, 22: 1222-1229. doi:10.1109/TPDS.2010.197

Gehrke J, Ginsparg P, Kleinberg J. 2003. Overview of the 2003 KDD Cup. ACM SIGKDD Explorations Newsletter, 5: 149. doi:10.1145/1081870.1081893

Goh KI, Kahng B, Kim D. 2001. Universal Behavior of Load Distribution in Scale-Free Networks. Physical Review Letters, 87: 278701. doi:10.1103/PhysRevLett.87.278701

Habiba YY, Berger-Wolf T, Saia J, 2010. Finding spread blockers in dynamic networks. Advances in Social Network Mining and Analysis, 8: 55-76. doi:10.1007/978-3-642-14929-0_4

Hadidjojo J, Cheong SA. 2011. Equal graph partitioning on estimated infection network as an effective epidemic mitigation measure. PloS one, 6: e22124. doi:10.1371/journal.pone.0022124

Hartvigsen G, Dresch JM, Zielinski AL, Macula AJ, et al. 2007. Network structure, and vaccination strategy and effort interact to affect the dynamics of influenza epidemics. Journal of Theoretical Biology, 246: 205-213. doi:10.1016/j.jtbi.2006.12.027

Hébert-Dufresne L, Allard A, Young J, et al. 2013. Global efficiency of local immunization on complex networks. Scientific Reports, 3: 6

Holme P, Kim BJ, Yoon CN, et al. 2002. Attack vulnerability of complex networks. Physical Review E Statistical, Nonlinear, and Soft Matter Physics, 65: 56-109

Hu K, Tang Y. 2012. Immunization for Complex Network Based on the Effective Degree of Vertex. International Journal of Modern Physics B, 26: 1250052. doi:10.1142/S021797921250052X

Kiss C, Bichler M. 2008. Identification of influencers - Measuring influence in customer networks. Decision Support Systems, 46: 233-253. doi:10.1016/j.dss.2008.06.007

Kitchovitch S, Liò P. 2011. Community structure in social networks: applications for epidemiological modelling. PloS one, 6: e22220. doi:10.1371/journal.pone.0022220

Leskovec J, Kleinberg J, Faloutsos C. 2005. Graphs over time. In: Proceeding of the Eleventh ACM SIGKDD International Conference on Knowledge Discovery in Data Mining - KDD '05. 177, ACM Press, New York, USA. doi:10.1145/1081870.1081893

Masuda N. 2009. Immunization of networks with community structure. New Journal of Physics, 11: 123018. doi:10.1088/1367-2630/11/12/123018

Miller JC, Hyman JM. 2007. Effective vaccination strategies for realistic social networks. Physica A: Statistical Mechanics and Its Applications, 386, 780-785. doi:10.1016/j.physa.2007.08.054

Mirzasoleiman B, Babaei M, Jalili M. 2012. Immunizing complex networks with limited budget. Europhysics Letters, 98: 38004. doi:10.1209/0295-5075/98/38004

Niu C, Li L, Xu D. 2009. A new network immunization strategy better than High Degree First. In: 2009 IEEE International Conference on Communications Technology and Applications. 821-825, doi:10.1109/ICCOMTA.2009.5349084

Opsahl T, Panzarasa P. 2009. Clustering in weighted networks. Social Networks, 31: 155-163. doi:10.1016/j.socnet.2009.02.002

Page L, Brin S, Motwani R, et al. 1999. The PageRank citation ranking: bringing order to the web. 1-17.

Peng C, Jin X, Shi M. 2010. Epidemic threshold and immunization on generalized networks. Physica A: Statistical Mechanics and Its Applications 389: 549-560. doi:10.1016/j.physa.2009.09.047

Restrepo J, Ott E, Hunt B. 2006. Characterizing the Dynamical Importance of Network Nodes and Links. Physical Review Letters, 97: 4. doi:10.1103/PhysRevLett.97.094102

Salathé M, Jones JH. 2010. Dynamics and control of diseases in networks with community structure. PLoS Computational Biology, 6: e1000736. doi:10.1371/journal.pcbi.1000736

Schneider CM, Mihaljev T, Havlin S, et al. 2011. Suppressing epidemics with a limited amount of immunization units. Physical Review E, 84: 061911. doi:10.1103/PhysRevE.84.061911

Schneider CM, Mihaljev T, Herrmann HJ. 2012. Inverse targeting —An effective immunization strategy. Europhysics Letters, 98: 46002. doi:10.1209/0295-5075/98/46002

Shavitt Y, Shir E. 2005. DIMES: Let the Internet Measure Itself. ACM SIGCOMM Computer Communication Review, 35: 10. doi:10.1145/1096536.1096546

Ventresca M, Aleman D. 2013. Evaluation of strategies to mitigate contagion spread using social network characteristics. Social Networks, 35: 75-88. doi:10.1016/j.socnet.2013.01.002

Vidondo B, Schwehm M, Bühlmann A, et al. 2012. Finding and removing highly connected individuals using suboptimal vaccines. BMC Infectious Diseases, 12: 51. doi:10.1186/1471-2334-12-51

Watts DJ, Strogatz SH. 1998. Collective dynamics of "small-world" networks. Nature, 393: 440-442. doi:10.1038/30918

Weerasinghe S. 2013. Analytical Frameworks of Social Network Analyses. In: Network Biology: Theories, Methods and Applications (WenJun Zhang, ed). 1-22, Nova Science Publishers, New York, USA

Yamada Y, Yoshida T. 2012. A comparative study of community structure based node scores for network immunization. 328-337, Active Media Technology. doi:10.1007/978-3-642-35236-2_33

Yoshida T, Yamada Y. 2012. Community structure based node scores for network immunization. PRICAI 2012: Trends in Artificial Intelligence, 7458: 1-6. doi:10.1007/978-3-642-32695-0_95

Zhang WJ. 2012a. Computational Ecology: Graphs, Networks and Agent-based Modeling. World Scientific, Singapore

Zhang WJ. 2012b. Several mathematical methods for identifying crucial nodes in networks. Network Biology, 2(4): 121-126

12

The predator-prey models for the mechanism of autocatalysis, pair wise interactions and movements to free places

Muhammad Shakil[1], **H. A. Wahab**[1], **Muhammad Naeem**[2], **Saira Bhatti**[3], **Muhammad Shahzad**[1]
[1]Department of Mathematics, Hazara University, Manshera, Pakistan
[2]Department of Information Technology, Hazara University, Manshera, Pakistan
[3]Department of Mathematics, COMSATS Institute of Information Technology, Abbottabad, Pakistan
E-mail: wahabmaths@yahoo.com,wahab@hu.edu.pk

Abstract
In this paper we aim to develop the modeled equations for different types of mechanism of the predator-prey interactions with the help of a quasi chemical approach while taking a special study case of foxes and rabbits, these mechanisms include autocatalysis mechanism, pair wise interactions and the mechanism of their movements to some free places. The chemical reactions representing the interactions obey the mass action law. The territorial animal like fox is assigned a simple cell as its territory. Under the proper relations between coefficients, this system may demonstrate globally stable dynamics.

Keywords predator prey interactions; autocatalysis mechanism; modeling of fox and rabbit interaction.

1 Introduction

We have introduced the quasi chemical approach to represent the different types of mechanisms between the predator and its prey by taking a case study of foxes and rabbits and in the second publication we determined the modeled equations for these mechanisms including the mechanism of circulation, mechanisms of attraction and repulsion and mechanism of sharing place.

The Lotka Voltera equations are a pair of first order non linear differential equations, and these are also known as the predator prey equations .i.e. when growth rate of one population is decreased and the other increased then these populations are said to be in a predator-prey situation. The Lotka–Volterra equations are frequently used to describe the dynamics of biological system in which two species interact, one as a predator and the other as a prey (Murray, 2003).

A biologist Humberto D'Ancona was doing a statistical study in the 1920's, on the numbers of each species sold at three main Italian ports. During his study from 1914-1923 of these fish species, he came across a surprising conclusion. He believed that these predator prey relationships between the sharks, rays, and fish were in their natural states outside of human interaction, namely before and after the war. D'Ancona asked his father in-law, Vito Volterra, a very successful mathematician, to analyze the data and draw some conclusions. Volterra spent weeks developing a series of models for interactions of two or more species, the first and simplest of these is the subject of this development (Volterra, 1926).

Alfred J.Lotka was an American mathematical biologist who discovered and development many of the same conclusions and models as Volterra and around the same time. His key study was on a species of plant and an herbivores species which relied heavily on that plant as a feeding source.

The Lotka Volterra predator prey models were originally introduced by Alfred J. Lotka in the theory of autocatalytic chemical reactions. This was effectively the logistic equation which was originally derived by Pierre Francois Verhulst. In 1920 Lotka extended the model to "organic systems" using a plant species and an herbivorous animal species as an example and in 1925 he utilized the equations to analyze predator-prey interactions in his book on biomathematics arriving at the equations that we know today (Lotka 1920).

In 1926 Vito Volterra (Volterra, 1926), made a statistical analysis of fish catches in the Adriatic independently investigated the equations. V. Voltera applied these equations to predator prey interactions; consist of a pair of first order autonomous ordinary differential equations.

Since that time the Lotka Voltera model has been applied to problems in chemical kinetics, population biology, epidemiology and neural networks. These equations also model the dynamic behavior of an arbitrary number of competitors. The Lotka–Volterra system of equations is an example of a Kolmogorov model, which is a more general framework that can model the dynamics of ecological systems with predator-prey interactions, competition, disease, and mutualism.

Lotka Volterra is the most famous predator prey model. According to Volterra if $x(t)$ is the prey population and $y(t)$ that of the predator at time t then Volterra's model is (Murray, 2003),

$$\frac{dx}{dt} = x(a - by); \tag{1}$$

$$\frac{dy}{dt} = y(cx - d). \tag{2}$$

Here a, b, c and d are positive constants.

The reaction diffusion system is a mathematical model which clarifies how the concentrations of one or more substances which are distributed in the space, change under the influence of two processes. In the local chemical reactions, the substances are transformed into each other and the diffusion occur which causes the substances to spread out over a surface in space. Such systems are naturally applied in chemistry, however; the system can also explain dynamical processes of non chemical nature. Mathematically, the reaction diffusion system takes the form of the semi-linear parabolic partial differential equation.

2 A Quasi-chemical Approach

In the existence of hundreds of different animals on the island, the modeling of the interaction among these animals is not an easy task. However, the interaction between two species is possible by taking one as a predator and the other as a prey. The same single population growth model can be used for predators (foxes), if prey (rabbits) is considered as an unlimited food supply to foxes. But Foxes are territorial animals, and in general each fox claims its own territory by marking their territories with signals that other foxes will recognize e. g., by leaving their droppings in prominent positions but they also pair up in winter season. So it will be difficult for the foxes to cover the whole island without defining their own territories.

It is a constant battle among the animals on a daily basis to survive. They have to find their food and avoid becoming a food. These species can be divide up on the basis that how they get their food, if they provide it they are producers (prey), if they need to find it they are consumers (predator) and if they breakdown dead material they are decomposers. This correlation between predator and prey intertwines into a complex food chain. In the ecosystem there are various species and every one play a significant role in maintaining populations near the carrying capacity and in keeping the system in balance. In the ecosystems Predator and prey assist them through particular adaptations to compete for food resources. Predator and prey populations are directly related and they cannot survive without each other. Here this relationship is illustrated by using foxes and rabbits.

In artificial world and in the social systems non-linear behaviors are the order of the day. Rabbits and foxes share some behaviors but vary in others and their interaction is non-linear in nature. The rabbits are proverbially very good in reproduction. In case when more rabbits are there, the more foxes will prey on them, so that the number of foxes will increase, when more foxes will predate a lot such that the number of rabbits start to decrease, swiftly it follows by falling fox numbers as there are a small amount of rabbits to sustain them therefore it is observed that the populations of rabbits and foxes are locked together in an interactive population 'dance.' This system is non-linear in its behavior. In mathematics non-linear simultaneous equations are used to represent these interactions for which there exist an infinite number of solutions.

Between the fox and rabbit populations here is a complicated and natural predator-prey relationship, since rabbits thrive in the absence of foxes and foxes thrive in the presence of rabbits. Foxes like to occupy a combination of forest and open fields. Foxes usually define their territory zones and use the transition zone or "edge" between these habitats as hunting areas. Foxes do not interfere in the defined areas of each others.

The cell-jump models may be considered as the proper diffusion models by themselves, the territorial animal like fox is given a simple cell as its territory. The sense in which the discrete equations for cells converge to the partial differential equations of diffusion is that the cell models give the semi-discrete approximation of the partial differential equations for diffusion which result in a system of ordinary differential equations in cells. There are jumps of concentrations on the boundary of cells. The system of semi discrete models for cells with "no-flux" boundary conditions has all the nice properties of the chemical kinetic equations for closed systems. Under the proper relations between coefficients, like complex balance or detailed balance, this system demonstrated globally stable dynamics. Here we specially refer this modeling by Gorban et al. (2011) who developed the idea and modeled diffusion equations for different mechanisms and proved the dissipation inequalities.

We have introduced the quasi chemical approach to represent the different types of mechanisms between the predator and its prey by taking a case study of foxes and rabbits and in the second publication we determined the modeled equations for these mechanisms including the mechanism of circulation, mechanisms of attraction and repulsion and mechanism of sharing place (Shakil et al., 2015). Here we briefly discuss the assumptions again for convenience of the readers.

Let us suppose that our space is divided on two territories, a system represented as a chain of territories each with homogeneous composition and elementary acts of transfers on the boundary (for us there are only two territories). On each territory we have some concentrations (number of foxes and rabbits) for these processes. Let F^{I} (fox/ foxes in first territory) and R^{I} (rabbit/ rabbits in first territory) is the vector concentration in the first territory I and F^{II} (fox/ foxes in second territory) and R^{II} (rabbit/ rabbits in second territory) is the vector of concentration in the second territory *II*. Then in our study case the following mechanisms are possible between the predator (foxes) and its prey (rabbits).

3 Pairwise Interaction

3.1 Here in pair wise interactions the movement of rabbits and foxes is being taken together. If a fox and a rabbit R_i are present in territory one and a fox and a rabbit R_j is in the territory two. As the foxes are territorial while rabbits move freely in nature and rabbits change their territories regularly while foxes do not do so. So if rabbit R_i moves from territory one to territory two and rabbit R_j moves from territory two to territory one then it is given by stoichiometric equations of the form:

$$(F^I + R_i^I) + (F^{II} + R_j^{II}) \rightarrow (F^I + R_j^I) + (F^{II} + R_i^{II}) \qquad (3)$$

This movement will not disturb the system and there will be a normal and usual interaction between the predator and its prey.

Let us suppose that c^I and c^{II} be the concentrations of foxes in the first and the second cell respectively, γ_F^I and γ_F^{II} be the stoichiometric vectors in the respective cells, w_r^I and w_r^{II} be the reactions rates in the respective cells and J_F be the total flux.

The mass action law is applicable here for the above mechanisms and we will calculate the stoichiometric vectors, reaction rates, vector of total flux and their modeled equations as,

$$\gamma_F^I = 0,\ \gamma_{R_i}^I = -1,\ \gamma_{R_j}^I = 1$$

$$w_r^I = kc^I c^{II} d_i^I d_j^{II},\ w_r^{II} = kc^I c^{II} d_j^I d_i^{II}$$

Total flux will be calculated as, $J = w_r^{II} - w_r^I$,

$$J = kc^I c^{II} d_j^I d_i^{II} - kc^I c^{II} d_i^I d_j^{II},$$

$$J = kc^I c^{II}\left[d_j^I d_i^{II} - d_i^I d_j^{II}\right],$$

$$J = kc^I\left(c^I + l\nabla c\right)\left[d_j^I\left(d_i^I + l\nabla d_i\right) - d_i^I\left(d_j^I + l\nabla d_j\right)\right],$$

First order approximation will be,

$$J = klc^2\left[d_j\nabla d_i - d_i\nabla d_j\right],$$

$$J = Dc^2\left[d_j\nabla d_i - d_i\nabla d_j\right], \tag{4}$$

In this approximation, $kl = D$, where l is the cell size.

Its modeled equation will be, $\frac{\partial c}{\partial t} = -div(J)$

$$\frac{\partial c}{\partial t} = -Dc^2\left[d_j\Delta d_i - d_i\Delta d_j\right]. \tag{5}$$

3.2 Now if rabbit from territory two moves to territory one then in this case both the rabbits will be present in territory one along with the fox of that territory, then for this situation the possible stoichiometric equation will be:

$$(F^I + R^I) + (F^{II} + R^{II}) \rightarrow (F^I + 2R^I) + F^{II} \tag{6}$$

For the above mechanism,

$$\gamma_F^I = 0,\ \gamma_R^I = 1.$$

$$w_r^I = kc^I c^{II} d^I d^{II},\ w_r^{II} = kc^I c^{II}\left(d^I\right)^2.$$

Total flux will be, $J = kc^I c^{II}\left(d^I\right)^2 - kc^I c^{II} d^I d^{II}$,

$$J = -kc^I c^{II}\left[d^I\left(d^{II} - d^I\right)\right],$$

$$J = -klc^I d^I\left(c^I + l\nabla c\right)\left[\frac{d^{II} - d^I}{l}\right],$$

$$J = -klc^I d^I\left(c^I + l\nabla c\right)\nabla d,$$

First order approximation will be,

$$J = -Dc^2 d\nabla d. \tag{7}$$

And its modeled equation will be,

$$\frac{\partial c}{\partial t} = Dc^2 d\Delta d. \tag{8}$$

3.3 Now if rabbit from territory one moves to territory two, then in this case both the rabbits will be present in territory two along with the fox of that territory, then for this situation the possible stoichiometric equation will be:

$$(F^I + R^I) + (F^{II} + R^{II}) \rightarrow F^I + (F^{II} + 2R^{II}). \tag{9}$$

For the above mechanism,

For the above mechanism,

$$\gamma_F^I = 0,\ \gamma_R^I = -1,\ w_r^I = kc^I c^{II} d^I d^{II},\ w_r^{II} = kc^I c^{II} \left(d^{II}\right)^2.$$

Total flux will be, $J = kc^I c^{II} \left(d^{II}\right)^2 - kc^I c^{II} d^I d^{II}$,

$$J = klc^I c^{II} d^{II} \left[\frac{d^{II} - d^I}{l}\right],$$

$$J = klc^I \left(c^I + l\nabla c\right)\left(d^I + l\nabla d\right)\nabla d,$$

Its first order approximation will be,

$$J = Dc^2 d\nabla d. \tag{10}$$

And its modeled equation will be,

$$\frac{\partial c}{\partial t} = -Dc^2 d\Delta d. \tag{11}$$

4 Movements to Free Places

4.1 Case study of foxes

If fox/foxes is/are in territory one/ two and due to some reasons its neighbouring territory *(Z)* is laying vacant, if this fox/foxes moves to that vacant territory (as it is stated that foxes are territorial animal and they stay in their own territory zones so it may or may not be possible in real) then for this situation it is given by stoichiometric equations of the form:

$$\begin{aligned} F^{I} + Z^{II} &\rightarrow F^{II} + Z^{II} \\ F^{II} + Z^{I} &\rightarrow F^{I} + Z^{I} \end{aligned} \tag{12}$$

For the mechanism $F^{I} + Z^{II} \rightarrow F^{II} + Z^{II}$

$$\gamma_F^I = -1,\ \gamma_z^I = 0,\ w_r^I = kc^I z^{II}.$$

Where z is supposed to be the concentration of free zones.

For the mechanism $F^{II} + Z^{I} \rightarrow F^{I} + Z^{I}$

$$\gamma_F^I = 1,\ \gamma_z^I = 0,\ w_r^{II} = kc^{II} z^{I}.$$

Total flux will be, $J_F = kc^{II} z^{I} - kc^{I} z^{II}$.

$$J_F = k\left[(c^I + l\nabla c)z^I - c^I\left(z^I + l\nabla z\right)\right] = kl\left[z\nabla c - c\nabla z\right].$$

$$J_F = D\left[z\nabla c - c\nabla z\right]. \tag{13}$$

Its modeled equation will be

$$\frac{\partial c}{\partial t} = D\left[c\Delta z - z\Delta c\right]. \tag{14}$$

4.2 Case study of rabbits

If rabbit/rabbits is/are in territory one/ two and due to some reasons its neighbouring territory is laying vacant, if this rabbit/rabbits moves to that vacant territory, then for this situation it is given by stoichiometric equations of the form:

$$\begin{aligned} R^{I} + Z^{II} &\rightarrow R^{II} + Z^{II} \\ R^{II} + Z^{I} &\rightarrow R^{I} + Z^{I} \end{aligned} \tag{15}$$

For the mechanism $R^I + Z^{II} \rightarrow R^{II} + Z^{II}$

$$\gamma_R^I = 1,\ \gamma_z^I = 0,\ w_r^{II} = kd^{II} z^I.$$

Total flux will be, $J_R = D\left[z\nabla d - d\nabla z\right].$ (16)

Its modeled equation will be, $\dfrac{\partial d}{\partial t} = D\left[d\Delta z - z\Delta d\right].$ (17)

4.3 Mixed study case for foxes and rabbits

If a fox/foxes and rabbit/rabbits are in territory one/ two and due to some reasons their neighbouring territory is laying vacant, if all the foxes/ fox and rabbit/rabbits moves to that vacant territory, then for this situation it is given by stoichiometric equations of the form:

$$\begin{aligned} &F^I + R^I + Z^{II} \rightarrow F^{II} + R^{II} + Z^{II} \\ &F^{II} + R^{II} + Z^I \rightarrow F^I + R^I + Z^I \end{aligned} \qquad (18)$$

For the mechanism $F^I + R^I + Z^{II} \rightarrow F^{II} + R^{II} + Z^{II}$.

$$\gamma_F^I = -1,\ \gamma_R^I = -1,\ \gamma_z^I = 0,\ w_r^I = kc^I d^I z^{II}.$$

For the mechanism $F^{II} + R^{II} + Z^I \rightarrow F^I + R^I + Z^I$,

$$\gamma_F^I = 1,\ \gamma_R^I = 1,\ \gamma_z^I = 0,\ w_r^I = kc^{II} d^{II} z^I.$$

Total flux will be, $J = k\left[c^{II} d^{II} z^I - c^I d^I z^{II}\right],$

$$J = k\left[\left(c^I + l\nabla c\right)\left(d^I + l\nabla d\right)z^I - c^I d^I\left(z^I + l\nabla z\right)\right],$$

$$J = kl\left[zc\nabla d + zd\nabla c - cd\nabla z + zl\nabla c\nabla d\right],$$

Its first order approximation will be,

$$J = D\left[c\left(z\nabla d - d\nabla z\right) + zd\nabla c\right]. \qquad (19)$$

Its modeled equation will be,

$$\frac{\partial c}{\partial t}=-D\left[c\left(z\Delta d-d\Delta z\right)+zd\Delta c\right]. \tag{20}$$

5 Autocatalysis Mechanisms

Autocatalysis is the procedure whereby a chemical is involved in its own production. Here in our study case we will see that how two foxes or two rabbits interact together to increase their populations.

5.1 Case study of foxes

As it is stated in the early lines that the foxes are the territorial animals and they define their own territories to prey, but they also pair up in winter seasons. when in winter seasons two foxes pair up in territory one/two together, let a male fox $F_{(m)}$ and a female fox $F_{(f)}$ interact, let n be the number of foxes which born as a result of that interaction, they all can be male foxes, females or a combination of both the sexes, this circle repeats itself. In this case the possible mechanism will be of the form,

$$\begin{aligned}&2F^{I}\rightarrow n_1F^{I},n_1>2\\&2F^{II}\rightarrow n_2F^{II},n_2>2\end{aligned} \tag{21}$$

For the mechanism $2F^{I}\rightarrow n_1F^{I}$

$$\gamma_F^I=n_1-2,\ w_r^I=k\left(c^I\right)^2.$$

For the mechanism $2F^{II}\rightarrow n_2F^{II}$

$$\gamma_F^I=0,\ w_r^{II}=k\left(c^{II}\right)^2.$$

Total flux will be, $J_F=k\left(c^{II}\right)^2-k\left(c^I\right)^2=k\left[\left(c^{II}-c^I\right)\left(c^{II}+c^I\right)\right].$

$$J_F=kl\left[\left(\frac{c^{II}-c^I}{l}\right)\left(c^I+l\nabla c+c^I\right)\right]=D\nabla c\left(2c+l\nabla c\right).$$

Its first order approximation will be, $J_F=2cD\nabla c.$ (22)

Its modeled equation for this case will be, $\frac{\partial c}{\partial t}=-2cD\Delta c.$ (23)

5.2 Case study of rabbits

Now in the study case of rabbits when two rabbits are in territory one /two, let a male rabbit $R_{(m)}$ and a female rabbit $R_{(f)}$ interact, let n be the number of rabbits which born as a result of that interaction, they all can be male rabbits, females or a combination of both the sexes, this circle repeats itself. In this case the possible mechanism will be of the form,

$$\begin{aligned} &2R^{I} \rightarrow n_3 R^{I}, n_3 > 2 \\ &2R^{II} \rightarrow n_4 R^{II}, n_4 > 2 \end{aligned} \tag{24}$$

For the mechanism $2R^{I} \rightarrow n_3 R^{I}$

$$\gamma_R^I = n_3 - 2, \; w_r^I = k\left(d^I\right)^2.$$

For the mechanism $2R^{I} \rightarrow n_4 R^{I}$

$$\gamma_R^I = n_4 - 2, \; w_r^{II} = k\left(d^{II}\right)^2.$$

Total flux will be, $J_R = 2dD\nabla c.$ (25)

Modeled equation for this case will be, $\dfrac{\partial d}{\partial t} = -2dD\Delta d.$ (26)

6 The Observations

Thus the discrete equations for cells converge result in a system of ODEs in cells. The jumps of concentrations on the boundary of cells are considered the discrete system of ODEs for cells produces the second order in the cell size approximation to the continuous diffusion equation These jumps are flows in both directions: from cell 1 to cell II and from cell II to cell I. Different types of mechanism were considered here and obtained the modeled equations for the predator prey interactions. These jumps or the integrations are supposed to obey the mass action law. The kinetic constants may be scaled with the cell size to keep $kl = d$ (k is the kinetic constant, l is the cell size, d is the diffusion coefficient for the particular mechanism). It seems a good hypothesis that for obtained modeled systems with convex Lyabunov functional and "no-flux" boundary conditions in bounded areas with smooth boundaries the cell model gives the uniform approximation to the solution of the correspondent PDE. The system of semi-discrete models for cells with "no-flux" boundary conditions has all the nice properties of the chemical kinetic equations for closed systems. Under the proper relations between coefficients, like complex balance or detailed balance, this system may demonstrate a globally stable dynamics. This global stability property can help with the study of the related PDE.

7 Future Work

In the future, we aim:

- To build up a concise study of complex biological systems and to tackle some key research questions concerning our study case by proposing some new techniques and algorithms that are inspired by those complex biological systems.
- To check the stability of all the mechanisms discussed in our study case.
- To find the solutions of the modeled equations determined for all these interactions.
- To check the existence and uniqueness of those solutions.

References

Gorban AN, Sargsyan HP, Wahab HA. 2011. Quasichemical models of multicomponent nonlinear diffusion. Mathematical Modelling of Natural Phenomena, 6(5): 184-262

Lotka AJ. 1920. Undamped oscillations derived from the law of mass action. Journal of American Chemical Society, 42: 1595

Murray JD. 2003. Mathematical Biology I: An Introduction. Springer-Verlag

Pedersen, A.B, Grieves, T, 2008. The interaction of parasites and resources cause crashes in a wild mouse population. Journal of Animal Ecology, 77: 370-377

Potts, J.R. et al. 2013. Quantifying behavioral changes in territorial animals caused by sudden population declines. American Naturalist, 182: 73-82

Voltera V. 1926.Variations and fluctuations of the number of the individuals in animal species living together. In: Animal Ecology (Chapman RN, ed). McGraw-Hill, New York, USA

Wahab HA, Shakil M, Khan T, et al. 2013.A comparative study of a system of Lotka-Voltera type of PDEs through perturbation methods, Computational Ecology and Software, 3(4): 110-125

Shakil M, Wahab HA, Naeem M, Bhatti S. 2014. A quasi chemical approach for the modeling predator-prey interactions, Network Biology, 4(3): 130-150

Shakil M, Wahab HA, Naeem M, Bhatti S, Shahzad M. 2015. Modeling of the predator-prey interactions, Network Biology, 5(2): 71-81

13

In silico prediction of three-dimensional structure and interactome analysis of Tubulin α subfamily of *Arabidopsis thaliana*

Jasmin Šutković, Mohamed Ragab Abdel Gawwad
Genetics and Bioengineering Department, International University of Sarajevo, Ilidza, 71220 Bosnia and Herzegovina
E-mail: mragab@ius.edu.ba

Abstract
Microtubules are essential components of cytoskeleton, rigid hollow rods approximately 25 nm in diameter. Microtubules are dynamic structures being continuously assembled and disassembled within the cell. The basic building blocks of microtubules are heterodimers of globular α- and β-tubulin subunits In *Arabidopsis thaliana* tubulin subunits are encoded by small gene families, six for α-tubulin and nine for β-tubulin.Both α- and β-tubulin bind GTP, which functions analogously to the ATP bound to actin to regulate polymerization. It is shown that tubulin alpha forms hydrogen bonds with the GTPase domain of b-tubulin. Multiple sequence alignment revealed highsimilarity between the family subunits. Due to the missing of three dimensional structuresin *A. thaliana, structural models were predicted and validated. Additionally, protein domains search revealed that all tubulin* α family subunits contain GTPase domain as the tubulin C terminal domain, confirming previous research. Finally the interactome analysis revealed several interactomes.AtTUA6 shows strong interaction with embryosac development arrest 10 protein (EDA10), involved in stimulating the exchange of guanyl nucleotides, enabling the replacement of GDP by GTP in association with a GTPases.

Keywords 3D structure; interactome; microtubules; tubulin α; functional annotation.

1 Introduction

Microtubules are dynamic cytoskeletal polymers essential for various cell functions such as intracellular organization, ordered vesicle transport, cell division and establishment of cell polarity. The basic building blocks of microtubules are heterodimers of globular α- and β-tubulin subunits. They are arranged in a head-to-tail fashion to form 13 protofilaments that constitute cylindrical microtubules with outer diameter around 25 nm. In *Arabidopsis thaliana* tubulin subunits are encoded by small gene families, six for α-tubulin (Schröder et al., 2001) and nine for β-tubulin (Snustad et al., 1992). It has been proposed that the function of microtubules is modulated by highly diverse posttranslational modifications of tubulin dimers (Blume et al., 2010). The alpha- and beta-tubulins are the major components of microtubules, while gamma-tubulin plays a major role in the nucleation of microtubule assembly. Microtubules are capable of performing various tasks during the life cycle of eukaryotic cells in spite of ahighly conserved basic structure. The different functions appear to be, at least in part, mediated by differentialaction of diverse associated proteins, including motor proteins (Schröder et al., 2001). However, the in vivo threedimensional structure of Tubulin alpha proteins is not resolved. In this study the in silicon three dimensional structure of all Tubulin α proteins in Arabidopsis *thaliana* will be predicted, phylogenetic relationships and Interactome for new functional annotations as well.

Microtubules

Microtubules play central roles in several of the most basic processes of eukaryotic cells: cell division, cell motility, intracellular transport, and the control of cell shape. In plant cells, rigid cell walls obviate the need for direct cytoskeletal maintenance of cell shape (Steven et al., 1987). Microtubules of the eukaryotic cytoskeleton perform essential and diverse functions and are composed of a heterodimer of alpha and beta tubulins. The genes encoding these microtubule constituents belong to the tubulin superfamily, which is composed of six distinct families. Genes from the alpha, beta and gamma tubulin families are found in all eukaryotes. The alpha and beta tubulins represent the major components of microtubules, while gamma tubulin plays a critical role in the nucleation of microtubule assembly. There are multiple alpha genes, which are highly conserved among species. This gene encodes alpha tubulin and is highly similar to mouse and rat TubA1 gene (Levilliers et al., 1998; Jolly et al., 2007).

About 13 of these protofilaments associate in parallel making the microtubule wall, and giving rise to a polymer with a well-defined polarity. Essential to the function of microtubules is their ability to switch stochastically between growing and shrinking phases (dynamic instability), a non-equilibrium behaviour of tubulin that is based on nucleotide binding and hydrolysis. Each tubulin monomer binds one molecule of GTP. The nucleotide bound to αtubulin, at the so called N-site, is non-exchangeable. The resulting metastable microtubule structure is thought to be stabilized by a cap of remaining GTP-tubulin subunitsat the ends, the loss of which results in rapid depolymerization (Lowe et al., 2001).As mentioned above,the alpha tubulin heterodimer is the structural subunit of microtubules. Tubulin alpha shares minimum 40% amino-acid sequence identity with tubulin beta , exist in several isotype forms, and undergo a variety of posttranslational modifications. The structures of alpha and beta tubulins are basically identical: each monomer is formed by a core of two beta-sheets surrounded by alpha-helices. The monomer structure is very compact, but can be divided into three regions (domains) based on function: the amino-terminal nucleotide-binding region, binds GTP , the carboxy-terminal region which probably constitutes the binding surface for motor proteins and an intermediate taxol-binding region (Nogales et al., 1998).

It is important to note that various isotypes of the two major microtubule subunits, Alpha and beta tubulin, can occur even within a single organism and that these may interact differentially with the proteins. Tubulin isotypes comprise post-translational modifications of a common progenitor and/or members of multigene families (MacRae and Carrie, 1989).In higher plants, as in higher animals, it is now accepted that tubulins are generally encoded by multigene families. Even in Arabidopsis, with its 'minimal' genome, families with at least six different α and nine β tubulin members have been found (Kopczak et al., 1992; Chu et al., 1993).

Tubulin α subfamily in Arabidopsis thalina

Tubulin α subfamily is the major constituent of microtubules. It binds two moles of GTP, one at an exchangeable site on the beta chain and one at a non-exchangeable site on the alpha-chain.This subfamilyincludessixtubulin α proteins, as show in Table 1 (Steven et al., 1987).

Alpha-tubulin 1 (TUA1) is encoded by the tua1 gene in *A. thaliana*.This protein is primarilyexpressed in stamens and mature pollen. TUA1 is one of the main constituent of cytoskeleton,involved in microtubule-based process. The location of TUA1 was measure in microtubules cytoskeleton, cytosol and plasma membrane (Ludwig et al., 1987, 1988).

Alpha-tubulin 2 (TUA2) proteinis also located in cytoskeleton and represent the main structural constituent of it.This protein is involved in cadmium transport (salt stress), microtubule-based processes and in the response to salt stress.TUA2 is located in cytoskeleton, cytosol and cell wall and cell membrane and in chloroplast. TUA2 is expressed 26 plant structure during 15 growth stage (Ludwig et al., 1987).

Alpha-tubulin 3 (TUA3)is involved is the function of cellular response to gravity.Like all other tubulin alpha proteins it is an important structural part of cytoskeleton.TUA3 is mainly expressed in mature pollen stage and during the germination stage of pollen. TUA3 shares the location with other tubulin alpha members (Ludwig et al., 1987; Saito et al., 2003).

Alpha-tubulin 4 (TUA4) functions in structural constituent of cytoskeleton.TUA4 is involved in response to cadmium ions, microtubule-based process, cellular response to gravity. This protein is located in cytosol, cell wall, plasma membrane and chloroplast. TUA4 is expressed in 29 plant structures during 15 growth stages (Ludwig et al., 1987, 1988; Saito et al., 2003).

Alpha-tubulin 5(TUA5), together with all tubulin alpha members, represent an important structural constituent of cytoskeleton.TUA5 regulates cadmium ion transport and microtubule-based processes.As atubulin complex it is located in cytosol, cell wall and cell membrane.It is expressed ion 6 plant structures (Mattingly et al., 2001; Saito et al., 2003; Abe and Hashimoto, 2005).

Alpha-tubulin 6 (TUA6) protein, like other tubulin alpha members it is a structural constituent of cytoskeleton, involved in response to salt stress, microtubule cytoskeleton organization and cellular response to gravity. Furthermore, it is involved protein polymerisation (Ludwig et al., 1987; Mattingly et al., 2001; Saito et al., 2003; Abe and Hashimoto, 2005; Ban et al., 2013).

2 Methodology

The fasta sequences of Tubulin α subfamily proteins (TUA1, TUA2, TUA3, TUA4, TUA5 and TUA6) were obtained from NCBI databases. The Gene accession numbers from NCBI and TAIR database are shown in Table 1.

Table 1 Tubulin α proteins and their NCBI accession IDs.

Tubulin α members	NCBI accession IDs	TAIR IDs
TUA1	NP_176654.1	AT1G64740
TUA2	NP_175423.1	AT1G50010
TUA3	NP_197478.1	AT5G19770
TUA4	NP_171974.1	AT1G04820
TUA5	NP_197479.1	AT5G19780
TUA6	NP_849388.1	AT4G14960

The multiple sequence alignment was performed with the ClustalW2 program (Larkin et al., 2007). Default parameters were applied and aligned sequences were executed using this software. Phylogenetic tree was constructed using Maximum likelihood method in Robust Phylogenetic Analysis software, where phylogenetic relationships between amino acids of TUA subunits were reconstructed and analysed (Dereeper et al., 2008).

In order to predict the three dimensional structure of Tubulin alpha subfamily in *Arabidopsis thaliana,*protein data bank (Pdb) files were obtained based on homology modelling using CPH models 3.2

servers (Nielsen et al., 2010). Additionally to validate the 3-D structures for all TUAs, by testing stereochemistry quality using Ramachandran plots obtained from RAMPAGE program (Lovell et al., 2003).

The protein's genetically mobile domain annotations were obtained using *Pfam* database analysis program (Finn et al., 2010). For the protein-protein interaction analysis, Arabidopsis interaction viewer was used (Popescu et al., 2009). Subcellular localization for each protein was predicted and confirmed by WoLF PSORT program (Horton, et al., 2007).

Arabidopsis Interactions Viewer V.2.0 was used for the protein- protein interaction in order to find the interactome for all AtRFC subunits (Popescu et al., 2009).

3 Results and Discussion

3.1 Modeling of 3D structure in Tubulin α subfamily

The three-dimensional (3-D) structure details of proteins are of major importance in providing insights into their molecular functions. 3-D structure of all six subunits showed highly structural similarities. Additionally outputs have been confirmed by Ramachandran plots (Fig. 1a and 1b). The modelled enzymes are monomers folded into α/β domains. AtTUA family consists of minimum 15 central stranded β sheet flanked by 11α helices. The similar structure is shared by other tubulin alpha enzymes from mammalian origins but the number of helices and β sheets differ from one species to another. The Ramachandran plots in Fig. 1a and 1b indicate the region of possible angle formations by w (phi) and c (psi) angles. The conventional terms represent the torsion angles on either side of alpha carbon in peptides. The plot was divided into three regions: favoured (91.5%), allowed (6%) and outlier (2.5%). This result is significant since the high percentage of residues in favoured region (90%). This indicates that the built model is of good quality, as shown in Fig. 1a and Fig. 1b.

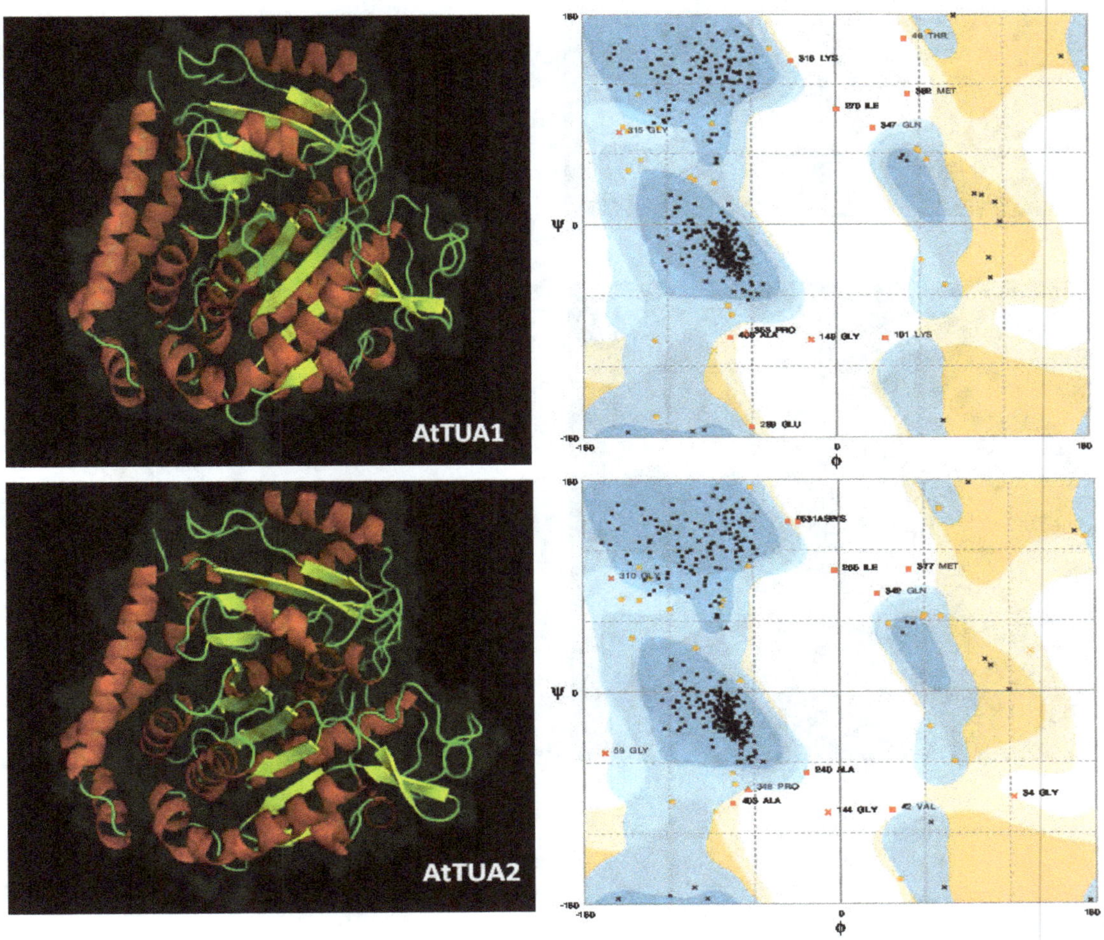

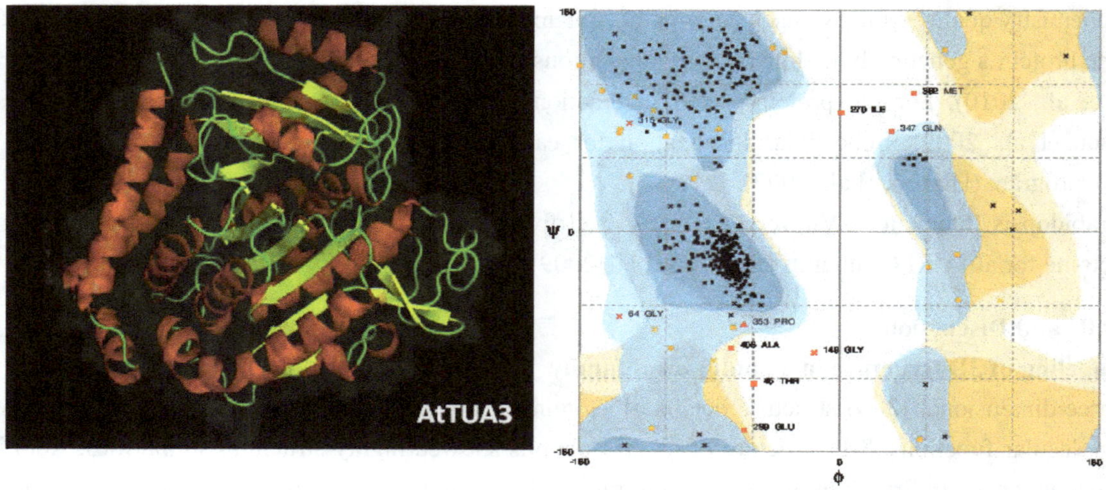

Fig. 1a 3D modelsand Ramachadran plots of AtTUA1/2/3.

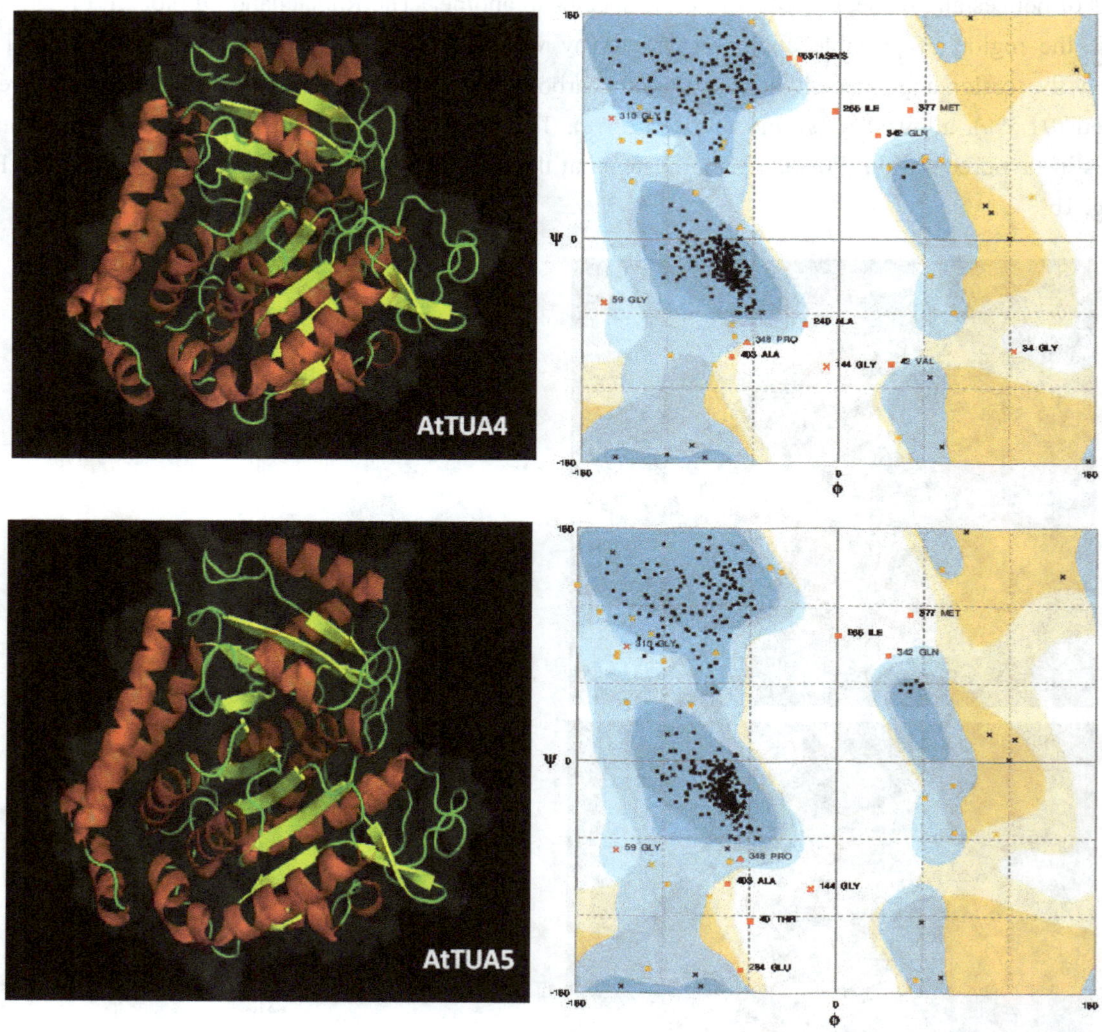

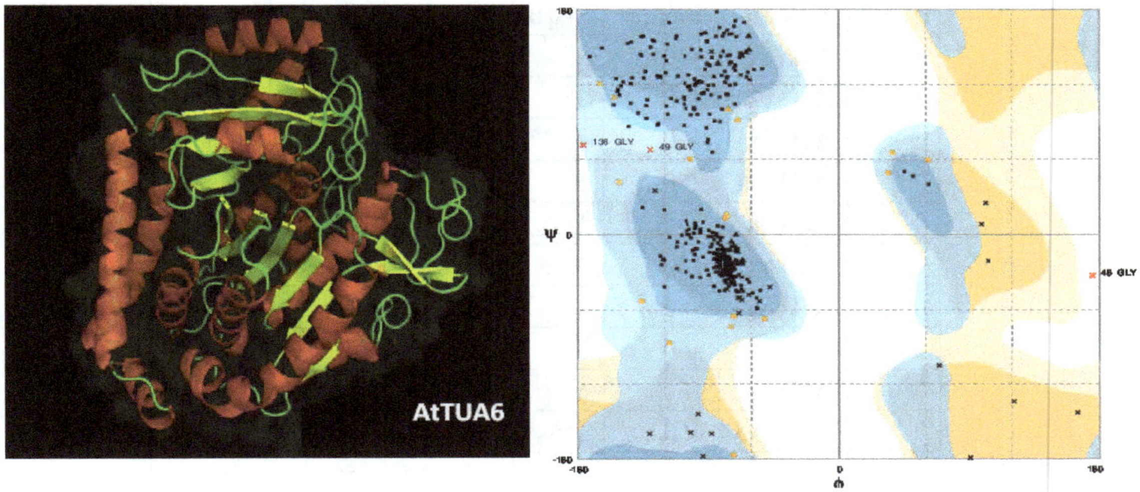

Fig. 1b 3D structure and Ramachadran plots of AtTUA4/5/6.

3.2 Multiple sequence alignment and phylogenetic tree analysis

Aligned sequences showed substantially significant homology, and constructed phylogenetic tree representing the already proven relationship between proteins in this family which play their major role as *GTPase* binding proteins.

The protein sequence of TUA1, TUA2, TUA3, TUA4, TUA5 and TUA6 were obtained from the NCBI sequence database (Table 1). The amino acid sequences are highly similar to the sequences of the homologous human TUA subunits, respectively, and also show amino acid sequence similarity to functionally homologous proteins of other organisms.

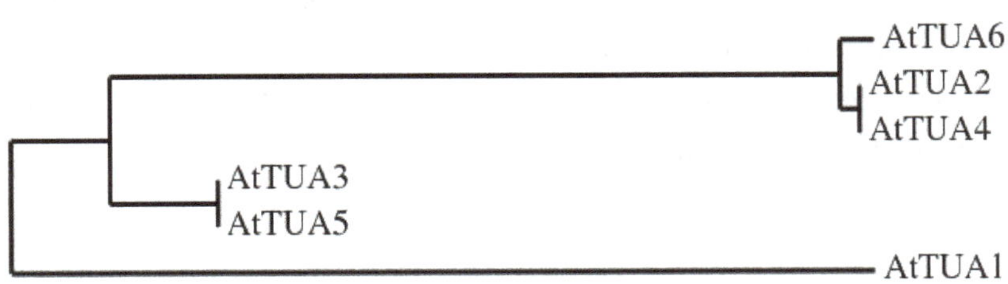

Fig. 2 Tubulin α subfamily phylogenetic tree.

The phytogenic relationships among the five subunits revealed that the two proteins (AtTUA3 and AtTUA5) are identical and that they diverged from AtTUA1. AtTUA2 and AtTUA4 evolved together with AtTUA6 with reserved evolutionary relationships (Fig. 2).This confirm the similarity in function of AtTUA subunits along the sequence alignment results. The sequence alignment scores in the TUA subfamily are shown in Table 2.

Table 2 ClustaW2 alignment result.

SeqA	Name	Length	SeqB	Name	Length	Score
1	AtTUA1	450	2	AtTUA2	450	88.6667
1	AtTUA1	450	3	AtTUA3	450	91.7778
1	AtTUA1	450	4	AtTUA4	450	88.6667
1	AtTUA1	450	5	AtTUA5	450	91.7778
1	AtTUA1	450	6	AtTUA6	450	88.4444
2	AtTUA2	450	3	AtTUA3	450	93.5556
2	AtTUA2	450	4	AtTUA4	450	100.0
2	AtTUA2	450	5	AtTUA5	450	93.5556
2	AtTUA2	450	6	AtTUA6	450	99.5556
3	AtTUA3	450	4	AtTUA4	450	93.5556
3	AtTUA3	450	5	AtTUA5	450	100.00
3	AtTUA3	450	6	AtTUA6	450	93.3333
4	AtTUA4	450	5	AtTUA5	450	93.5556
4	AtTUA4	450	6	AtTUA6	450	99.5556
5	AtTUA5	450	6	AtTUA6	450	93.3333

3.3 Domain search in Pfam database

Pfam domain search shows that all members in the TUA subfamily contains the Tubulin GTPase domain and the Tubulin C-terminal domain, confirming the sequence alignment results. Tubulin is homologous to FtsZ protein, responsible for bacterial cell division (alpha-beta tubulin structure). FTsZ and tubulins are GTPases, having the GTPase domain. FtsZ can polymerise into tubes, sheets, and rings in vitro and is ubiquitous in bacteria and archaea. FtsZ is found at the C-terminal domain (Nogales et al., 1998).

In all AtTUA the ATPase domain extends from 49 to 246 amino acid residues, while C-terminal domain extends from 248 to 393 residues, shown in Table 3.

All six subunits show conserved regions characteristic of Tubulin/FtsZ family GTPase binding domain and the Tubulin C-terminal domain (Nogales et al., 1998).

Table 3 AtTUA family domains.

Domains	Tubulin/FTSZfamily, GTPase domain		Tubulin c-terminal domain	
Proteins	*Start*	*End*	*Start*	*End*
TUA1	49	246	248	393
TUA2	49	246	248	393
TUA3	49	246	248	393
TUA4	49	246	248	393
TUA5	49	246	248	393
TUA6	49	246	248	393

3.4 Subcellular localization

The Subcellular localizations of AtTUAs are analysed with using WoLF PSORT program. Results showed good and significant reliabilities that all Tubulin alpha members are located in cytoskeleton and/or cytosol. TAIR software confirmed those localization results (Table 4).

Table 4 AtRFC family localization results.

Protein	Subcellular localization
AtTUA1	Cytostol
	Cytoskeleton
AtTUA2	Cytostol
AtTUA3	Cytoskeleton
	Cytostol
AtTUA4	Cytostol
	Cytoskeleton
AtTUA5	Cytostol
	Cytoskeleton
AtTUA6	Cytostol
	Cytoskeleton

3.5 Interactions

According to data obtained from the Arabidopsis Interactions Viewer (TAIR) we found several interactions belonging to all six subunits of AtTUA family. The interactome network of TAIR analysis revealed that AtTUA1, AtTUA4 and AtTUA6 subunits showed strongly interaction with Tubulin folding cofactor B (Fig. 3). Additionally AtTUA4 and AtTUA6interact with Prefoldin 6 protein and EDA10 protein, whereas only AtTUA6 interacts with Prefoldin 3 protein.

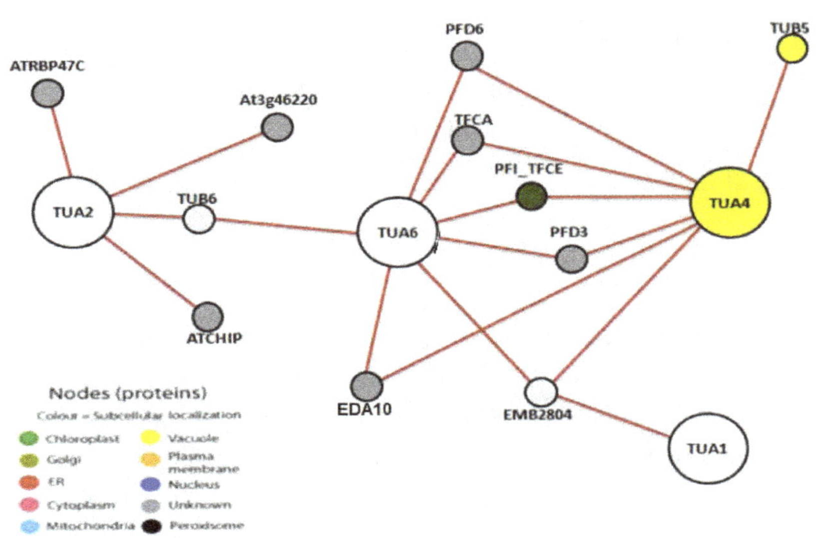

Fig. 3 AtTUAfamilyinteraction network (only siginifcant interactions are shown).

Tubulin cofactor B is involved in embryo development and cell division (Du et al., 2010). Prefoldin 3 and 6, members of the PFD complex, are thought to function along with the TCP ring complex to properly fold microtubule proteins. Additionally, Prefoldin 3 is shown to play essential role in Arabidopsis tolerance is salt stress (Rodríguez-Milla and Salinas, 2009). Embryo sac development arrest10 protein (EDA10) is involved in guanyl-nucleotide exchange factor activity by stimulating of the exchange of guanyl nucleotides associated with a GTPase (Pagnussat et al., 2005).

The common partner for AtTUA2 and AtTUA4 are tubulin folding factors A, E and B. Tubulin folding factors are mostly involved in the control of the correct balance of α- and β-tubulin monomers (Victor et al., 2002). AtTUA3 and AtTUA5 show no significant interactions (Table 5 and Fig. 3).

Table 5 Interactomes of AtRFC subunits.

AtRFCs	Interactom ID	Interactome name and function
TUA1	EMB2804	Tubulin folding cofactor B
TUA2	ATRBP47C	RNA-binding protein 47C
TUA2	ATCHIP	Carboxyl terminus of HSC70-interacting protein
TUA2	At3g46220	Unknown name and function
TUA2	TUB6	Tubulin beta-6 chain
TUA4	EDA10	Embryo sac development arrest 10
TUA4	TUB5	Tubulin beta-5 chain
TUA4	PFD6	Prefoldin 6 chain
TUA4	PFI_TFC E	Tubulin folding cofactor E Pfifferling (PFI)
TUA4	TFCA	Tubulin folding cofactor A (KIESEL)
TUA4	PFD3	Prefoldin 3 chain
TUA4	EMB2804	Tubulin folding cofactor B
TUA6	PFD6	Prefoldin 6
TUA6	PFI_TFC E	Tubulin folding cofactor E / Pfifferling (PFI)
TUA6	TFCA	KIS_tubulin folding cofactor A (KIESEL)
TUA6	EMB2804	Tubulin folding cofactor B
TUA6	EDA10	Embryo sac development arrest 10
TUA6	TUB6	Tubulin beta-6 chain
TUA6	PFD3	Prefoldin 3

4 Conclusions

Tubulin family is a multi protein complex consisting of several subunits. Tubulin α subfamily is the major constituent of microtubules. It binds two moles of GTP, one at an exchangeable site on the beta chain and one at a non-exchangeable site on the alpha-chain. This subfamily includes six tubilin α proteins. The 3D structure and inteactome analysis confirmed their similarity and functions.Additinally, through the strong interaction between AtTUA6 and Embryo sac development arrest10 (EDA10), AtTUA may additionaly functions in the exchange of guanyl nucleotides associated with a GTPase. It was expected that AtTUA1, AtTUA3 and AtTUA4 interacts with various tubulin coding factors, confirming previous research. Interestingly, AtTUA4 and AtTUA6 are shown to interact with Prefoldin2 and Profoldin 6 protein that supports the proper folding of microtubules and have important role in Arabidopsis tolerance to salt stress.

References

Abe T, Hashimoto T. 2005. Altered microtubule dynamics by expression of modified alpha-tubulin protein causes right-handed helical growth in transgenic Arabidopsis plants. Plant Journal, 43(2): 191-204

Ban Y, Kobayashi Y, Hara T, et al. 2013. Alpha tubulin is rapidly phosphorylated in response to hyperosmotic stress in rice and Arabidopsis. Plant and Cell Physiology, 54(6): 848-858

Blume Y, Yemets A, Sheremet Y, et al. 2010. Exposure of beta-tubulin regions defined by antibodies on an *Arabidopsis thaliana* microtubule protofilament model and in the cells. BMC Plant Biology, 10(29): 1471-2229

Chu BY et al. 1993. Alteration of P-Tubulin gene expression during low-temperature exposure in leaves of *Arabidopsis thaliana*. Plant Physiology, 103: 371-377

Dereeper A, Guignon V, Blanc G, et al. 2008. Robust phylogenetic analysis for the non-specialist. Nucleic Acids Research, 36(l2): 465-469

Du Y, Cui M, Qian D, et al. 2010. AtTFC B is involved in control of cell division. Frontiers in Bioscience, 1(2): 752-63

Finn RD, Mistry J, Tate J, et al. 2010. The Pfam protein families database: Bateman. Nucleic Acids Research, 38: 211-222

Horton P, Park KJ, Obayashi T, et al, 2007. WoLF PSORT: protein localization predictor. Nucleic Acids Research, 35: 585-587

Jolly C, Mitar I, Sattentau QJ. 2007. Requirement for an intact T-cell actin and tubulin cytoskeleton for efficient assembly and spread of human immunodeficiency virus type 1. Journal of Virology, 81(11): 5547-5560

Kopczak SD, Haas NA, Hussey PJ, et al. 1992. The small genome of Arabidopsis contains at least six expressed alpha-Tubulin genes. Plant Cell, 4: 539-547

Larkin MA, Blackshields G, Brown NP, et al. 2007. Clustal W and Clustal X version 2. Bioinformatics, 23(21): 2947-2948

Levilliers J, Crozet F, Chaib H, et al. 1998. Sequence characterization of a newly identified human alpha-tubulin gene (TUBA2). Genomics, 47(1): 125-130

Lowe J, Amos LA.1998. Crystal structure of the bacterial cell-division protein FtsZ. Nature, 391: 203-206

Lowe J, Li H, Downing KH, et al. 2001. Downing and E. Nogales. Refined structure of αβ-Tubulin at 3.5 A resolution. Journal of Molecular Biology, 313(5): 1045-1057

Lovell SC, Davis IW, Arendall WB, et al. 2003. Structure validation by Calpha geometry: phi, psi and Cbeta deviation. Proteins, 50(3): 437-450

Ludwig SR, Oppenheimer DG, Silflow CD, et al. 1987. Characterization of the alpha-tubulin gene family of *Arabidopsis thaliana*. Proceedings of the National Academy of Sciences of USA, 84(16): 5833-5837

Ludwig SR, Oppenheimer DG, Silflow CD, et al. 1988. The α1-tubulin gene of Arabidopsis thaliana: primary structure and preferential expression in flowers. Plant Molecular Biology, 10(4): 311-321

MacRae TH, Carrie ML. 1989. Tubulin synthesis, structure, and function: what are the relationships? Biochemistry and Cell Biology, 67(11-12): 770-790

Mattingly KS, Beaty BJ, Mackie RS, et al. 2001. Molecular cloning and characterization of a metal responsive Chironomus tentans alpha-tubulin cDNA. Aquatic Toxicology, 54(3-4): 249-260

Nielsen M, Lundegaard C, Lund O, et al. 2010. CPHmodels-3.0 - Remote homology modeling using structure guided sequence profiles. Nucleic Acids Research, 38

Nogales E, Downing KH, Amos LA, et al. 1998.Tubulin and FtsZ form a distinct family of GTPases. Nature Structural and Molecular Biology, 5: 451-458

Nogales E, Sharon GW, Kenneth HD. 1998. Structure of the αβtubulin dimer by electron crystallography. Nature, 391:199-203

Pagnussat GC,Yu HJ, Ngo QA, et al. 2005.Genetic and molecular identification of genes required for female gametophyte development and function in Arabidopsis. Development, 132(3): 603-614

Popescu SC, Popescu GV, Bachan S, et al. 2009. MAPK target networks in *Arabidopsis thaliana* revealed using functional protein microarrays. Genes and Development, 23: 80-92

Rodríguez-Milla MA, Salinas J.2009. Prefoldins 3 and 5 play an essential role in Arabidopsis tolerance to salt stress. Molecular Plant, 2(3): 526-534

Saito Y, Soga K, Wakabayashi K, et al. 2003. Increase in expression level of alpha-tubulin gene in Arabidopsis seedlings under hypergravity conditions. Biological Sciences in Space, 17(3): 177-180

Schröder J, Stenger H, Wernicke W.2001. Alpha-tubulin genes are differentially expressed during leaf cell developmentin barley (*Hordeum vulgare* L). Plant Molecular Biology, 45: 723–730

Snustad DP, Haas NA, Kopczakand SD, et al. 1992. The small genome of Arabidopsis contains at least nine expressed P-tubulin genes. Plant Cell, 4: 549-556

Steven RL, David GO, Carolyn DS, et al. 1987. Characterization of the α tubulin gene family of *Arabidopsis thaliana*. Proceedings of the National Academy of Sciences of USA, 84: 5833-5837

Victor K, Paul EG, Jaideep M, et al. 2002. The Arabidopsis Tubulin-folding Cofactor A gene is involved in the control of the α/β-Tubulin monomer balance. Plant Cell, 14(9): 2265-2276

14

Removal of electromyography noise from ECG for high performance biomedical systems

Navdeep Prashar, Jyotsna Dogra, Meenakshi Sood, Shruti Jain
Department of Electronics and Communication Engineering, Jaypee University of Information Technology, Solan, India
E-mail: nav.prashar@gmail.com, jyotsnadogra1989@gmail.com, meenakshi.sood@juit.ac.in, shruti.jain@juit.ac.in

Abstract
This paper presents the review of the biomedical system which consists of an energy source, signal processing, signal conditioning and signal transmission. These blocks are designed by various optimization techniques to achieve high operating speed, compressed area and minimum energy consumption. These techniques are mainly divided in to four aspects: (a) increasing the longevity of device using energy harvesting approaches; (b) reducing the delay to enhance the operating frequency; (c) reducing the data storage using data compression; (d) increasing the data rate transmission with reduced power consumption. This review paper briefly summarizes the various techniques and device performance achieved by these techniques. To attain these high performance systems input played a vital role. This paper also presents the different low pass IIR filter approximation method techniques to remove Electromyography noise from ECG input signal. For this purpose, we have taken MIT-BIH Arrhythmia database. We have calculated signal to noise ratio and power spectral density. On comparing their performance parameters of different low pass IIR filters, Elliptic filter has found best suited to remove this type of noise.

Keywords energy harvesting; implantable devices; inductive coupling; SNR; PSD; ECG; IIR.

1 Introduction

For the last sixty years, biomedical implantable devices are available for consumers. In 1957, for the first time Earl Bakken included transistor in biomedical implantable device for cardiac pacemaker (Rhees, 2009). These transistors enable biomedical devices to monitor and diagnose different types of signals such as electroretiuogram (ERG), electrocardiogram (ECG) and electromyography (EMG) of human body (Dogra et al., 2018; Zhang, 2018). The main focus of research is on patient safety and comfort in recent years. Therefore, the main aim of the research is to ensure efficient energy transfer and to the implanted device to reduce the power consumption (Amandeep et al., 2017; Jain et al., 2015; Jain, 2016).

With the advancement in integrated circuits (IC) technology and revolution in circuit design technique replicates the complex biomedical operations such as speech spectral analysis, image processing and digital filtering etc which enhance the device performances (Chandrakasan et al., 2008; Wong et al., 2004, and Paradiso et al., 2005). In Electronic Biomedical Devices, reduction in power consumption can be achieved by continued device scaling and integration. For example, an implanted medical device comprises of low-power general purpose processor which consumes almost 10 mW of energy having 3 days of battery operation. Alternately, a dedicated solution is obtained by implementing a specialized low power technique, which consumes approximately (Wise et al., 2004; Mishra et al., 2016). Generally, implanted devices were powered by batteries, but limited life time of batteries can lead to many challenges. Such issues are resolved by wireless–telemetry bio devices (WTB) which have been powered by radio frequency (RF) signals. In WTB, transmission of power has been performed via inductive coupling coil which is converted to DC voltage. Many other energy harvesting sources has been utilized to power the implanted devices such as knee implants that use the vibration of piezoelectric materials or body motion (Chen et al., 2009; Mishra et al., 2016; Zeng et al., 2011).

For biomedical devices a system level approach is needed which reduces the power consumption to μW (Jain, 2017). Advanced architecture and circuits for biomedical applications should be used to lessen the energy dissipation. This review focuses on the different specialized techniques which lead to the design to be compatible for high performance biomedical applications that require reduced power consumption, compress chip area and speed up operating frequency (Gyselinckx et al., 2005; Cadei et al., 2014).

Different Infinite impulse response (IIR) and Finite impulse response (FIR) filters are used to remove the noise from input ECG signal. FIR filters are simplest filter as they do not require any feedback system. These non–recursive filters are characterized by equation

$$y(m) = \sum_{k=1}^{N} ak\, y(m-K) + \sum_{k=o}^{M} bk\, x(m-K) \qquad (1)$$

IIR filters are known as recursive filters as their output depends on past inputs and past outputs. These filters are designed by using different approximation methods like Butterworth, Chebyshev I and Elliptic. IIR filters are determined by equation

$$y[n] = \sum_{k=0}^{N-1} h(k).X(n-k) \qquad (2)$$

In this paper, high frequency noise from ECG signal has been removed by designing different low pass IIR filter using Butterworth, Chebyshev I and Elliptic approximation methods.

2 Electronic Biomedical System

The generic biomedical system is shown in Fig. 1. The main components of this biomedical system includes energy source, sensors/transducers; signal conditioning, data storage, data transmission, control system and display equipment. A reliable biomedical system has been obtained by optimizing each and every block.

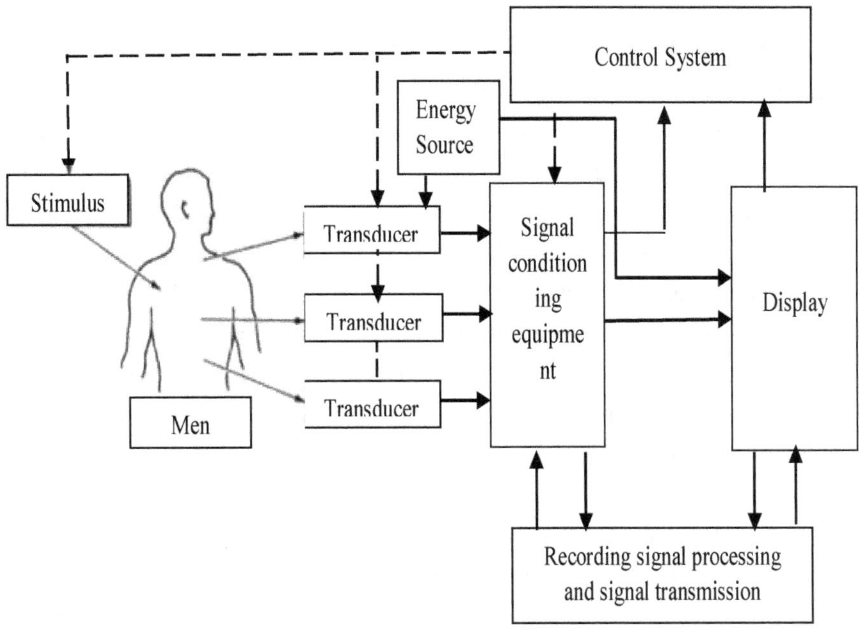

Fig. 1 Generic block diagram of Biomedical System (Cromwell, 2004).

Few existing biomedical devices such as pacemaker, hearing-aid, analog cochlear processor, retinal stimulator, neural recording and body-area monitoring with their performance parameters are shown in Table 1 (Chandrakasan et al., 2008).

Table 1 Review of existing applications of Biomedical Devices (Chandrakasan, 2008).

Energy Source	Power	Processors	Application
10 year lifetime battery	<10μW	1 KHz DSP	Pacemaker
1-week lifetime rechargeable battery	100-2000 μW	32 kHz-1MHzDSP	Hearing-Aid
1-week lifetime rechargeable battery	200 μW	Analog DSP	Analog cochlear processor
Inductive Power	$25X10^{4}\mu W$	No embedded DSP	Retinal stimulator
Inductive Power	$(1 X10^{3}-10X10^{3}\mu W$	-	Neural recording
Battery	140 μW	<10 MHz DSP	Body-area Monitoring

The reviews of the various components of Biomedical System are discussed in the following sections:

2.1 Energy source

In biomedical devices, the main source of energy is battery. Batteries can be easily replaced in non-implanted device but in case of implanted devices that cannot be possible without surgery. Therefore, an energy harvesting techniques has been introduced that provide continuous power to implanted devices. Paradiso et al.

(2005) described various energy harvesting sources such as thermoelectric, light, vibration, piezoelectric, electrostatic generators, near-field inductive energy transfer, far-field electromagnetic energy transfer. Cadei et al. (2014) explained thermoelectric generators that generate few hundred μ watts by creating temperature difference between outer and inner parts. Cascading of large thermocouples has been done to raise the output power. Glynne et al. (2001) and Kim et al. (2011) discussed the phenomenon of conversion of kinetic energy in to electric energy in piezoelectric generators to generate the power in mW. However, requirement of bulky movement made it not suitable for many implant devices. Kymissis et al. (1998) explained a piezoelectric transducer that produced the output power of 1W while attached to the shoe heel. Cheng et al. (1989) explained electrostatic energy harvesting based on the interference of coulomb forces. Wireless electromagnetic energy transfer is most efficient technique to impart energy to implanted biomedical devices. These techniques involve transmission of electromagnetic energy in to the body and gather the same through coil and antenna. Weiland et al. (2005) and Huang et al. (1998) explained low-frequency energy transfer using near-field electromagnetic induction process for many biomedical applications like radio-frequency identification (RFID), retinal prosthetics and neuro stimulators. To transfer the energy at far distance, Kurs et al. (2007) and Karthaus et al. (2003) presented a quantitative model of non-radiative power transfer over distances up to 8 times the radius of coil via strongly coupled magnetic resonances. High frequencies above hundreds of megahertz are another method to achieve far field communication. However, channel losses and low penetration to the skin are some of the limitations that keep the efficiency low. For implanted devices micro-electronic mechanical system (MEMS) technology has been introduced which fabricate mechanical system in to microchip. Amirtharajah et al. (2000) presented a fabricated MEMS energy harvester for low-power biomedical system. To transfer the energy continuously some other technologies like optical charging, ultrasonic transducers has been introduced which sent the energy through body tissue optically and mechanically (Murakawa et al., 1999). Table 2 shows the energy harvesting approaches for implanted biomedical devices (Amar et al., 2015).

2.2 Biomedical signal processing

Biomedical signal processing includes operation like filtering, data computation and data compression. Digital signal processing is preferred over analog due to its large noise margin and robust to distortion (Sood et al., 2017, Rana et al., 2016a; Bhusri et al., 2016a; Sharma et al. 2017). The limitation of analog circuits is that they require additional hardware for conversion in to digital signals.

Table 2 Energy harvested approaches for implanted biomedical devices (Amar et al., 2015).

Energy Harvesting Approaches	Generated Power (μW)
Thermoelectric	180 $\mu W/cm^2$
Piezoelectricity	0.33 μW
Electrostatic	80 μW
Electromagnetic	400 μW
Optical charging	22×10^3 μW /cm
Ultrasonic transducer	15×10^2 $\mu W /cm^2$
Inductive coupling	6150 μW

2.2.1 Filtering

Different methods have been implemented to remove the unwanted signal. Bhusri et al. (2016b), Mohamed (2016), and Rana et al. (2016b) presented least-square finite impulse response filtering technique to the ECG input signal to filter out low frequency noise. Implementation of least-square linear phase FIR filter (LLFE) design on field programmable gate array (FPGA) achieved accuracy of 97.8% with reduction in resource utilization.

2.2.2 Data computation

Biomedical devices required fast computation with greater accuracy and low power consumption along with reduced device area. Pardhu et al. (2014) designed a new hybrid CMOS design for one-bit full adder which operates at low voltage. This adder is then realized in 4-2, 5-2, 5-3, 7-2, 11-2, 15-4, 31-5 compressors which is used in multipliers to reduce the delay but design complexity leads to high power consumption. Ajay and Laurde (2015) designed an efficient booth–encoded wallace tree (BEWT) multiplier architecture for fast computation using fast fourier transform (FFT). In this architecture partial products have been reduced by booth encoder and then added by Wallace tree. Output accuracy can be increased by increasing the significant digits of twiddle factor by compromising with the increased area of circuit .Circuit analyzation gave an accuracy of ± 0.1% for twiddle factor =0.7071. BEWT multiplier architecture consist of five blocks- 2's complement, partial product generator, booth –encoder, Wallace tree and carry look-ahead (CLA) adder as shown in Fig 2.

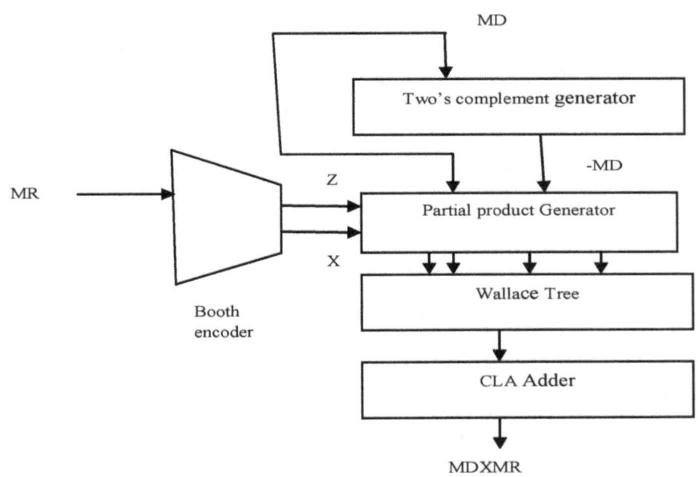

Fig. 2 Architecture of BEWT multiplier (Ajay and Laurde, 2015).

2.2.3 Data compression

In signal processing, data compression encodes the information in fewer bits than its original representation to reduce the storage (Sood et al., 2014). In this context, many new reliable and efficient data compression techniques have been developed by researchers. Ranjith and Muniraj (2012) presented a dictionary based data compression technique for bio signal processors that leads to significant reduction in storage. On reviewing the results this technique showed 28% saving of data storage. Also the previous research challenges such as bitmask selection and don't care resolution of previous test data compression technique has been overcome by dictionary based data compression technique.

2.3 Signal transmission

In implanted biomedical devices, signal transmission has been done through modulation technique. In this technique, carrier variation can be done by Amplitude shift keying (ASK), Frequency shift Keying (FSK) and Phase shift Keying (PSK). Different types of modulation techniques with their performance parameters for bio medical devices are shown in Table 3. These modulation techniques allow high data transmission, low power consumption, intensive data security and greater accuracy (Ziemer et al., 2010; Zhu et al., 2010). ASK modulation also called as on-off keying (OOK) of the simplest technique to be implemented in implanted biomedical devices. However, low data rate, high noise sensitivity, coupling variations have some of the limitations of ASK. To overcome these limitations, two more modulation techniques known as FSK and PSK modulations along with their derivatives have been used that provide high date rate and more immunity to noise and interference but complexity and high power consumptions are some of their issues. To overcome these issues, Trigui et al. (2016) introduced a new technique called quad-width carrier modulation based on pulse width modulation for data transmission that allow high data rate and minimum power consumption with reduced complexity.

Table 3 Performance analysis of different modulation techniques.

Author name and reference no.	Type of modulation	Technology (µm)	Carrier (MHz)	Data rate (Mbps)	Power Consumption (µW)	Supply voltage (V)
Zhu, 2010	FSK	0.35	6.459	0.45	20	2.5
Trigui, 2016	QCWM	0.13	27.12	10.85	35.5	1.2
Kao, 2009	ASK	0.18	2	1	396	1.8
Lee et al., 2011	PSK	0.35	0.250	9.536743e-7	31.5	5
Li et al., 2010	BPSK	0.18	10	2	485.4	3.3
Deng et al., 2006	QPSK	0.18	13.56	4	750	1.8
Lu and Sawan, 2008	OQPSK	0.18	13.56	8	-	1.8
Kiani and Ghovanloo, 2012	PDM	0.35	50	13.56	-	1.8
Kiani and Ghovanloo, 2013	PHM	0.35	66.6	20	-	1.8

3 Response of Different IIR Filters for Denoising ECG Signal

Denoising is the process to reconstruct the signal from noisy one. Its aim to remove noise and preserve useful information. ECG is an important tool to measure health and disease detection. Due to many noise sources, this signal has to be denoised and presented in a clear waveform.ECG signal have frequency range between 0.5 Hz –100 Hz. An electrocardiogram demonstrates the electrical activity in the heart, and may be analyzed in characteristic parts, named P, Q, R, S, and T waves. In ECG Low frequency components are P and T waves, QRS resides at higher frequency. When an ECG is recorded, it would be corrupted with various kinds of noise. These noises are classified as

1. Baseline wanders (low frequency range)
2. Power line interference (Medium frequency range)
3. Electromyography noise (high frequency range)
4. Burst noise
5. Electrode contact noise
6. Muscle contraction
7. Motion artifacts Noise in ECG signal

This paper mainly stresses on the removal of high frequency noise particles called as Electromyography noise by designing different low pass infinite impulse response filters using different approximation methods (Butterworth, Chebyshev I and Elliptic). Taking the input MIT-BIH Arrhythmia from Physio Bank ATM, signal type v5, length 10 sec, 100 Records. We have considered 1454 input samples. The input ECG signal is sampled at sampling frequency f_s= 360Hz.In this input MIT-BIH Arrhythmia are fed to different low pass IIR filters to remove high frequency noise. By implementing these filters, signal time response, frequency response, power spectral density (PSD) and signal to noise ratio (SNR) has been observed. We have considered the cut-off frequency (f_c =100 Hz) for low pass filters. Order of the filter N is 7.

Normalized Frequency (f_n) can be calculated as

$$\omega_n = f_n * f_s \quad (3)$$

$$2\pi(100) = f_n * 360$$

$$f_n = 0.55\pi \text{ rad/sample} \quad (4)$$

4 Results and Discussion

In this section, results of different low pass IIR filters have been analysed through signalTime response, Frequency response and by calculating the value of diffrerentperformance parameters such as SNR and PSD. Signal to noise ratio of a signal determines the signal strength relative to noise. As the signals have wide dynamic range SNR is expressed in Decibles

$$SNR_{db} = 10 \log_{10} (P_{signal}) - 10 \log_{10} (P_{noise}) \quad (5)$$

Power spectral density descibes the signal power distribution over the frequency and is expressed in dB/Hz. PSD is determined by equation

$$PSD = \frac{1}{fs\, N} \left| \sum_{n=1}^{N} x_n e^{-j\left(\frac{2\pi f}{F}\right)n} \right|^2 \quad (6)$$

PSD has been estimated by Burg's method is an autoregressive power spectral density estimate pxx, of a discrete time signal x. Order of autoregressive (AR) model taken is 4. Here, FFT of a signal gave complex numbers for paricular frequency therfore PSD computed for a signal is negative.

Time domain output waveform of ECG signal filtered by IIR Low Pass filter using Butterworth approximation method is shown in Fig. 3(a) and corresponding Frequency spectrum of an ECG filtered signal is shown in Fig. 3(b).

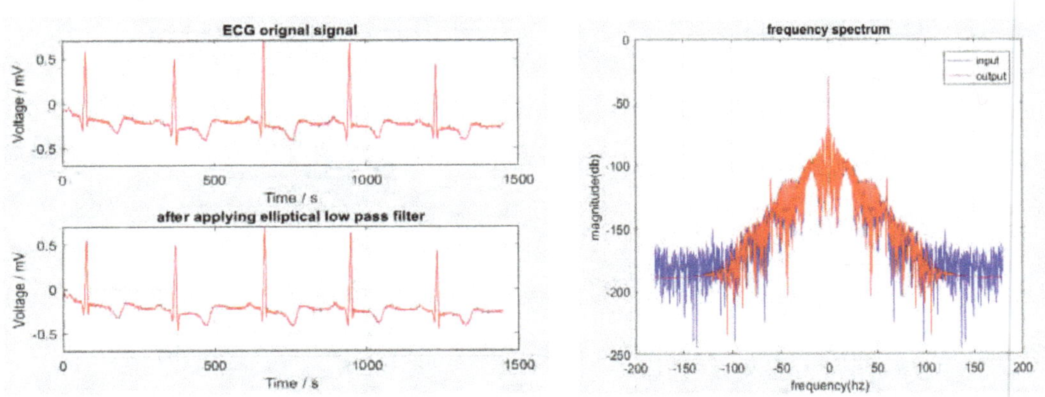

(a) (b)

Fig. 3 (*a*) Time domain analysis of ECG signal filtered by IIR filter using Butterworth approximation. (*b*) Frequency spectrum of an ECG signal filtered by IIR filter using Butterworth approximation.

Time domain output waveform of ECG signal filtered by IIR Low Pass filter using Chebyshev I approximation method is shown in Fig. 4(a) and corresponding Frequency spectrum of an ECG filtered signal is shown in Fig. 4(b).

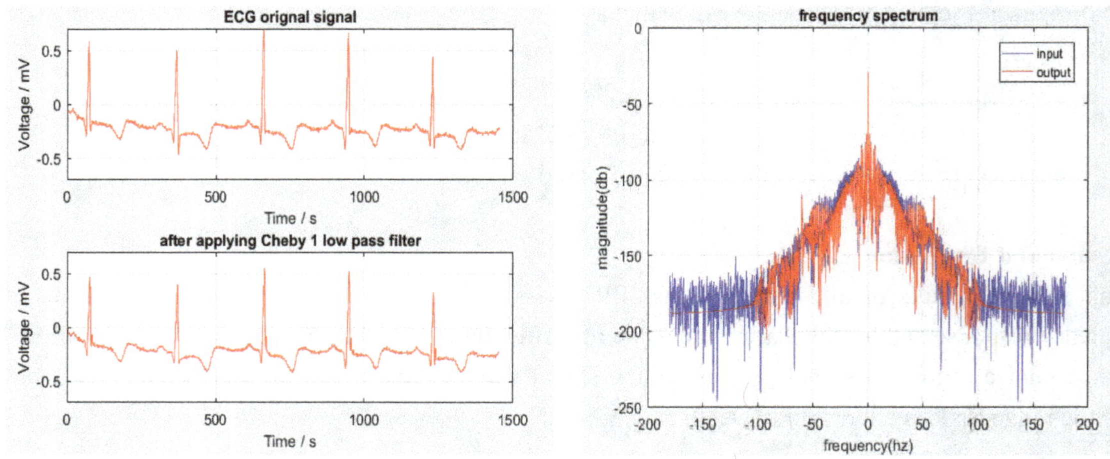

(a) (b)

Fig. 4 (*a*) Time domain analysis of ECG signal filtered by IIR filter using Chebyshev I approximation. (*b*) Frequency spectrum of an ECG signal filtered by IIR filter using Chebyshev I approximation.

Time domain output waveform of ECG signal filtered by IIR Low Pass filter using Elliptic approximation method is shown in Fig 5(a) and corresponding Frequency spectrum of an ECG filtered signal is shown in Fig 5 (b).

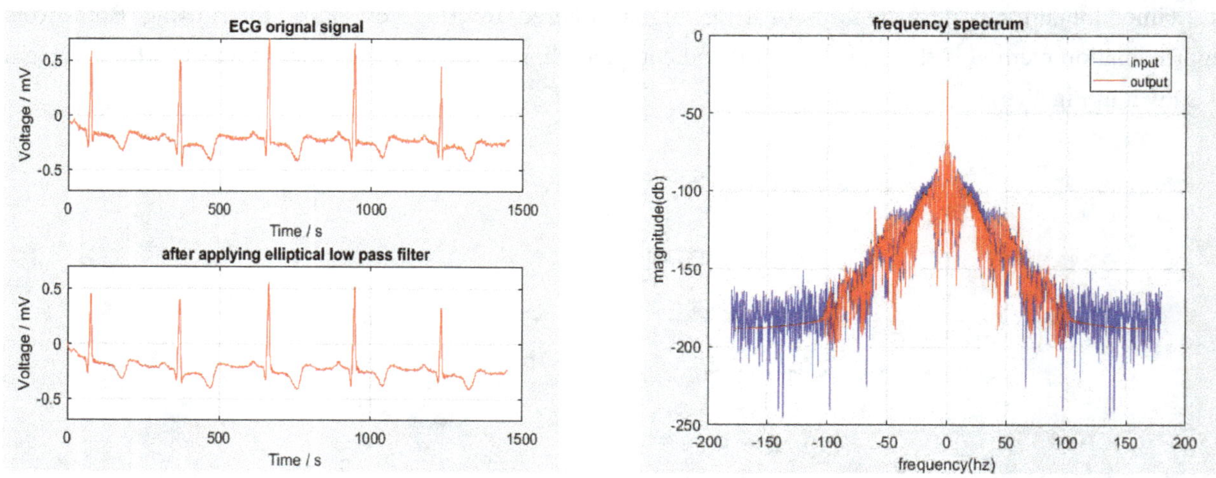

(a) (b)

Fig. 5 (*a*) Time domain analysis of ECG signal filtered by IIR filter using Elliptic approximation. (*b*) Frequency spectrum of an ECG signal filtered by IIR filter using Elliptic approximation.

4.1 Comparison of different low pass IIR filters

Comparison between different low pass IIR filters has been done on the basis of the performance metric of SNR and PSD. Filter with greater SNR and reduced PSD is well suited for removing the noise from ECG signal.

SNR and PSD of input ECG signal before filtering and after filtering by digital low pass IIR filter with specified approximation method has been calculated and shown in Table 4 and Table 5 respectively.

Table 4 SNR comparison of different low pass IIR filters.

IIR Filter with Approximation Method	SNR before filtering(dB)	SNR after filtering(dB)
Butter worth Filter	-5.5313	11.8139
Chebyshev I Filter	-5.5313	9.0951
Elliptic Filter	-5.5313	12.2930

Table 5 PSD comparison of different low pass IIR filters.

IIR Filter with Approximation Method	PSD before filtering (dB/Hz)	PSD after filtering (dB/Hz)
Butter worth Filter	-54.8834	-68.5366
Chebyshev I Filter	-54.8834	-70.3884
Elliptic Filter	-54.8834	-70.9543

4.2 Comparison of proposed work with previous reference paper

Comparison of proposed work with previous reference papers has been done. For proper comparative analysis of proposed work, we have chosen the orders of different filter same as mentioned in previous reference papers. Comparison of proposed work with previous reference paper is shown in Table 6.

Table 6 Comparison of proposed work with previous Reference papers.

Type of Filter	Order of filter	Author name	SNR after filtering [Existing work]	SNR after filtering [Proposed work]	PSD after filtering [Existing work]	PSD after filtering [Proposed work]
Butterworth Filter	5	(Singh et al., 2013)	-	-	-56.0660	-68.4152
Butterworth Filter	1	(Sarvankumar et al., 2015)	5.70	24.38	-	-
Chebyshev 1 Filter	2	(Sarvankumar et al., 2015)	5.37	7.61	-	-
Elliptic Filter	3	(Sarvankumar et al., 2015)	5.37	13.90	-	-
Butterworth Filter	10	(Bhogeshwar et al., 2014)	1.12	9.39	-	-
Chebyshev 1 Filter	10	(Bhogeshwar et al., 2014)	1.10	3.58	-	-
Elliptic Filter	10	(Bhogeshwar et al., 2014)	1.73	2.10	-	-
Butterworth Filter	4	(Mishra and Mehra, 2014)	-	-	-81	-68.2
Chebyshev 1 Filter	4	(Mishra and Mehra, 2014)	-	-	-82	-70.10

5 Conclusion

This paper presents the review of various existing techniques and their characteristics to develop a reliable Biomedical System. It has been observed that reliability of these devices depend on energy harvesting, signal filtering, data compression and modulation techniques. However, researchers have initial challenge to denoised the biomedical signals. In context to this problem, different low pass IIR filter has been designed to remove the high frequency noise artifacts from an ECG signal. On comparing their performance parameters of different low pass IIR filters, Elliptic filter has found best suited to remove this type of noise.

References

Amandeep, Jain S, Bhusri S. 2017. CAD for two class classification of lung cancer using statistical features. International Journal of Pharmaceutical Sciences Review and Research, 45(2): 55-60

Amar AB, Kouki AB, Cao H. 2015. Power approaches for implantable medical device. Sensors, 15(11): 28889-28914

Amirtharajah R, Meninger S, Mur-Miranda J, Chandrakasan A, Lang J. 2000. A micropower programmable DSP powered using a MEMS-based vibration-to-electric energy converter. IEEE International Solid-State Circuits Conference Dig. Tech. Pap, 362-363, 469

Ajay A, Laurde RM. 2015. VLSI Implementation of an improved multiplier for FFT Computation in Biomedical Applications. IEEE Computer Society Annual Symposium on VLSI. Montpellier, France

Bhogeshwar SS, Soni MK, Bansal D. 2014. Design of Simulink Model to denoise ECG signal using various IIR & FIR filters, 2014 International Conference on Reliability, Optimization and Information Technology - ICROIT 2014, India

Bhusri S, Jain S, Virmani J. 2016a. Classification of breast lesions using the difference of statistical features. Research Journal of Pharmaceutical, Biological and Chemical Sciences, 7(4): 1365-1372

Bhusri S, Jain S, Virmani J. 2016b. Breast Lesions Classification using the Amalagation of morphological and texture features. International Journal of Pharma and BioSciences, 7(2)B: 617-624

Cadei A, Dionisi A, Sardini E, Serpelloni, M. 2014. Kinetic and thermal energy harvesters for implantable medical devices and biomedical autonomous sensors. Measurement Science and Technology, 25(1): 012003-0120016

Chandrakasan AP, Verma N, Daly DC. 2008. Ultralow-power electronics for biomedical applications. Annual Review of Biomedical Engineering, 10: 247-274

Chen H, Liu M, Chen J, Wang Z. 2009. Power harvesting using PZT ceramics embedded in orthopedic implants. IEEE Transactions on Ultrasonics, Ferroelectrics, and Frequency, 56: 2010-2014

Cheng DK. 1989. Field and Wave Electromagnetics. Addison-Wesley, New York, USA

Deng S, Hu Y, Sawan M. 2006. A high data rate QPSK Demodulator for inductively powered electronics implants. Proceedings of the IEEE International Symposium on Circuits and Systems.21-24, Kos, Greece

Dogra J, Prashar N, Jain S, Sood M. 2018. Improved methods for analyzing MRI brain images. Network Biology, 8(1): 1-11

Glynne-Jones P, Beeby S, White N. 2001. Towards a piezoelectric vibration-powered micro generator, IEE Proceedings - Science, Measurement and Technology, 148: 68-72

Gyselinckx B, VanHoof C, Ryckaert J, Yazicioglu R, Fiorini P, Leonov V. 2005. Human++: autonomous wireless sensors for body area networks. Proceedings of the IEEE 2014 Custom Integrated Circuits Conference, 13-19

Huang Q, Oberle M. 1998. A 0.5-mW passive telemetry IC for biomedical applications. IEEE Journal of Solid-State Circuits, 33(7): 937-946

Jain S, Naik PK, Bhooshan SV, 2011. Nonlinear Modeling of cell survival/ death using artificial neural network. In: Proceedings of International Conference on Computational Intelligence and Communication Networks. 565-568, Gwalior, India

Jain S, Chauhan DS. 2015. Implementation of fuzzy system using different voltages of OTA for JNK pathway leading to cell survival/ death. Network Biology, 5(2): 62-70

Jain S. 2016. Regression analysis on different mitogenic pathways. Network Biology, 6(2): 40-46

Jain S. 2017. Regression modeling of different proteins using linear and multiple analysis. Network Biology. 7(4): 80-93

Kao CH, Tang KT. 2009.Wireless Power and data transmission with ASK demodulator and power regulator for a biomedical implantable SOC. Proceedings of the IEEE Life Science Systems and Applications Workshop. 179-182, Lissa Bethesda, MD, USA

Karthaus U, Fischer M. 2003. Fully integrated passive UHF RFID transponder IC with 16.7-μWminimumRF input power. IEEE Journal of Solid-State Circuits, 38(10): 1602-1608

Kim HS, Kim JH, Kim J. 2011. A review of piezoelectric energy harvesting based on vibration. International Journal of Precision Engineering and Manufacturing, 12: 1129-1141

Kurs A, Karalis A, Moffatt R, Joannopoulos JD, Fisher P, Soljacic M. 2007. Wireless power transfer via strongly coupled magnetic resonances. Science, 317(5834): 83-86

Kymissis J, Kendall C, Paradiso J, Gershenfeld N. 1998. Parasitic Power Parvesting in Phoes,Wearable Computers. Proceedings of the IEEE Second International Symposium on Digest of Papers. 132-129, Pittsburg, PA, USA

Cromwell L. 2004. Biomedical Instrumentations and Measurements (2nd Edition). PHI Publications, India

Lee SY, Cheng CJ, Liang MC. 2011.A low-power bidirectional telemetry device with a near-field charging feature for a cardiac micro stimulator. IEEE Transactions on Biomedical Circuits and Systems, 5: 357-367

Li CP, Wu ZH, Li B. 2010. A new integrated low-power BPSK demodulator for wireless implantable neural recording system. Proceedings of the IEEE International Conference on Electron Devices and Solid-State Circuits. 1-4, Hong Kong

Lu Z, Sawan M. 2008. An 8 mbps data rate transmission by inductive link dedicated to implantable devices. Proceedings of the IEEE International Conference on Biomedical Circuits and Systems. 3057-3060, Seattle, WA, USA

Kiani M, Ghovanloo M. 2013. A 20-mb/s pulse harmonic modulation transceiver for wideband near-field data transmission. IEEE Transactions on Circuits and Systems II: Express Briefs, 60(7): 382–386.

Kiani M, Ghovanloo M. 2012. Pulse delay modulation (pdm) a new wideband data transmission method to implantable medical devices in presence of a power link. IEEE Biomedical Circuits and Systems Conference, 256-259

Mishra B, Mehra R. 2014. Design and simulation of low pass IIR filter for ECG interference reduction. International Journal of Engineering Research and Technology, 3(3): 1322-1326

Mishra R, Jain S, Durga Prasad C. 2015. A review on piezoelectric material as a source of generating electricity and its possibility to fabricate devices for daily uses of army personnel. International Journal of Systems, Control and Communications, 6(3): 212-221

Mishra R, Jain S, Durgaprasad C, Sahu S. 2015. Vibration energy harvesting using drum harvester. International Journal of Applied Engineering Research, 10(14): 34995-35001

Mishra R, Jain S, Durgaprasad C. 2016. Analysis of piezoelectric buzzers as vibration energy harvesters. 10th INDIACom: 3rd International Conference on Computing for Sustainable Global Development. 2476-2480, BVICAM, New Delhi, India

Mishra R, Jain S, Thakur B, Verma YP, Durgaprasad C. 2016. Performance analysis of piezoelectric drum transducers as shoe-based energy harvesters. International Journal of Electronics Letters, 1-15

Murakawa K, Kobayashi M, Nakamura O, Kawata S. 1999. Aireless near-infrared energy system for medical implants. IEEE Engineering in Medicine and Biology Magazine, 18: 70-72

Singh N, Ayub S, Saini JP. 2013. Design of Digital IIR Filter for Noise Reduction in ECG Signal. CICN '13 Proceedings of the 2013 5th International Conference on Computational Intelligence and Communication Networks, 171-176

Paradiso JA, Starner T. 2005. Energy scavenging for mobile and wireless electronics. IEEE Pervasive Computing, 4(1): 18-27

Pardhu T, Manusha S, Sirisha K. 2014. Low powerVLSI compressor for biomedical applications. CS&IT-CSCP, 111-120

Penner A. 2013. Acoustically Powered Implantable Stimulating Device. U.S. Patent US8577460 B2,5 November

Rana S, Jain S, Virmani J. 2016a. Classification of focal kidney lesions using wavelet-based texture descriptors. International Journal of Pharma and Bio Sciences, 7(3)B: 646-652

Rana S, Jain S, Virmani J. 2016b. SVM-Based Characterization of Focal Kidney Lesions from B-Mode Ultrasound Images. Research Journal of Pharmaceutical, Biological and Chemical Sciences, 7(4): 837

Ranjith J, Muniraj NJR. 2012. VLSI implementation of efficient code compressor for bio-signal processors. Proceedings on Engineering, 30: 953-996

Rhees DJ. 2009. From Frankenstein to the Pacemaker: A profile of the Bakken Museum. IEEE Engineering in Medicine and Biology Magazine, 28: 78

Sharma S, Jain S, Bhusri S, 2017. Two class classification of breast lesions using statistical and transform domain features. Journal of Global Pharma Technology, 9(7): 18-24

Sood M. 2017, Performance analysis of classifiers for seizure diagnosis for single channel EEG data. Biomedical and Pharmacology Journal, 10(2): 795-803

Sood M, Bhooshan SV. 2014. Design and development of prediction model to detect seizure activity utilizing higher order statistical features of EEG signals. Research Journal of Pharmaceutical, Biological and Chemical Sciences, 5(3): 1129-1145

Trigui A, Ali M, Ammari AC, Savaria Y, Sawan M. 2016. Quad-Level Carrier Width Modulation demodulator for microimplants. 14th IEEE International New Circuits and Systems Conference (NEWCAS) Conference. Vancouver, Canada

Tsai NC, Sue CY. 2007. Review of mems-based drug delivery and dosing systems. Sensors and Actuators A: Physical, 134: 555-564

Weiland JD, Liu W, Humayun MS. 2005. Retinal prosthesis. Annual Review of Biomededical Engineering, 7: 361-401

Wise KD, Anderson DJ,Hetke JF, Kipke DR, Najafi K. 2004. Wireless implantable Microsystems: High-density electronic interfaces to the nervous system, Proceedings of IEEE, 92(1): 76-97

Wong L, Hossain S, Ta A, Edvinsson J, Rivas D, Naas H. 2004. A very low-power CMOS mixed-signal IC for implantable pacemaker applications. IEEE Journal of Solid-State Circuits, 39(12): 2446-2456

Zeng H, Zhao Y. 2011. Sensing movement: Microsensors for body motion measurement. Sensors, 11: 638-660

Zhang WJ. 2018. Fundamentals of Network Biology. World Scientific, Singapore

15

Robustness of plant-plant networks with different levels of habitat modification and interaction intimacy

Walter Santos de Araújo
Department of General Biology, Center of Biological Sciences and Health, Universidade Estadual de Minas Gerais, 39401-089, Montes Claros, Brazil
E-mail: walterbioaraujo@yahoo.com.br

Abstract
Anthropogenic modification of natural environments is the main causes of species extinction in the globe, which directly leads to loss of interspecific links and modifies the structure of ecological networks. The objective of present study is to evaluate the effect of human-induced habitat modification on the connectivity and robustness of ecological interaction networks composed by plant-epiphyte and plant-parasite interactions. In total were analyzed eight distinct binary networks of plant-plant interactions in Brazil, being three epiphyte networks and five parasite networks occurring both in conserved and anthropized habitats. The results show that the human-induced habitat modification influences the connectance of plant-plant networks, since networks of anthropized habitats had greater connectance than the networks of conserved habitats. In addition, the results showed higher values of robustness in the plant-parasite networks when compared to plant-epiphyte networks, and these differences were mainly observed in anthropized habitats. This study presented a new approach for studies of plant-plant ecological interactions, because is the first to compare the effect of human-induced habitat modification on the plant-plant network robustness.

Keywords Brazil; coextinctions; ecological networks; habitat loss; Neotropical interactions.

1 Introduction

Anthropogenic habitat modification is the main causes of species extinction in the globe (Smart et al., 2006), which directly affects the loss of interspecific links, disrupting ecological networks and compromising the functioning of ecosystems (Tylianakis et al., 2010). Plants are the organisms more frequently studied concerning to ecological interaction networks in terrestrial ecosystems. There are a very large bibliography about the effects of habitat disturbance on networks involving plant-animal interactions (review in Hagen et al., 2012), but there are still relatively few studies focused on plant-plant interactions. In this sense, an issue unexplored is how plant-plant networks are affected by human-induced habitat modification.

There is a relative consensus that human-induced habitat modification has negative consequences on native plant species richness (Ellis et al., 2012). These negative effects are directly linked to deforestation of natural vegetation cover and to colonization with exotic plant species (Didham et al., 2007). Thereby, habitat-specialized native plant species tend to be more vulnerable to habitat degradation and exotic competition than generalist native plants. In this sense, while natural habitats are composed exclusively by native plant species (including a great proportion of specialized species), anthropized habitats tend to be dominated by generalist native plant species and exotic plant species. Because the generalist and exotic plants tend to have many connections in the network, it is expected that networks of anthropogenic habitats be more connected and robust than networks of natural habitats.

Ecological interactions between plant species can vary from antagonistic (e.g., interactions between host-plants and parasite plants) to comensalistic interactions (e.g., interactions between host-plants and epiphyte plants) (Montesinos, 2015). The level of intimacy of interaction and degree of biological association between partners can influence the structure and dynamic of interactions within a network (Pires and Guimarães Jr, 2013). For example, the parasite life-form lead the plants to have some level of specialization in the use of host-plants, because they are intimate linked to their host-plants. Parasite plants can be hemiparasites (partially parasitic), presenting leaves for photosynthesis and also absorbing nutrients from the host-plant through its haustoria, or holoparasites (totally parasitic) use exclusively the host resources (Thompson, 1994; Norton and Lange, 1999). All these two types of parasites depend on their hosts and therefore, extinction of host species can lead to loss from parasitic species (i.e., secondary extinctions). On the other hand, epiphyte plant species live on host-plants as inquilines in a commensalistic way, without parasitism (Silva et al., 2010). Therefore, plant-epiphyte communities tend to have more occasional links between the plants, resulting in networks with high connectivity and nestedness (Burns, 2007). Since specialized plant species are more prone to local extinction due to the habitat modification (Didham et al., 2007), plant-parasite networks are expected to have lower levels of connectivity and robustness to secondary extinctions than plant-epiphyte networks.

The objective of present study is to evaluate the effect of habitat modification on the structure and robustness of plant-plant networks with distinct levels of ecological association (plant-epiphyte versus plant-parasite interactions). To describe the plant-plant networks were used the connectance and the network robustness to secondary extinctions. Given the biological homogenization and dominance of generalist plants in human-modified habitats, it is tested the hypothesis that plant-plant interaction networks in anthropic habitats have higher connectance and robustness than networks of conserved habitats. Moreover, since plant-parasite links lead to more intimate interactions between host- and hosted-plants than commensalistic links, it is expected to find a lower level of connectance and robustness in plant-parasite networks.

2 Material and Methods

2.1 Network compilation

Plant-plant interaction networks were compiled from data available in the Google Scholar data base (<https://scholar.google.com.br/>). The search was focused on studies investigating the association between epiphytes (e.g., orchids, bromeliads, etc.) or parasites (e.g., mistletoes) and their host-plants at local level. In order to allow network structure analysis, were used only studies that listed the plant species involving in the interactions and were composed by at least three host-plant species and three hosted plant species (epiphytes or parasites). In total were compiled eight distinct binary networks of plant-plant interactions, being three epiphyte networks and five parasite networks (Table 1; Fig. 1).Networks were characterized concerning the habitat conservation status in "conserved habitats" (i.e., habitats characterized by natural vegetation with little or no anthropic effect, such as primary forest) and "anthropized habitats"(i.e., habitat with secondary or planted vegetation under strong anthropic influence, such as urban gardens) according the original description of authors.

Table 1 Networks of parasite and epiphyte interactions compiled from literature data for Brazil. Legend: S_{host} = richness of host-plant species; S_{hosted} = richness of hosted plant species (epiphytes or parasites); S_{total} = total species richness; C = connectance (pure value).

Network	Habitat conservation status	Network type	S_{host}	S_{hosted}	S_{total}	C	Latitude
Azevedo (2010)	Anthropized	Epiphyte	3	14	17	0.548	-23.067
Fontoura et al. (2009)	Conserved	Epiphyte	35	20	55	0.141	-22.783
Leal et al. (2006)	Anthropized	Parasite	6	3	9	0.722	-25.430
Moura et al. (2009)	Anthropized	Parasite	26	4	30	0.413	-8.054
Pereira (2014) – net 1	Conserved	Parasite	11	13	24	0.154	-2.751
Pereira (2014) – net 2	Conserved	Parasite	15	10	25	0.200	-2.751
Rotta (2001)	Anthropized	Parasite	27	3	30	0.457	-25.418
Sousa & Colpo (2017)	Conserved	Epiphyte	3	7	10	0.571	-23.819

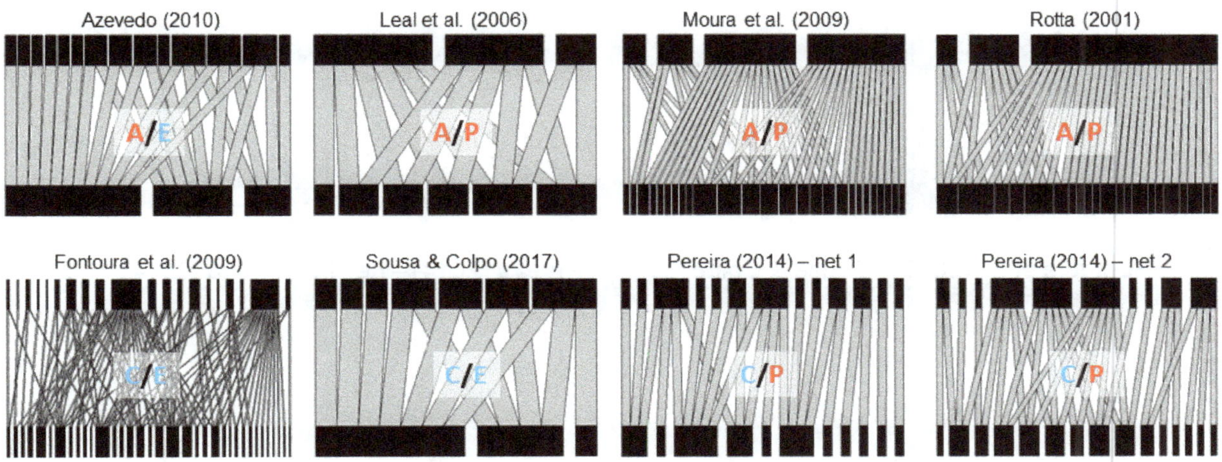

Fig. 1 Structure of the plant-plant networks analyzed in this study. For each network, lower bars represent host plant species and upper bars represent hosted plant species (parasites or epiphytes). Bar thickness is proportional to the number of interactions of each species (drawn at different scales). Networks were ordered by habitat conservation status (A = Anthropized habitats; C = conserved habitats) and network type (E = epiphyte network; P = parasite network).

2.2 Evaluating the network structure

In order to describe the structure of plant-plant networks were used the network connectance, and two measures of robustness to secondary extinctions (random and realistic). Connectance was evaluated because is the most commonly used descriptor of topological structure of bipartite binary networks (review in Dormann et al., 2009), and can influence the fragility of ecological networks to coextinctions (Dunne et al., 2002). The connectance is a measure of proportion of potential interactions that are realized in the network (Zhang, 2011, 2012).The connectance is negatively influenced by network size (i.e., total number of species) (Dunne et al., 2002, Araújo et al., 2015), and to control this effect, the connectance was calculated as the residuals from a linear regression between the number of realized interactions and the number of potential interactions (both log-transformed) (hereafter “residual connectance”). Residual connectance therefore allows the comparison of different-sized networks in terms of higher or lower connectance than expected based on their size (positive and negative residuals, respectively) (Araújo et al., 2015). In the present study, connectance (pure value) was used only to describe the networks, while residual connectance was used in the analyses.

Robustness is a measure of the resistance degree of the network to coextinctions (Dormann et al., 2009; Zhang, 2016, 2018; Araújo et al., 2017). To characterize the network robustness was used the exponent of the curve generated by the proportion of remaining hosted plant species in function of the proportion of primary extinctions of host-plant species (Fig. 2). In this sense, was calculates the robustness considering two different scenarios. In the scenario 1 the primary extinctions are random (i.e., all host-plant species have the same probability of extinction), which was called "random robustness". In the scenario 2, the chance of extinction is inversely proportional to the number of host-plant links (i.e., the host species with the less links are extinct first). Because this is a more realistic scenario this measure was called "realistic robustness". We performed all network analyses using Bipartite package (Dormann et al., 2008) from R program (R Development Core Team, 2015).

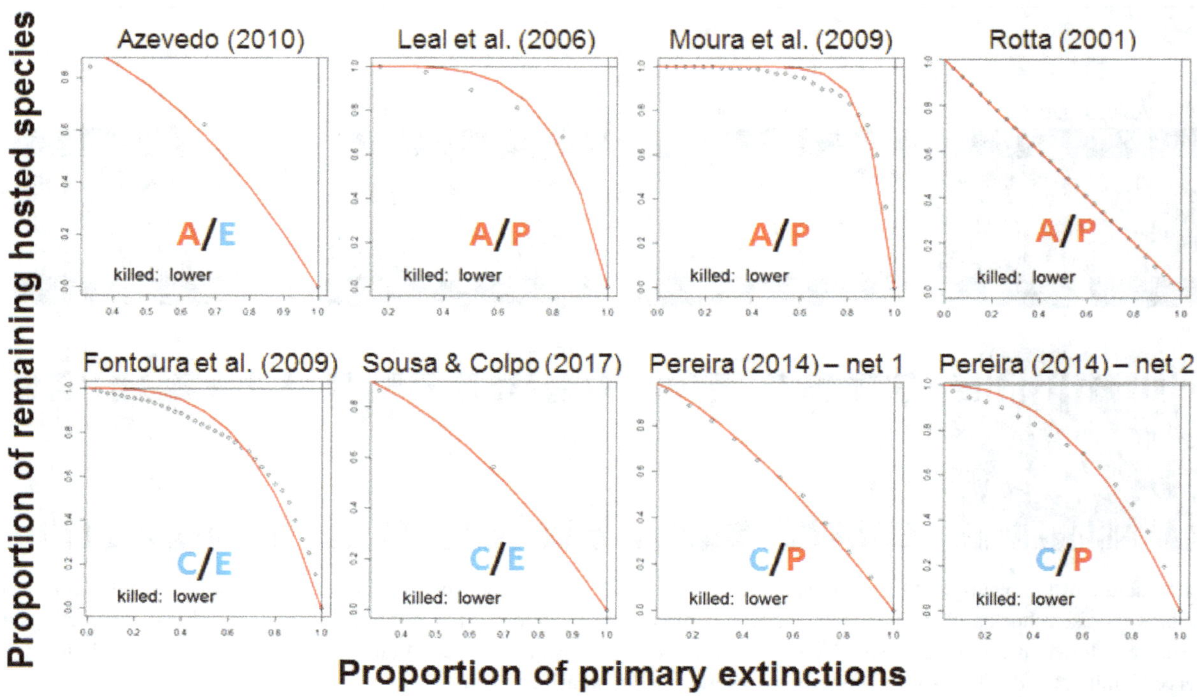

Fig. 2 Robustness to secondary extinctions of hosted species (epiphytes or parasites) resulting from random primary species loss (host-plants) in different plant-plant networks in Brazil. Networks were ordered by habitat conservation status (A = Anthropized habitats; C = conserved habitats) and network type (E = epiphyte network; P = parasite network). A less inclined curve pattern, indicating greater robustness, is found for parasite networks in anthropized habitats, such as the networks of Leal et al. (2006) and Moura et al. (2009).

2.3 Data analyses

Because the plant-plant networks compiled ranged in the location and sampling effort, the sampled taxonomic range was used as additional explanatory variables, in order to control for the confounding effects. Taxonomic range is a measure of level of taxonomic inclusion in the study because different studies ranged in the number of sampled plant families. For example, some authors often choose to limit the sampling procedure within a taxon, including only plants from a given family (e.g., Rotta, 2001, Sousa and Colpo, 2017), while some other studies sampled several families of plants (e.g., Fontoura et al., 2009). The sampled taxonomic range was calculated as the number of host-plant families (log-transformed) summed to number of hosted plant families (log-transformed) sampled in each study. Although latitude is an important geographic variable related to possible general climatic differences among areas, it was not included in the final analyses because it did not influence any of the variables of interest (all p values > 0.05). To test the effects of habitat type (conserved versus anthropized habitats), network type (parasite versus epiphyte networks) and sampled taxonomic range on the connectance, random robustness and realistic robustness of plant-plant networks were used generalized linear models (glm's).

3 Results

Connectance of plant-plant networks was moderately high, with a mean value of C = 0.40 ± 0.21 (mean ± SD), reaching a maximum value of C = 0.72 in the Leal et al. (2006) network (Table 1). Connectance values were higher for networks of anthropized habitats (C = 0.53 ± 0.13) than for networks of preserved habitats (C = 0.26 ± 0.20). Comparing the connectance values between the different network types, was found C = 0.42 ± 0.24 for epiphyte networks and C = 0.38 ± 0.22 for parasite networks.

The statistical analyses showed that the residual connectance differs significantly between plant-plant networks of conserved and anthropized habitats ($F_{(1,3)}$ = 24.48, p = 0.01, Table 2). In general, networks of anthropized habitats were more connected than expected by network size, while networks of conserved habitats were less connected than expected by size (Fig. 3).

Table 2 Generalized linear models of the effects of habitat type (anthropized versus conserved), network type (epiphytes versus parasites) and sampled taxonomic range on the residual connectance, random robustness and realistic robustness of plant-plant interaction networks in Brazil.

Network variable	Explanatory variable	*F*	*p*
Residual connectance	Intercept	0.239	0.659
	Habitat type	24.48	0.016
	Network type	3.792	0.147
	Network type*Habitat type	4.929	0.113
	Sampled taxonomic range	0.050	0.837
Random robustness	Intercept	261.37	0.001
	Habitat type	6.914	0.078
	Network type	17.92	0.024
	Network type*Habitat type	29.70	0.012
	Sampled taxonomic range	3.684	0.151
Realistic robustness	Intercept	781.99	<0.001
	Habitat type	2.376	0.221
	Network type	47.35	0.006
	Network type*Habitat type	87.37	0.003
	Sampled taxonomic range	23.58	0.017

The network robustness based on random robustness (Table 2) was significantly influenced by network type ($F_{(1,3)}$ = 17.92, p = 0.02), and interaction between network type and habitat type ($F_{(1,3)}$ = 29.70, p = 0.01). Plant-plant networks composed by parasitic interactions were more robustness to random primary extinctions than epiphyte networks, which was mainly observed in anthropized habitats (Fig. 4). A similar pattern was observed to realistic robustness (Table 2), where also were found significant effects of network type ($F_{(1,3)}$ = 47.35, p< 0.01) and interaction between network type and habitat type ($F_{(1,3)}$ = 87.37, p< 0.01)on the network robustness. In this sense, parasite networks were more robust to primary extinction of more rare species than the epiphyte networks, mainly in the anthropized habitats (Fig. 5). Additionally, the realistic robustness also was influenced by sampled taxonomic range ($F_{(1,3)} = 23.58$, $p = 0.01$), being that networks with larger sample range were more robust than networks that sampled few taxa.

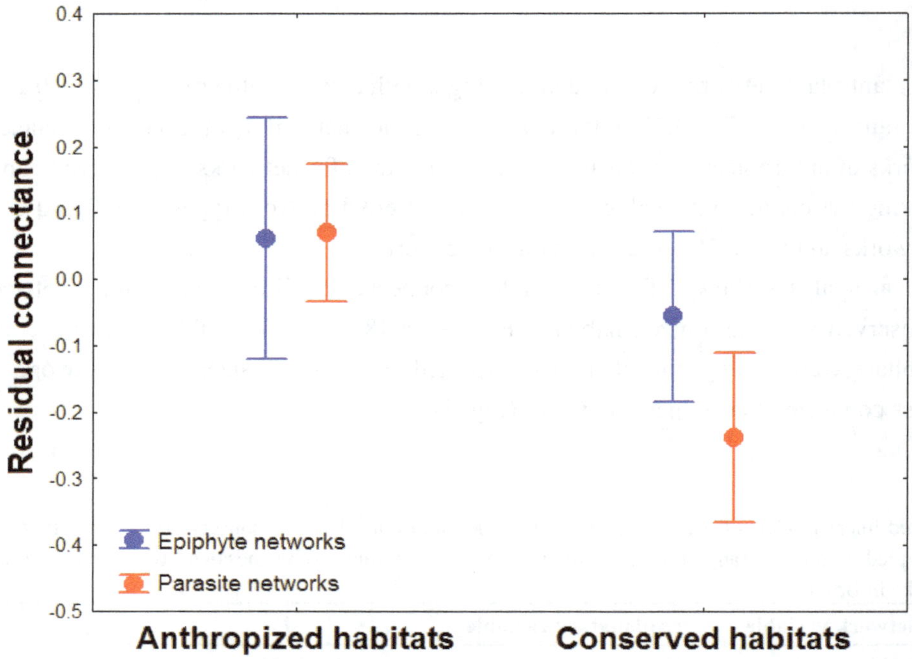

Fig. 3 Comparison between residual connectance between the different habitat type (anthropized versus conserved) and network type (epiphytes versus parasites).

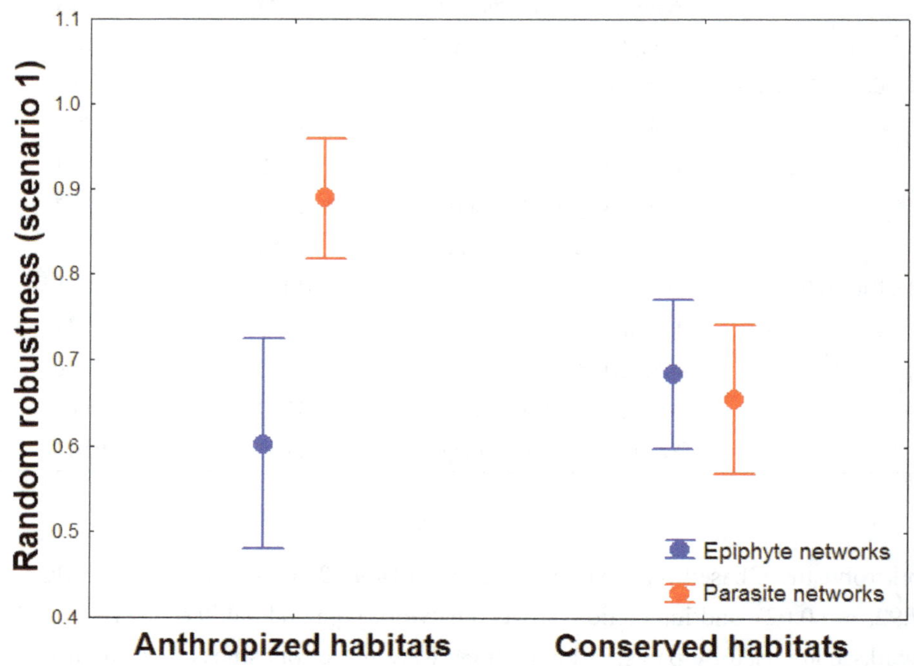

Fig. 4 Comparison between random robustness (scenario 1) between the different habitat type (anthropized versus conserved) and network type (epiphytes versus parasites).

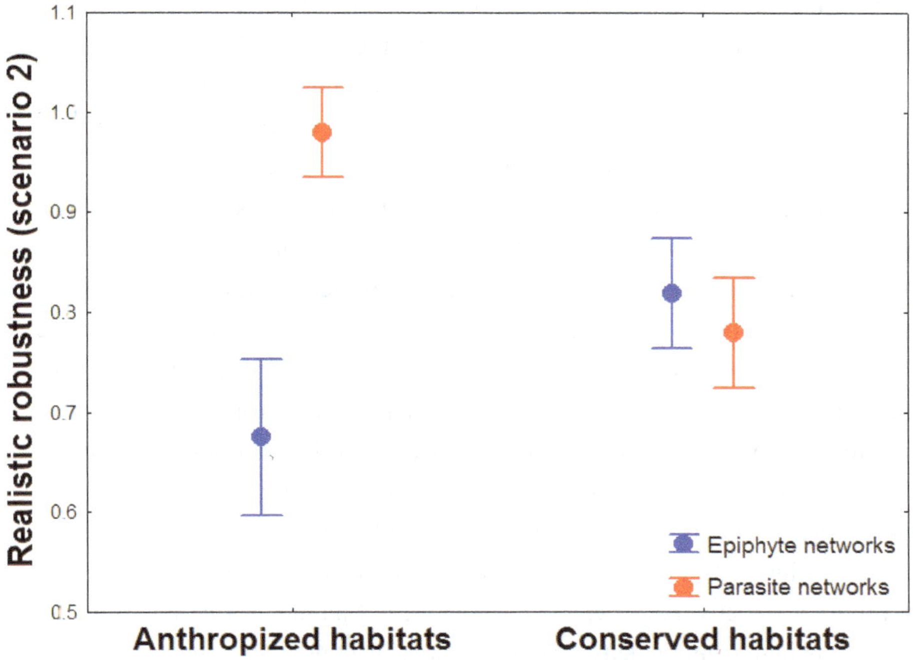

Fig. 5 Comparison between realistic robustness (scenario 2) between the different habitat type (anthropized versus conserved) and network type (epiphytes versus parasites).

4 Discussion

The results show that the human-induced habitat modification influences the connectance of plant-plant networks. As expected, the networks of anthropized habitats had greater residual connectance than the networks of conserved habitats. This result corroborates previous studies that point out that habitat modification can influence the connectivity of ecological networks (Hagen et al., 2012; Araújo et al., 2015). Contrary to residual connectance, the robustness of plant-plant networks was significantly influenced by network type (epiphytes versus parasites) and by interaction between network type and habitat type. To contrary of expectative, the results show that parasite interaction networks were more robust (considering both scenarios of primary extinctions) than epiphyte networks, mainly in anthropized habitats.

In addition to the anthropogenic habitats have less plant species richness (Ellis et al., 2012) they present microclimatic characteristics and species composition different of natural environments. The opening up and the homogenization of natural vegetation, for instance, have been shown to cause a series of changes in environmental conditions, such as increasing microclimatic oscillations and reducing the natural buffer against pollution (Smart et al., 2006). Besides that, exotic plant species are usually more common in human-disturbed habitats due both to direct introduction and to habitat invasion (Didham et al., 2007; Parker et al., 2010). In this context, the anthropic habitats are characterized by species that are few selective in their ecological links, which results in a great connectivity inside the networks in these environments. On the other hand, no effect of the type of interaction on the residual connectance was found, contrary to what was expected. This result may be due to the strength exerted by the habitat type, especially in anthropic habitats, limiting the species more specialized, generating a high degree of connectance in the network (independently of the type of interaction).

The results showed higher values of robustness in the plant-parasite networks when compared to plant-epiphyte networks, and these differences were mainly observed in anthropized habitats. This result contrary the initial expectation that plant-parasite networks would have lower robustness because are characterized by more intimate links, as compared to plant-epiphytes networks that have more occasional links.

The main explanation for this unexpected result can be related to degree of host specificity of parasite plants (Norton and Carpenter, 1998). In a thematic review, Arruda et al. (2012) point out that much of the Neotropical parasitic plants are host-generalists, and that extreme host specialization is rare. Despite this, there is evidence that the specificity of parasitic plants may be dependent on the scale of observation, since they may have a local preference for a particular host (Arruda et al., 2006; Grenfell and Burns, 2009). In this sense, the level generalization or local specialization of the parasite plants can be dependent on multiple factors related to the dispersant animals of their seeds, as well as the characteristics of the host-plants (Arruda et al., 2012). At network level, a large variation in the specialization range of species, from many specialists to generalists, can generate a nested pattern of interactions, which is usually very robust (Piazzon et al., 2011).

This study presented a new approach for studies of plant ecological interactions, because is the first to compare the effect of habitat modification on the plant-plant network robustness. Combining data available from the literature to build networks with different types of plant interactions, the results shown that habitat modification influences the network connectance, which was higher in networks of anthropized habitats than in conserved habitat networks. In addition, the results indicate that the level of ecological association (interaction intimacy) influences in the network structure and dynamic, since networks composed of parasite links were more robust than networks of epiphytic links. It is findings can be interpreted as general patterns that emerged, despite the high heterogeneity among the plant-plant networks analyzed.

Acknowledgements
I thank to Universidade Estadual de Montes Claros by logistical support. This research did not receive any specific grant from funding agencies.

References

Araújo WS, Grandez-Rios JM, Bergamini LL, Kollár J. 2017. Exotic species and the structure of a plant-galling network. Network Biology, 7: 21-32

Araújo WS, Vieira MC, Lewinsohn TM, Almeida-Neto M, 2015. Contrasting effects of land use intensity and exotic host plants on the specialization of interactions in plant-herbivore networks. PLoSOne, 10: e0115606

Arruda R, Fadini RF, Carvalho LN, Del-Claro K, Mourão FA, Jacobi CM, Teodoro GS, Berg E, Caires CS, Dettke GA. 2012. Ecology of neotropical mistletoes: an important canopy-dwelling component of Brazilian ecosystems. Acta Botanica Brasilica, 26: 264-274

Arruda R, Carvalho LN, Del-Claro K. 2006. Host specificity of a Brazilian mistletoe, *Struthanthus* aff. *polyanthus* (Loranthaceae), in cerrado tropical savanna. Flora, 201: 127-134

Azevedo DB. 2010. Epífitas Vasculares Ocorrentes Em Três Espécies De Forófitos na Área Urbana Da Ilha Da Marambaia - Mangaratiba/RJ. Monograph, Universidade Federal Rural do Rio de Janeiro, Brazil

Burns KC. 2007. Network properties of an epiphyte metacommunity. Journal of Ecology, 95: 1142–1151

Didham RK, Tylianakis JM, Gemmell NJ, Rand TA, Ewers M. 2007. Interactive effects of habitat modification and species invasion on native species decline. Trends in Ecology and Evolution, 22: 489-496

Dormann CF, Gruber B, Fründ J. 2008. Introducing the Bipartite Package: Analysing Ecological Networks. R News, 8: 8-11

Dormann CF, Fründ J, Blüthgen N, Gruber B. 2009. Indices, graphs and null models: analyzing bipartite ecological networks. The Open Ecology Journal, 2: 7-24

Dunne JA, Williams RJ, Martinez ND. 2002. Network structure and biodiversity loss in food webs: robustness increases with connectance. Ecology Letters, 5: 558-567

Ellis EC, Antill EC, Kreft H. 2012. All is not loss: plant biodiversity in the Anthropocene. PLoS One, 7: e30535

Fontoura T, Rocca MA, Schilling AC, Reinert F. 2009.Epífitas da floresta seca da Reserva Ecológica Estadual de Jacarepiá, sudeste do Brasil: relações com a comunidade arbórea. Rodriguésia, 171-185

Hagen M, Kissling WD, Rasmussen C, De Aguiar MA, Brown LE, Carstensen DW, Guimaraes Jr PR. 2012. Biodiversity, species interactions and ecological networks in a fragmented world. Advances in Ecological Research, 46: 89-210

Grenfell M, Burns KC. 2009. Sampling effects and host ranges in Australian mistletoes. Biotropica, 41: 656-658

Leal L, Bujokas WM, Biondi D. (2006). Análise da infestação de erva-de-passarinho na arborização de ruas de Curitiba, PR. Floresta, 36(3): 323-330

Montesinos D. 2015. Plant–plant interactions: from competition to facilitation. Web Ecology, 15: 1-2

Moura RM, Pereira RDCA, Cano MOO. 2009. Enxerto–De–Passarinho: Ameaça A Árvores Frutíferas E Ornamentais No Recife e Municípios Vizinhos. Anais da Academia Pernambucana de Ciência Agronômica, 5: 320-328

Norton DA, Lange PJ. 1999. Host specifcity in parasitic mistletoes (Loranthaceae) in New Zealand. Functional Ecology, 13: 552-559

Norton DA, Carpenter MA. 1998. Mistletoes as parasites: host specifcity and speciation. Trends in Ecology and Evolution, 13: 101-105

Parker JD, Richie LJ, Lind EM, Maloney KO. 2010. Land use history alters the relationship between native and exotic plants: the rich don't always get richer. Biological Invasions, 12(6): 1557-1571

Pereira JL. 2014. Prevalência De Infecção E Padrões De Uso De Hospedeiros Por Ervas-De-Passarinho (Loranthaceae e Viscaceae) Em Espécies De Árvores Tropicais De Interesse Madeireiro. Dissertation, Universidade Federal do Oeste do Pará, Brazil

Piazzon M, Larrinaga AR, Santamaría L. 2011. Are nested networks more robust to disturbance? A test using epiphyte-tree, comensalistic networks. PLoS One, 6(5): e19637

Pires MM, Guimarães JR PR. 2013. Interaction intimacy organizes networks of antagonistic interactions in different ways. Journal of the Royal Society Interface, 10: 20120649

R Development Core Team. 2015. *R*: A language and environment for statistical computing. R Foundation for Statistical Computing, Vienna, Austria

RottaE. 2001. Erva-De-Passarinho (Loranthaceae) Na Arborização Urbana: Passeio Público De Curitiba, Um Estudo De Caso. PhD Thesis, Universidade Federal do Paraná, Brazil

Silva IA, Ferreira AW, Lima MI, Soares JJ. 2010. Networks of epiphytic orchids and host trees in Brazilian gallery forests.Journal of Tropical Ecology, 26(2): 127-137

Sousa MM, Colpo KD. 2017. Diversity and distribution of epiphytic bromeliads in a Brazilian subtropical mangrove. Anais da Academia Brasileira de Ciências, 89(2): 1085-1093

Smart SM, Thompson K, Marrs RH, Le Duc MG, Maskell LC, Firbank LG. 2006. Biotic homogenization and changes in species diversity across human-modified ecosystems. Proceedings of the Royal Society of London B: Biological Sciences, 273(1601): 2659-2665

Thompson JN. 1994. The Coevolutionary Process. The University of Chicago Press, Chicago, USA

Tylianakis JM, Laliberté E, Nielsen A, Bascompte J. 2010. Conservation of species interaction networks.

Biological conservation, 143(10): 2270-2279

Zhang WJ. 2011. Constructing ecological interaction networks by correlation analysis: hints from community sampling. Network Biology, 1(2): 81-98

Zhang WJ. 2012. Computational Ecology: Graphs, Networks and Agent-based Modeling. World Scientific, Singapore

Zhang WJ. 2016. Network robustness: implication, formulization and exploitation. Network Biology, 6(4): 75-85

Zhang WJ. 2018. Fundamentals of Network Biology. World Scientific Europe, London

16

A comparative analysis on computational methods for fitting an ERGM to biological network data

Sudipta Saha[1], **Munni Begum**[2]

[1]Dalla Lana School of Public Health, University of Toronto, Toronto, ONM5S 2J7, Canada

[2]Department of Mathematical Sciences, Ball State University, Muncie, IN47306, USA

E-mail: sudip.saha@mail.utoronto.ca, mbegum@bsu.edu

Abstract

Exponential random graph models (ERGM) based on graph theory are useful in studying global biological network structure using its local properties. However, computational methods for fitting such models are sensitive to the type, structure and the number of the local features of a network under study. In this paper, we compared computational methods for fitting an ERGM with local features of different types and structures. Two commonly used methods, such as the Markov Chain Monte Carlo Maximum Likelihood Estimation and the Maximum Pseudo Likelihood Estimation are considered for estimating the coefficients of network attributes. We compared the estimates of observed network to our random simulated network using both methods under ERGM. The motivation was to ascertain the extent to which an observed network would deviate from a randomly simulated network if the physical numbers of attributes were approximately same. Cut-off points of some common attributes of interest for different order of nodes were determined through simulations. We implemented our method to a known regulatory network database of *Escherichia coli (E. coli)*.

Keywords biological networks; regulatory networks; exponential random graph models; Monte Carlo maximum likelihood estimation; maximum pseudo likelihood estimation; *E. coli*.

1 Introduction

Over the last decade, there has been a growing interest in the study of biological interaction networks at the macro and micro molecular levels (Zhang 2012). Identifying basic structural relationships among micro components is the main goal in the field of systems biology (Li and Zhang, 2013). A formal basis for handling such complex networks includes computational tools to support the modelling and simulation through methods developed in mathematical biology and bioinformatics. Since 1960s, with some notable precursors in the preceding decades, a variety of mathematical formalisms have been proposed to describe this kind of complex networking. During the last few years, modelling efforts targeted several distinct types of networks at the molecular level, such as gene regulatory networks (Pavlopoulos et al., 2011; Mason and Verwoerd, 2007), metabolic networks (Ideker et al., 2001), signal transduction networks (Stock, 1990) or protein-protein interaction networks (Pavlopoulos et al., 2011), transcription regulatory networks (Begum et al., 2014). Networks of interactions that are not restricted to a cell (intercellular communications) or take place at an altogether different level of detail (immunological networks, ecological networks) are also of immense interest.

In this paper, we considered a transcription regulatory network for the model organism *Escherichia coli* K-12 (*E. coli*) from *RegulonDB* (Salgado et al., 2006) version 7.4 (http://regulondb.ccg.unam.mx/). The *RegulonDB* contains information on transcription initiation and the regulatory network of *E. coli*. Downloadable experimental datasets are available on the regulatory network interactions *RegulonDB*. The transcription factor (TF) - transcription factor (TF) interaction network data are considered in this work. A transcriptional unit is defined as a set of one or more genes within an operon transcribed as a set through the utilization of a single promoter. In the original dataset (represented as a table) of *E.coli* in the *RegulonDB* website, there are four columns. The first column is the name of the Transcription Factor (TF), the second column is TF regulated by TF, third column is Regulatory effect of the TF on the regulated gene (+ activator, - repressor, +- dual, ? unknown) and the fourth column is the evidence of support of the existence of the regulatory interaction. The first two columns are considered and it created the TF-TF interaction network. The observed TF-TF network, which is a directed network with loops, is given in Fig. 1.

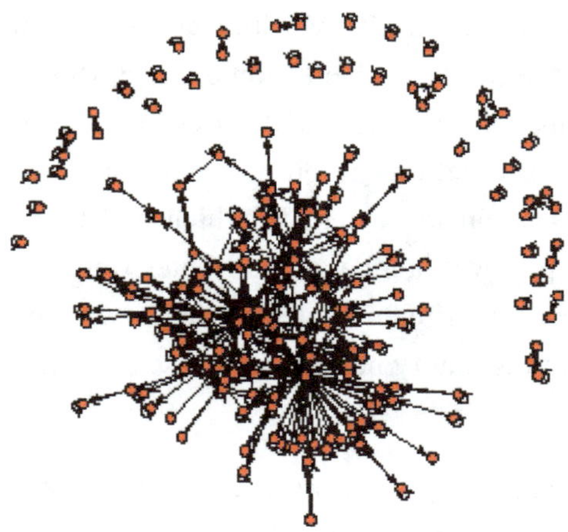

Fig. 1 Observed TF-TF network.

Each vertex is a TF and an edge between two TFs represents a regulation. An edge from a TF to another TF represents that the first TF regulates the second. We explored this observed network and counted the number of several network attributes i.e. edge, triangle and stars. In this observed network, there are *387* edges, *114* triangles, twenty *3-ostars*, thirty-four *3-istars*, ten *5-ostars*, and nine *5-istars* and the network has two big clusters and several small clusters. The basic definitions of some network attributes are given below,

Edges or arcs: This term adds one network statistic that is equal to the number of *edges* in the network. For undirected networks, an *edge* is same as*k-star (1)* [see below] whereas for directed networks, an *edge* represents both *ostar (1)* and *istar (1)* (Morris et al., 2008).

Triangles: This term adds one statistic to the model that is equal to the number of *triangles* in the network. For an undirected network, a *triangle* is defined to be any set $\{(i,j),(j,l),(l,i)\}$of three edges. For a directed network, a *triangle* is defined as any set of three edges $\{(i \rightarrow j) and (j \rightarrow l)\}$ and either $(l \rightarrow i)$ or $(l \leftarrow i)$(Morris et al., 2008).

k-star: This term adds one statistic when there exists ties between one node and *k* number of other nodes. For a directed network the star statistics are replaced by outgoing stars (*k-ostar*) and incoming stars (*k-istar*) (Morris et al., 2008).

2 Exponential Random Graph Model (ERGM)

An Exponential Random Graph Model (ERGM) models the probability distribution (mass function / density function) for a given class of graphs. Given an observed graph and a set of local features of that graph, the probability distribution of the graph is estimated. The distribution provides a concise summary of the class of graphs to which the observed graph belongs, i.e. the probability distribution can be used to calculate the probability that any given graph is drawn from the same distribution as the observed graph (Fronczak, 2012; Robins et al., 2007; Saul and Filkov, 2007; Wasserman and Pattison, 1996).

ERGMs represent the generative process of tie formation in networks with two basic types of processes namely dyadic dependence and dyadic independence. A *dyad* refers to a pair of nodes and the relations between them. Dyadic dependent processes are those in which the state of one dyad depends stochastically on the state of other dyads. Dyadic independent processes exhibit no direct dependence among dyads. This distinction between these two types of processes affects the specification, estimation, and behaviour of ERGMs. Models with only dyadic independent terms have a likelihood function that simplifies to a form that can be maximized using standard logistic regression models. In contrast, models for processes with dyadic dependence require computationally intensive estimation and imply complex forms of feedback and global dependence that confound both intuition and estimation (Handcock et al., 2003; Hunter and Handcock, 2006).

Although an ERGM presents a flexible means to model complex networks, the likelihood function for parameter estimation involves a mathematically intractable normalizing constant. ERGMs generalize the Markov random graph models (Frank and Strauss, 1986), and edge and dyadic independence models. Several statistical computational methods had been proposed to address this difficulty in parameter estimation in an ERGM. These are the Markov chain Monte Carlo maximum likelihood estimation (MCMCMLE) method and the Maximum pseudo likelihood estimation (MPLE) method (Handcock et al., 2003; Robins et al., 2007; Snijders, 2002). We briefly discuss the general ERGM, which is also known asp^* model, to layout the theoretical background of such models.

The general log-linear form of p^* model is expressed as,

$$P(X = x) = \frac{\exp[\boldsymbol{\theta}'\mathbf{z}(\mathbf{x})]}{\kappa(\boldsymbol{\theta})} \tag{1}$$

here $\boldsymbol{\theta}$ is a vector of model parameters, $\mathbf{z}(\mathbf{x})$is a vector of network statistics, and $\kappa(.)$is a normalizing constant which is hard to compute for large networks. In order to simplify the estimation process of the model parameters, the log-linear model form of the p^* model can be re-expressed as a logit model. In particular, as per (Wasserman and Pattison, 1996), $\boldsymbol{X}_{ij}^{+}$ denotes an adjacency matrix where a tie from $i \rightarrow j$ is forced to be present. That is $\boldsymbol{X}_{ij}^{+} = \{X_{kl}, with\, X_{ij} = 1\}$. $\boldsymbol{X}_{ij}^{-}$ denotes an adjacency matrix where a tie from $i \rightarrow j$ is forced to be absent. That is $\boldsymbol{X}_{ij}^{-} = \{X_{kl}, with\, X_{ij} = 0\}$. Finally, $\boldsymbol{X}_{ij}^{c}$ denotes an adjacency matrix with complement relation for the tie from $i \rightarrow j$. That is, $\boldsymbol{X}_{ij}^{c} = \{X_{kl}, with\; (kl) \neq (i,j)\}$. The p^*model in Equation (1) can be turned to a logistic regression model by considering a set of binary random variables$\{X_{ij}\}$, where$X_{ij} = 1$ implying a tie from $i\; to\; j$ as follows.

$$\log\left\{\frac{P(X_{ij} = 1|\mathrm{X}_{ij}^{c})}{P(X_{ij} = 0|\mathrm{X}_{ij}^{c})}\right\} = \omega_{ij} = \boldsymbol{\theta}'[\mathbf{z}(\mathbf{x}_{ij}^{+}) - \mathbf{z}(\mathbf{x}_{ij}^{-})] \tag{2}$$

$$\omega_{ij} = \boldsymbol{\theta}'\boldsymbol{\delta}(\boldsymbol{x}_{ij}) \quad (3)$$

Here $\boldsymbol{z}(\boldsymbol{x}_{ij}^{+})$ is defined as vector of network statistic x_{ij}^{+}, $\boldsymbol{z}(\boldsymbol{x}_{ij}^{-})$ is the vector of network statistic x_{ij}^{-} and $\boldsymbol{\delta}(\boldsymbol{x}_{ij})$ is the vector of difference statistics obtained from the network statistics $\mathbf{z}(.)$ when the variable X_{ij} changes from *1* to *0*. The model in Equation (3) is referred to as the $logit\ p^*$ model for single binary relation. One can work with either the log-linear form of p^* model given in Equation (1) or the logit form given in equation (3).

3 Computational Methods

There are two methods commonly used to estimate the maximum likelihood fit to exponential random graph models. These are the maximum pseudo-likelihood estimation (MPLE) and the Markov chain Monte Carlo maximum likelihood estimation (MCMCMLE) (Handcock et al., 2003; Robins et al., 2007; Snijders, 2002). The pseudo likelihood function is simply the product of the probabilities of X_{ij} with each probability conditional on the rest of the data. The method avoids the technical difficulty inherent in the maximum likelihood approach. The maximum pseudo likelihood estimator (MPLE) for an ERGM, which maximize the pseudo likelihood, may easily be found (at least in principle) by using logistic regression as a computational device. However, when the ERGM in question is not a dyadic independence model, the statistical properties of pseudo likelihood estimators for a network are not well understood (Hunter and Handcock, 2006).

Monte Carlo maximum likelihood estimation (MCMCMLE) is preferred for dyadic dependent p^* models. The MCMCMLE of the parameter vector θ is obtained by maximizing the approximate likelihood. The MCMCML estimation algorithm is implemented to the software package *statnet* (Handcock et al., 2003) under the statistical computational environment R. We use these two packages *statnet* and *ergm* (Handcock et al., 2008) to fit the exponential random model given in equation (1).

4 Simulation Study

We conduct a simulation study for generating random network under varying conditions. We choose conditions by assigning different number of nodes and network statistics. The primary reason behind conducting the simulation is to determine the cut-off points for different number of nodes for specific attributes and also to compare our simulated models with an observed model. For the comparison part, we create two networks by imposing the same number of network attributes to the models and then compare the results of estimates with the TF-TF interaction network of *E. coli* by fitting ERGM.

We consider various network statistics such as *arc, stars*, and *triangles*. A *k-star* is defined where there exist ties between one node and *k* number of other nodes. For a directed network the star statistics are replaced by outgoing stars (*k-ostar*) and incoming stars (*k-istar*). In particular *arc, 5-ostar, 5-istar, 6-ostar, 6-istar,* and *triangles* are considered as our network attributes. We physically impose these attributes into the simulated network by keeping approximately the same number of attributes as the observed network. We also observe that if we simulate *triangles, ostars, istars*, and *arcs* are automatically created. We randomly assign these attributes to the simulated networks for different number of nodes (*n=20, 50, 100*) and determine the conditions for these statistics to become insignificant. The cut-off points for single attributes and for a combination of attributes are assessed. However, due to the convergence issues, we were unable to obtain the cut-off points for some cases. A cut-off point is defined as the value where network attributes become significant to insignificant and vice versa. The rationale is that if the biological network behaves almost the same as the random network, then if we have an observed network with different number of nodes, we can determine up to which point (approximately) certain statistics become insignificant.

We explore the TF-TF interaction network of *E. coli* from the *RegulonDB* database and found that there are ten*5-ostars*, nine*5-istars*, ten*6-ostars*, eight*6-istars* and *114 triangles*. The network contains *175* nodes with density *0.012*. Once we determine the number of attributes in the observed network, then we mimic this network and randomly simulate two networks. Then we consider different combinations of attributes (*ostars, istars* and *triangles*) and fit the models by ERGM. We fit the same models for the observed data by using ERGM and then compare the estimates of ERGM for both MCMCMLE and MPLE method.

We begin with networks with small number of nodes and move toward networks with higher number of nodes. With only *20* nodes, we consider reasonably smaller magnitude of network attributes such as *arcs, 3-ostars, 3-istars* and *triangles* as our attributes of interest and then fit the models with ERGM to get the estimates and also to determine the cut-off points. We start with smaller number of attributes, two*3-ostars*, two 3-*istars*, and two *triangles*. We increment each attribute one at a time to determine the cut-off points. Simulated network with 77 triangles, twelve*3-istars* and fourteen*3-ostars* is presented in Fig. 2.

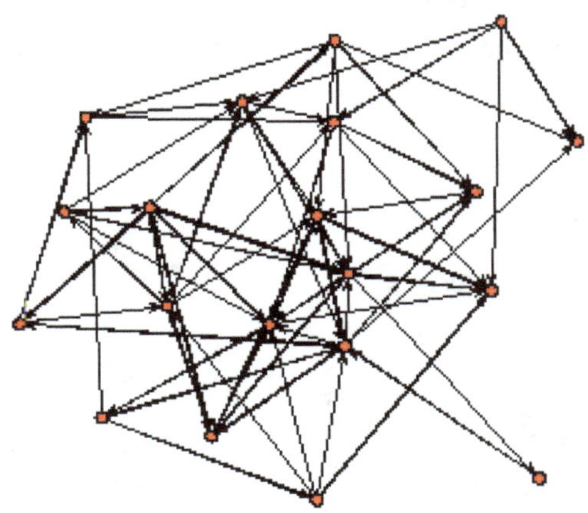

Fig. 2 Simulated network for *n=20.*

Next we increased the number of nodes to 50 and 100. A summary of the simulated networks with nodes 20, 50, and 100including the cut-off points for each network statistics is presented in Table 1.

Table 1 Summary of simulation studies for different numbers of nodes.

	For *n*=20			**For *n*=50**			**For *n*=100**		
	Trian-gles	3-Ostar	3-Istar	Trian-gles	3-Ostar	3-Istar	Trian-gles	3-Ostar	3-Istar
Lower cut-offs	-	7	6	-	-	3	-	4	5
% of*n* (apps)[a]	-	35%	30%	-	-	6%	-	4%	5%
Higher cut-offs	76-80	17	17	-	35	-	-	64	64
% of*n* (apps)	390%	85%	85%	-	70%	-	-	64%	64%

[a]% of n (apps) means that the lower cut-offs are the percentage of n (i.e. node). For example, for n=20, lower cut-off of 3-Ostar is 7 which is 35% of n=20.

It is to be noted that the cut-off points for *3-ostar* and *3-istar* are quite similar, although we could not find any conclusive answer when the number of nodes is *50*. For *3-ostar* and *3-istar*, we can say that, the cut-off points spread out with the increase in the number of nodes. That is if we move toward higher number of nodes, the lower cut-off points become smaller and the higher cut-off points become smaller. For *n=20*, the total spread of insignificant region is close to *(85-35) = 50%* and which is approximately *60%* for *n=100.* For triangles, cut-off points should be bigger than the number of nodes *n*. In summary, we can say that, for network data if we increase the order of the nodes, the spread of the insignificant region gradually becomes larger for any specific attributes. To determine the exact percentage of cut-off points, we have to do similar study for different other nodes, and then we can generalize the idea.

4.1 Comparisons of results under simulation schemes

In our observed TF-TF model, we have *175* nodes, *114* triangles, ten *5-ostars*, nine 5-*istars*, ten 6-*ostars* and eight *5-istars*. An *R*-script is written to count the number of attributes in the model. Then we randomly simulate two different network models to compare the estimates of these network attributes with the observed network. In both cases, we have very close estimates of network attributes from the simulated models compared to the actual model. In Fig. 3 and 4, we represent the observed TF-TF network with and without loops.

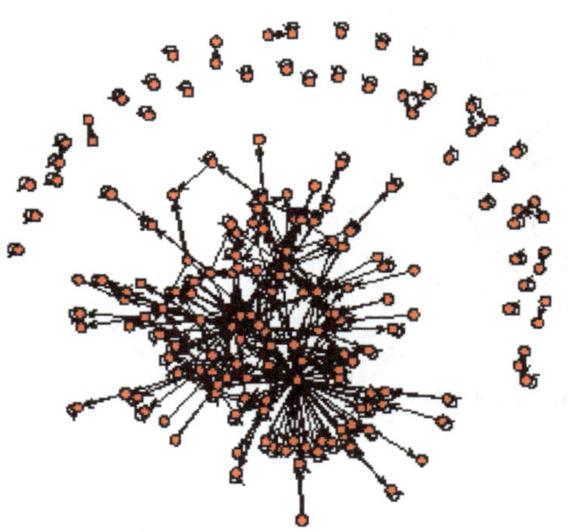

Fig. 3 Observed TF-TF network with looping.

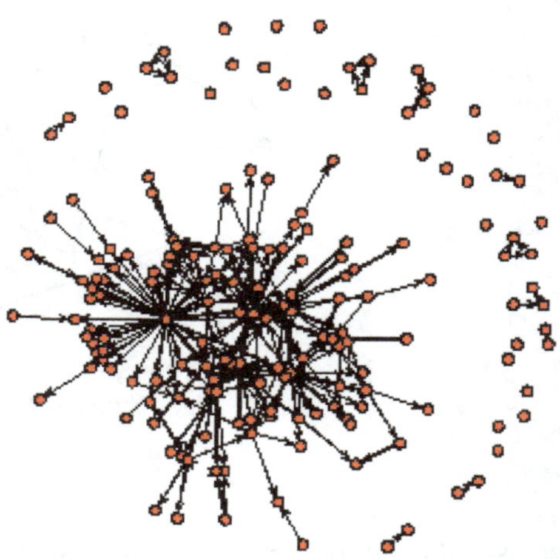

Fig. 4 Observed TF-TF network without looping.

The estimates of attributes of the observed network are presented in Table 2.

Table 2 Estimates of the observed network.

Network Attributes	MCMCMLE Estimates	MPLE Estimates
Edges	-5.3500647	-5.35
Triangle	0.9355000	9.355e-01
5-Ostar	0.0003851	1.564e-05
5-Istar	0.0022043	2.204e-01
6-Ostar	7.797e-05	1.676e-06
6-Istar	1.034e-01	1.034e-03

To compare the estimates of network attributes between the observed and simulated network, we randomly simulate two networks by imposing the same number of attributes as TF-TF, one with *5-ostars, 5-istars*, and *triangles* (network-1) and the other with same number of *6-ostars, 6-istars*, and *triangles* (network-2). The summary table of the number of network attributes is presented in Table 8.It is to be noted that we found very similar estimates for the common network attributes edge and triangle from network-1 and network-2. In Tables 3 and 4, we presented the estimates of the observed and simulated networks for both MCMCMLE and MPLE methods.

Table 3 Estimates from observed versus simulated networks with MCMC MLE.

Network Attributes	Estimates from observed networks	Estimates from simulated networks
Edges	-5.3500647	-5.73286
Triangle	0.9355000	2.90743
5-Ostar	0.0003851	-0.01720
5-Istar	0.0022043	-0.08434
6-Ostar	7.797e-05	-1.342e-01
6-Istar	1.034e-01	-8.702e-04

Table 4 Estimates from observed versus simulated networks with MPLE.

Network Attributes	Estimates from observed networks	Estimates from simulated networks
Edges	-5.35	-5.675632
Triangle	9.355e-01	2.905757
5-Ostar	1.564e-05	-0.016141
5-Istar	2.204e-01	-0.083797
6-Ostar	1.676e-06	-0.1343137
6-Istar	1.034e-03	-0.0006957

From Tables 3 and 4, we conclude that except triangles the rest of the estimates of network attributes are very close for both MCMCMLE and MPLE method. Therefore, from the biological point of view, if the observed network is available and the numbers of certain network attributes are known, then it behaves almost same as the random model for most of the cases. However, to generalize the case we need more experiment and more exploration among higher order of species. The simulated networks (1 & 2)are presented in Figs 5 and 6.

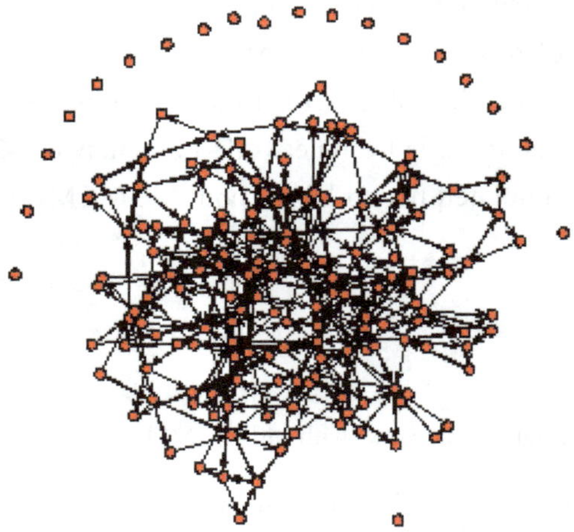

Fig. 5 Simulated network-1.

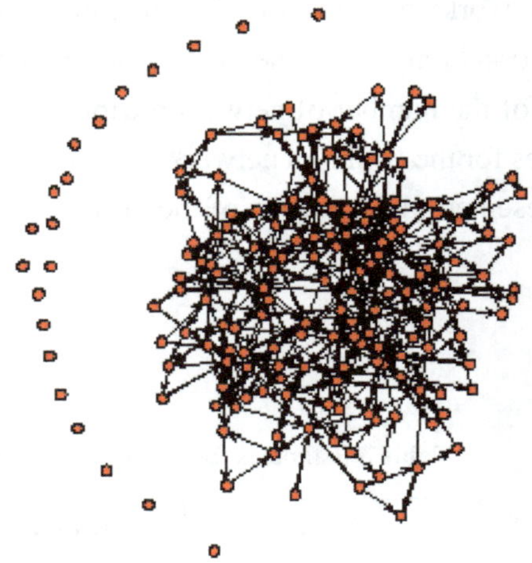

Fig. 6 Simulated network-2.

From this experiment, we observe that if we want to simulate a biological data, then one way would be to explore the observed data and count the number of statistics that we are interested and then physically impose the number of statistic and then compare. There are several other ways to simulate network models using several packages on R. The simplest one is to take the density of the observed model and simulate it using binomial distribution. Also, once a model is fitted by using ERGM package, it can be simulated from the fitted model. ERGM takes the estimates of the network attributes and simulates a similar type of model. However, in such a case the physical number of attributes differs substantially. Again, we can also simulate networks by using Erdos-Renyi model. The comparison of networks obtained using different simulation approaches is presented in the following section.

4.2 Comparison over simulation methods

In this section, we simulate several networks by the existing simulation schemes. We simulated a network by using Erdos-Renyi modelling scheme where we consider *175* nodes to create similarity with our observed TF-TF network and then consider the density of the TF-TF model. The summary of the estimates that we obtain under different approaches, are provided in Tables 5, 6, and 7 (for both MCMCMLE and MPLE).

Table 5 Estimates from Erdos-Renyi model

Network Attributes	MCMCMLE Estimates	MPLE Estimates
Edges	-4.438846	-4.42969
Triangle	-0.058951	-0.06431
5-Ostar	-0.007336	-0.00546
5-Istar	-0.120974	-0.15166
6-Ostar	-0.01302	-0.01366
6-Istar	-0.75202	-0.84257

Table 6 Estimates from Binomial simulated model

Network Attributes	MCMCMLE Estimates	MPLE Estimates
Edges	-4.33561	-4.30465
Triangle	-0.06395	-0.08794
5-Ostar	-0.01032	-0.04404
5-Istar	-0.07059	-0.07491
6-Ostar	-0.10063	-0.25368
6-Istar	-0.31184	-0.29370

Table 7 Estimates from fitted ERGM models

Network Attributes	MCMCMLE Estimates	MPLE Estimates
Edges	-5.3318479	-5.332e+00
Triangle	0.7194116	7.194e-01
5-Ostar	0.0001207	6.297e-06
5-Istar	0.0016440	1.644e-03
6-Ostar	1.484e-05	5.333e-07
6-Istar	-5.887e-02	-5.887e-02

We notice that as long as we consider the same network, estimates of certain attributes are always similar. Although some of the estimates we obtain in this simulation study are very close, the physical numbers of statistics differ substantially. As the simulation scheme takes the fitted estimates into account, the physical number of different attributes should be close to the observed model. It is important since the exact numbers of network statistics might have a significant influence on the overall process. The simulated networks using Erdos-Renyi modelling scheme, binomial density, and *ergm* package in *R* are presented in Figs 7, 8, and 9.

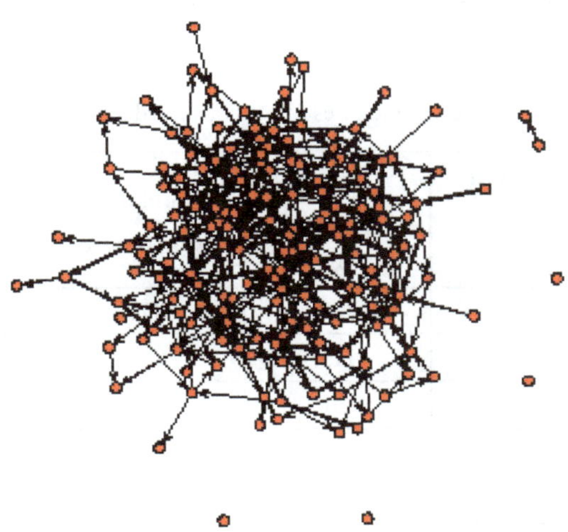

Fig. 7 Simulated from Erdos-Renyi model.

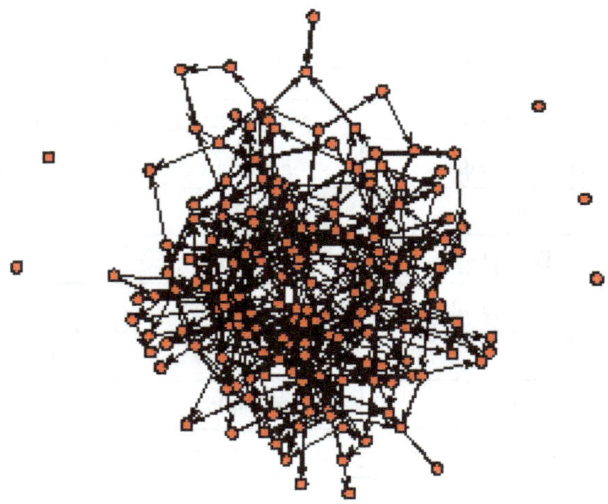

Fig. 8 Simulated network using binomial probability.

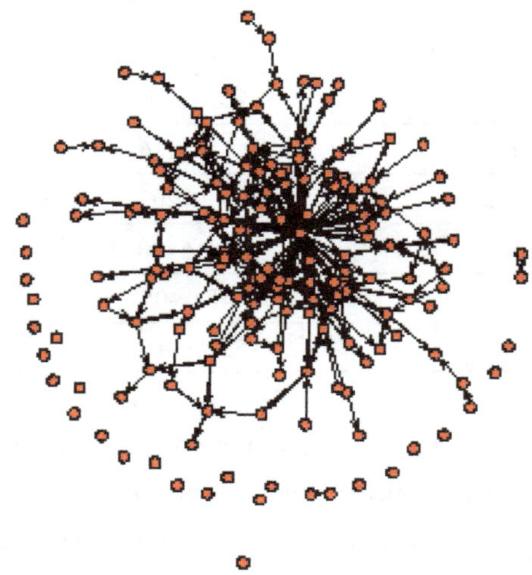

Fig. 9 Simulated network from fitted ERGM model.

The numbers of network attributes for different simulation models are presented in Table 8.

Table 8 Summary table of estimates observed versus simulated networks.

Network Attributes	Observed TF-TF network	Our Simulated network	Simulation using density	Erdos-Renyi simulation	ERGM fitted simulation
Edges	263	327	377	375	247
Triangle	114	115	12	9	82
5-Ostar	10	10	17	6	7
5-Istar	9	9	12	8	2
6-Ostar	10	10	6	2	1
6-Istar	8	8	3	1	3

From Table 8, we can see that the network attributes are different under all the simulation schemes. In our process as we are physically imposing the attributes, it is very close to the observed model. The only difference in the attributes is for the triangles which differ by just *1*. From Table 8, we can say that, in terms of number, the ERGM simulated network generates close result. However, the numbers of triangles substantially differ from the original observed model. For the simple binomial simulation, the edges do not even come close and the other attributes also significantly differ. We find similar characteristic for Erdos-Renyimodelling scheme. The reason behind this could be that both the binomial and Erdos-Renyi consider the density only while simulation. Thus, the number of attributes along with the edges is very close. However, other attributes such as the number of *5-istars* or *5-ostars* are not very close. In our random simulation, we emphasize on the number of attributes because a biological process is a very complicated process. A single edge might have significant influence over the entire process. Therefore, for biological simulation, we should always keep in mind the physical number of attributes that we are interested in.

5 Conclusions

The number of commonly used network attributes such as *k-istar, k-ostar* and *triangles* in the TF-TF regulatory network of *E. coli* is determined. These networks attributes statistically serve as the significant local structures for the *E. coli* regulatory network. An observed regulatory network of the model organism *E. coli* was exploredin terms of finding statisticallysignificant local structure in this study. Simulation of two network models, network-1 and network-2, and comparison of the estimates of the observed and simulated models are presented. In both cases, the estimates we obtain are very similar with the observed TF-TF network except for *triangles*. Networks simulated using existing methods are compared in terms of these estimates as well. At the end, our models provide close results and same number of network attributes, which is very important for biological network data. Therefore, it can concluded that for the*E. coil* regulatory network, the network can be reproduced by taking the counts for different attributes, and the simulated network will behave as the observed network.

Simulation of different networks with different number of nodes and network attributes were performed. The cut-off points were determined for a number of attributes at which point specific attributes become significant to insignificant, or vice versa. We observed that for smaller numbers of network attributes, the estimates usually become significant. If the number of attributes increases in a given model, the attributes become insignificant.

We also observe that the models in ERGM do not always converge. Addressing the convergence issue would be a desirable upgrade for the computational method. For the several models considered, convergence failure occurred while estimating parameters for any of the methods. For example, for our observed network, the model with *edges, 4-istars, 4-ostars* and *triangles* did not converge. Also, due to the convergence issue, cut-off points could not be determined for several network attributes. In addition, computation for networks with self loops demonstrates convergence problems. Therefore, while the ERGM provides flexible methodology, these issues remain in need of further analysis.

References

Begum M, Bagga J, Blakey A, Saha S. 2014. Network motif identification and structure detection with graphical models. Network Biology, 14(4): 155-169

Frank O, Strauss D. 1986. Markov graphs. Journal of the American Statistical Association, 81(395): 832-842

Fronczak A. 2012. Exponential Random Graph Models. ArXiv e-prints. Available at http://arxiv.org/pdf/1210.7828.pdf

Handcock MS, Hunter DR, Butts CT, et al. 2003. Statnet: Software Tools for the Statistical Modeling of Network data. Statnet Project. Available at http://statnetproject.org/

Handcock MS, Hunter DR, Butts CT, et al. 2008. ergm: A package to fit, simulate and diagnose exponential-family models for networks. Journal of Statistical Software, 24. Available at http://www.ncbi.nlm.nih.gov/pmc/articles/PMC2743438/

Hunter DR, Handcock MS. 2006. Inference in curved exponential family models for networks. Journal of Computational and Graphical Statistics, 15(3): 565-583

Ideker T, Thorsson V, Ranish JA, et al. 2001. Integrated genomic and proteomic analysis of a systematically perturbed metabolic network. Science, 292(5518): 929-934

Li JR, Zhang WJ. 2013. Identification of crucial metabolites/reactions in tumor signaling networks. Network Biology, 3(4): 121-132

Mason O, Verwoerd M. 2007. Graph theory and networks in biology. IET Systems Biology, 1(2): 89-119

Morris M, Handcock MS, Hunter DR. 2008. Specification of exponential-family random graph models: Terms and computational aspects. Journal of Statistical Software, 24(4): 1-24

Pavlopoulos GA, Secrier M, Moschopoulos CN, et al. 2011. Using graph theory to analyze biological networks. BioData Mining, 4(10): 1-27

RobinsGL,Pattison PE,Kalish Y, LusherD. 2007. An introduction to exponential random graph (p*) models for social networks. Social Networks, 29(2): 173-191

Salgado H, Gama-Castro S, Peralta-Gil M, et al. 2006. RegulonDB (version 5.0): Escherichia coli K-12 transcriptional regulatory network, operon organization, and growth conditions. Nucleic Acids Research, 34(1): 394-397

Saul ZM, Filkov V. 2007. Exploring biological network structure using exponential random graph models. Bioinformatics, 23(19): 2604-2611

Snijders TAB. 2002. Markov Chain Monte Carlo Estimation of Exponential Random Graph Models.Journal of Social Structure, 3: 1-40.

Stock JB, Stock AM, Mottonen JM. 1990. Signal Transduction in Bacteria. Europe PubMed Central, 344(6265): 395-400

Wasserman S, PattisonPE. 1996. Logit models and logistic regressions for social networks: I. An introduction to Markov graphs and p*. Psychometrika, 61(3): 401-425

Zhang WJ. 2012. Computational Ecology: Graphs, Networks and Agent-based Modeling. World Scientific, Singapore

17

Compedium model using frequency / cumulative distribution function for receptors of survival proteins: Epidermal growth factor and insulin

Shruti Jain
Department of Electronics and Communication Engineering, Jaypee University of Information Technology, Solan-173234, India
E-mail: jain.shruti15@gmail.com

Abstract
In this paper I used the frequency and cumulative distribution functions to make a best model of the receptors of the survival proteins i.e. Epidermal Growth Factor Receptor (EGFR) and Insulin Receptor (IRS) using ten concentrations combination of TNF, EGF and Insulin. It has been revealed that survival and apoptosis signals induced by the receptors of EGF and insulin are temporarily separated and this is reflected in my model by the differences between the values of the parameters used. I conducted the analysis using KS-d, KS, AD stat, AD p-value, chi square, chi square p-value and chi square df for different distribution functions for EGFR and IRS. The frequency and cumulative distribution curves for different distribution techniques like exponential, log-normal, normal, gamma, chi-square etc are plotted using chi- square tests.

Keywords EGF; insulin; receptors; frequency distribution; cumulative distribution.

1 Introduction
The programmed cell death (Jain, 2012; Suzzane, 2005; Weiss, 2001), Tumor Necrosis Factor α (TNF-α) (Brockhaus, 1990; Janes, 2005; Jain, 2011a, 2011b; Thoma, 1990), functions as apoptosis cues, whereas growth factors such as EGF (Arteaga, 2003; Libermann, 1984; Normano, 2006; Ullirich, 1990) and insulin (Lizcano, 2002; White, 2003; Jain, 2009, 2010, 2011a, 2010c, 2015a, 2015b) exert survival effects. These factors in single or in combination activate various key players in the network pertaining to cell survival/ apoptosis. Many proteins involved in this process that interact systematically regulating a specific pathway or cross talk with other proteins of different pathways. As a result many pathways activated simultaneously leading to many biochemical and physiological changes inside the cell. The final outcome of whether a cell dies or survives depends in the concentrations of key players among the pathways.

The epidermal growth factor (EGF) binds with its receptor to form EGF receptor (EGFR) which further binds with Src homology 2 (SH2) (Jain, 2015a; Toma, 1990) leading to the activation of RAS/extracellular signal regulated kinase (ERK) pathway (Kyriakis, 1996), the phosphatidylinositol 3 kinase(PI3K) pathway (Jain, 2014, 2015c, 2015d, 2016a, 2016b), and the activator of transcription (JAK/ STAT) pathway.

Insulin is a hormone that binds to its receptor (the insulin receptor) on cell membranes and initiates signal transduction leading to cell survival/ apoptosis. Binding of insulin to its receptor induces phosphorylation of tyrosine residue on the inner part of the receptor (White, 2003; Jain, 2015b). The phosphorylated tyrosine residues allow other intracellular proteins to bind to the intracellular domain of the receptor, and become phosphorylated.

In this paper I calculated KS, AD stat, AD p-value, chi square, chi square p-value and chi square df for different survival proteins. Later I plotted frequency distribution and cumulative distribution function curves for different distribution functions for EGFR and IRS.

2 Material and Methods

For mathematical modeling we have used frequency and cumulative distribution functions for cell survival/ cell death. Different parameters are calculated with the different concentrations of the TNF, EGF and Insulin (Jain, 2011b, 2012).

Cumulative distribution function specifies the distance of multivariate random variables X. The real valued random variables X with a given probability distance will found to have a value less than or equal to *x*. There are different test performed on different distribution functions: Anderson darling test, Kolmogorov–Smirnov test and chi- square test (Jain, 2016b).

a) The Kolmogorov–Smirnov (K-S/ KS) test is an equality test using nonparametric. It can also be used as a goodness of fit test. One sample/one dimensional K-S test is used to compare a sample with a probability function while two sample/ 2-D test is used to compare two samples. The K–S statistic for a given cumulative distribution function *F*(*x*) is

$$D_n = \sup_x \left| F_n(x) - F(x) \right| \quad ..(1)$$

where sup *x* is the supremum value of the distances.

b) The Anderson Darling (AD) test/ Shapiro Wilk test is a statistical test and is based on the distance

$$A = n \int_{-\infty}^{\infty} \frac{\left(F_n(x) - F(x)\right)^2}{F(x)\left(1 - F(x)\right)} dF(x) \quad ..(2)$$

c) A chi-squared test (χ^2) test are used to determine the difference between observed and expected frequency in one or more categories.

3 Results

Table 1, Table 2 and Table 3 shows the KS-d, KS, AD stat, AD p-value, chi square, chi square p-value and chi square df values for different tests (KS test, chi square test, AD test) for different distribution functions for EGFR. The frequency distribution curves and cumulative distribution curves for different distribution techniques are shown in Fig. 1 and Fig. 2 respectively.

Table 1 Different distribution function values using Kolmogorov-Smirnov Test for EGFR.

	K-S d	K-S	AD Stat	AD p-value	Chi-sq	Chi-sq p-value	Chi-sq df
Gaussian Mixture(Mixing.Coef.1,Mean 1, Std.Dev 1, Mixing Coef.2,...)	0 026799	0 978501	0 1503	0 998562	6 067	0 108415	3
Normal (location,scale)	0 183078	0 000000	18 5049	0 000000	193 533	0 000000	7
Log Normal (scale,shape)	0 231571	0 000000	22 6154	0 000000	266 733	0 000000	7
Half Normal (scale)	0 236718	0 000000	23 2578	0 000000	279 467	0 000000	7
Rayleigh (scale)	0 210974	0 000000	25 5090	0 000000	232 733	0 000000	6
Weibull (scale,shape)	0 568500	0 000000	117 2716	0 000000	1424 267	0 000000	8
General Pareto (scale,shape)	0 640743	0 000000	144 1607	0 000000	1907 267	0 000000	8
Triangular(min,max,mode)	0 867025	0 000000	532 6983	0 000000	1555 267	0 000000	7

Table 2 Different distribution function values using chi square test for EGFR.

	K-S d	K-S	AD Stat	AD p-value	Chi-sq	Chi-sq p-value	Chi-sq df
Gaussian Mixture	0 026799	0 978501	0 1503	0 998562	6 067	0 108415	3
Normal (location,scale)	0 231571	0 000000	22 6154	0 000000	266 733	0 000000	7
Log Normal (scale,shape)	0 236718	0 000000	23 2578	0 000000	279 467	0 000000	7
Half Normal (scale)	0 640743	0 000000	144 1607	0 000000	1907 267	0 000000	8
Rayleigh (scale)	0 568500	0 000000	117 2716	0 000000	1424 267	0 000000	8
Weibull (scale,shape)	0 183078	0 000000	18 5049	0 000000	193 533	0 000000	7
General Pareto (scale,shape)	0 867025	0 000000	532 6983	0 000000	1555 267	0 000000	7
Triangular(min,max,mode)	0 210974	0 000000	25 5090	0 000000	232 733	0 000000	6

Table 3 Different distribution function values using Anderson darling test for EGFR.

	K-S d	K-S	AD Stat	AD p-value	Chi-sq	Chi-sq p-value	Chi-sq df
Gaussian Mixture	0 026799	0 978501	0 1503	0 998562	6 067	0 108415	3
Weibull (scale,shape)	0 183078	0 000000	18 5049	0 000000	193 533	0 000000	7
Normal (location,scale)	0 231571	0 000000	22 6154	0 000000	266 733	0 000000	7
Log Normal (scale,shape)	0 236718	0 000000	23 2578	0 000000	279 467	0 000000	7
Triangular(min,max,mode)	0 210974	0 000000	25 5090	0 000000	232 733	0 000000	6
Rayleigh (scale)	0 568500	0 000000	117 2716	0 000000	1424 267	0 000000	8
Half Normal (scale)	0 640743	0 000000	144 1607	0 000000	1907 267	0 000000	8
General Pareto (scale,shape)	0 867025	0 000000	532 6983	0 000000	1555 267	0 000000	7

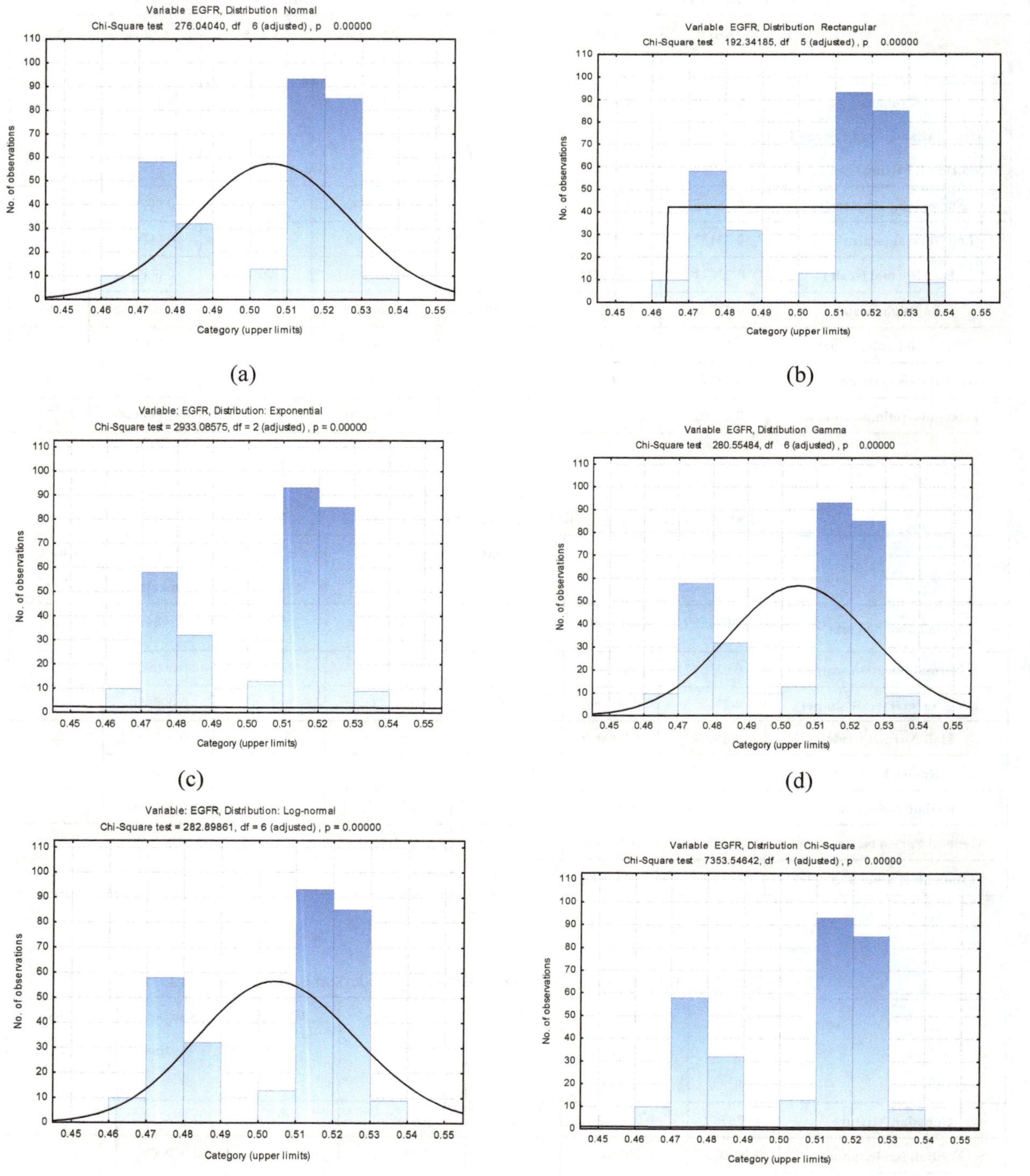

Fig. 1 Frequency Distribution curves for EGFR using different distribution techniques (a) Normal, (b) Rectangular, (c) Exponential, (d) Gamma, (e) Log-normal, (f) Chi-square.

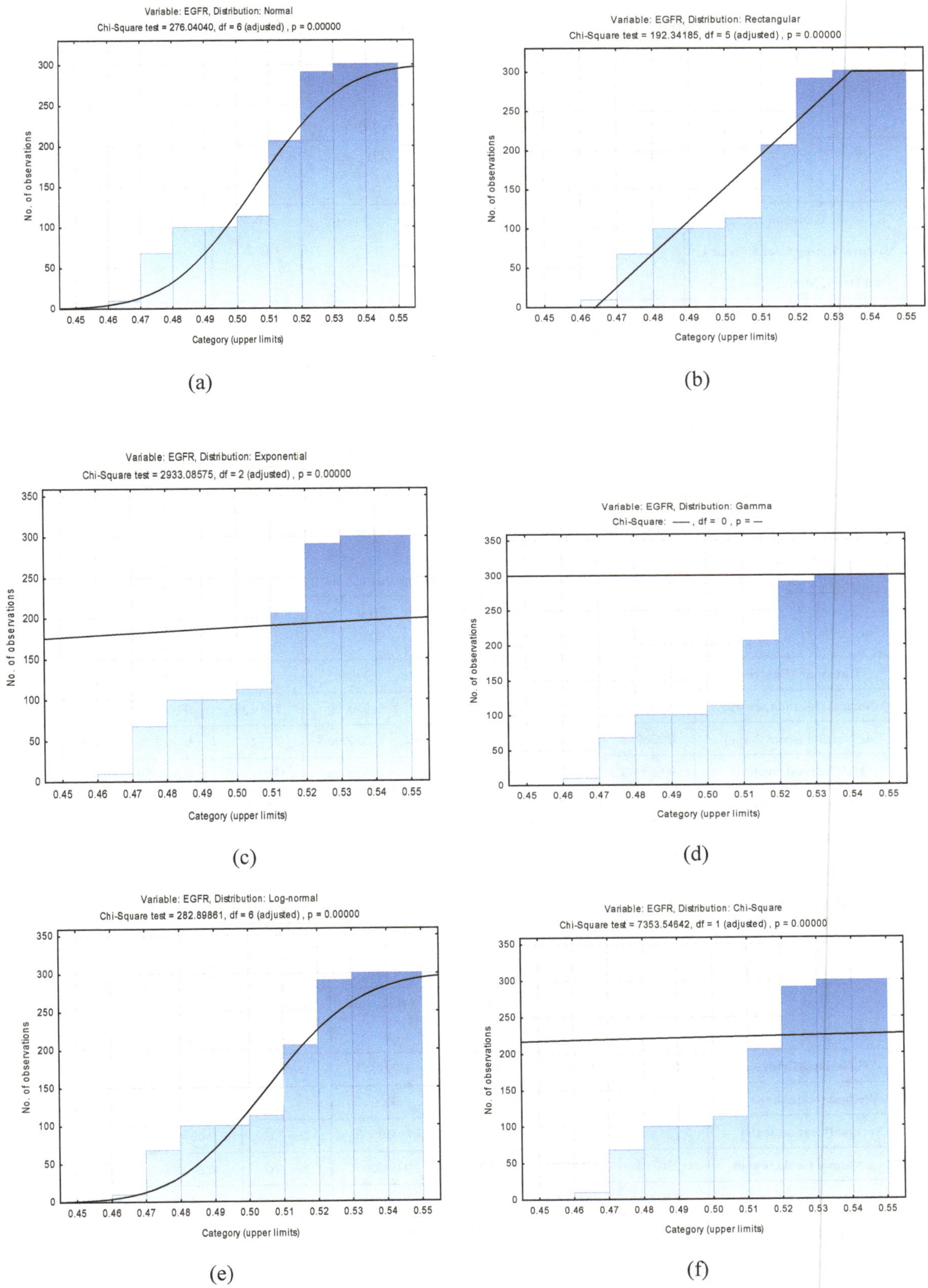

Fig. 2 Cumulative Distribution curves for EGFR using different distribution techniques (a) Normal, (b) Rectangular, (c) Exponential, (d) Gamma, (e) Log-normal, (f) Chi-square.

Table 4, Table 5 and Table 6 shows the KS-d, KS, AD stat, AD p-value, chi square, chi square p-value and chi square df for different distribution functions for IRS. The frequency distribution curves and cumulative distribution curves for different distribution techniques are shown in Fig. 3 and Fig. 4 respectively.

Table 4 Different distribution function values using Kolmogorov-Smirnov Test for IRS.

	K-S d	K-S	AD Stat	AD p-value	Chi-sq	Chi-sq p-value	Chi-sq df
Gaussian Mixture(Mixing.Coef.1,Mean 1, Std.Dev 1, Mixing Coef.2,...)	0 026799	0 978501	0 1503	0 998562	6 067	0 108415	3
Normal (location,scale)	0 183078	0 000000	18 5049	0 000000	193 533	0 000000	7
Log Normal (scale,shape)	0 210974	0 000000	25 5090	0 000000	232 733	0 000000	6
Half Normal (scale)	0 231571	0 000000	22 6154	0 000000	266 733	0 000000	7
Rayleigh (scale)	0 236718	0 000000	23 2578	0 000000	279 467	0 000000	7
Weibull (scale,shape)	0 568500	0 000000	117 2716	0 000000	1424 267	0 000000	8
General Pareto (scale,shape)	0 640743	0 000000	144 1607	0 000000	1907 267	0 000000	8
Triangular(min,max,mode)	0 867025	0 000000	532 6983	0 000000	1555 267	0 000000	7

Table 5 Different distribution function values using chi square test for IRS.

	K-S d	K-S	AD Stat	AD p-value	Chi-sq	Chi-sq p-value	Chi-sq df
Gaussian Mixture	0 026799	0 978501	0 1503	0 998562	6 067	0 108415	3
Normal (location,scale)	0 231571	0 000000	22 6154	0 000000	266 733	0 000000	7
Log Normal (scale,shape)	0 236718	0 000000	23 2578	0 000000	279 467	0 000000	7
Half Normal (scale)	0 640743	0 000000	144 1607	0 000000	1907 267	0 000000	8
Rayleigh (scale)	0 568500	0 000000	117 2716	0 000000	1424 267	0 000000	8
Weibull (scale,shape)	0 183078	0 000000	18 5049	0 000000	193 533	0 000000	7
General Pareto (scale,shape)	0 867025	0 000000	532 6983	0 000000	1555 267	0 000000	7
Triangular(min,max,mode)	0 210974	0 000000	25 5090	0 000000	232 733	0 000000	6

Table 6 Different distribution function values using Anderson darling test for IRS.

	K-S d	K-S	AD Stat	AD p-value	Chi-sq	Chi-sq p-value	Chi-sq df
Gaussian Mixture	0 026799	0 978501	0 1503	0 998562	6 067	0 108415	3
Weibull (scale,shape)	0 183078	0 000000	18 5049	0 000000	193 533	0 000000	7
Normal (location,scale)	0 231571	0 000000	22 6154	0 000000	266 733	0 000000	7
Log Normal (scale,shape)	0 236718	0 000000	23 2578	0 000000	279 467	0 000000	7
Triangular(min,max,mode)	0 210974	0 000000	25 5090	0 000000	232 733	0 000000	6
Rayleigh (scale)	0 568500	0 000000	117 2716	0 000000	1424 267	0 000000	8
Half Normal (scale)	0 640743	0 000000	144 1607	0 000000	1907 267	0 000000	8
General Pareto (scale,shape)	0 867025	0 000000	532 6983	0 000000	1555 267	0 000000	7

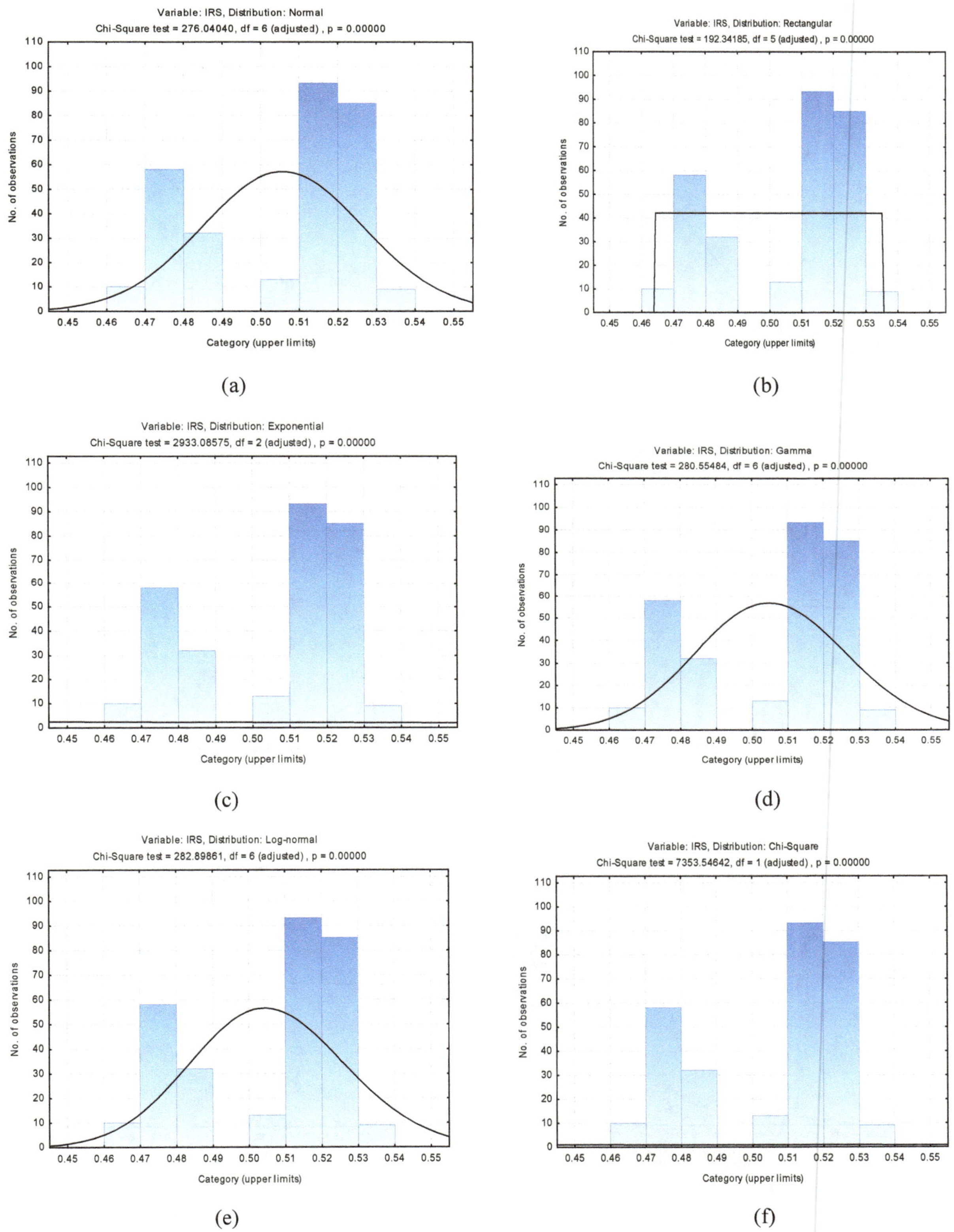

Fig. 3 Frequency Distribution curves for IRS using different distribution techniques (a) Normal, (b) Rectangular, (c) Exponential, (d) Gamma, (e) Log-normal, (f) Chi-square.

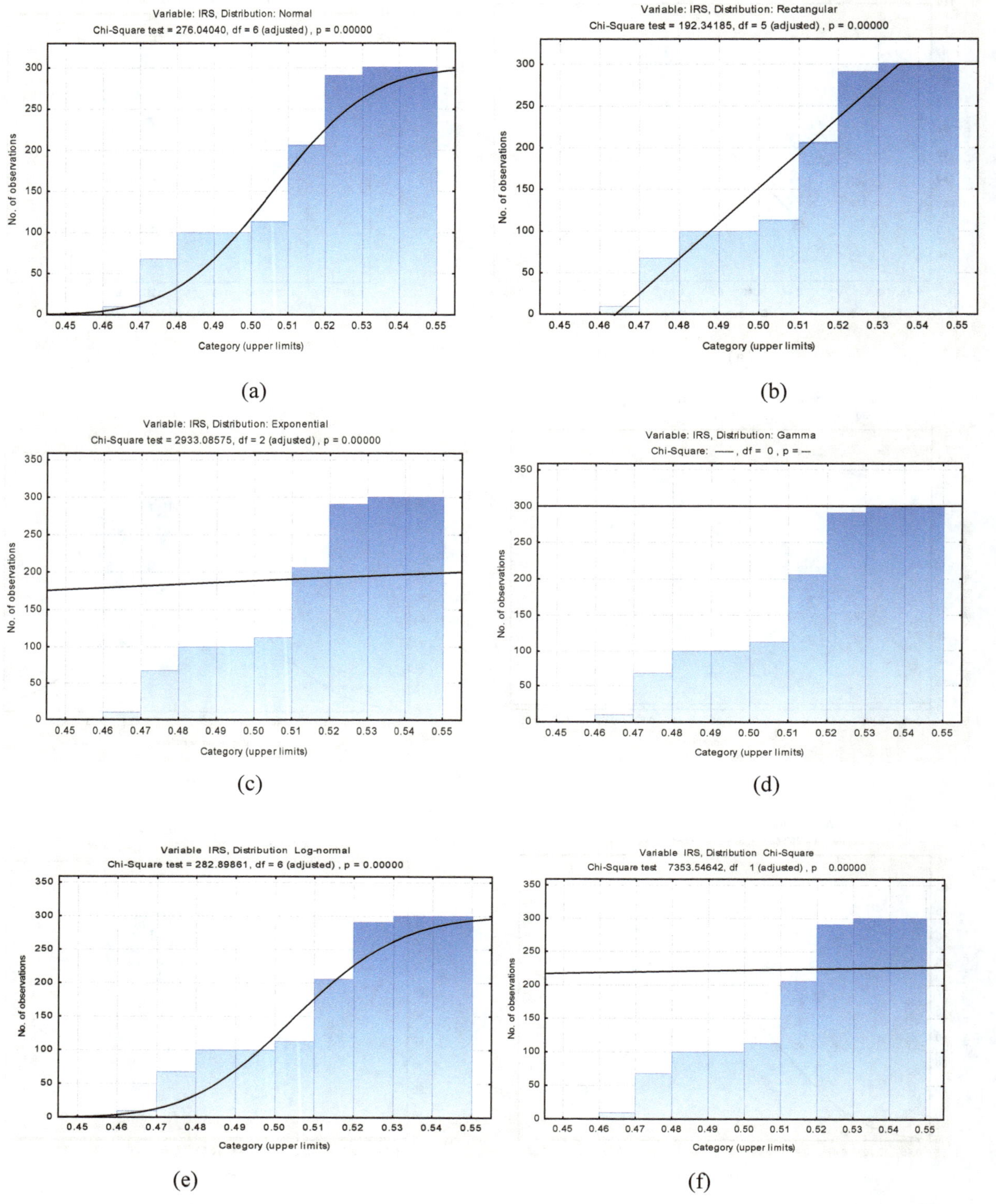

Fig 4 Cummulative Distribution curves for IRS using different distribution techniques (a) Normal, (b) Rectangular, (c) Exponential, (d) Gamma, (e) Log-normal, (f) Chi-square.

4 Conclusion

In this paper we have find the KS-d, KS, AD stat, AD p-value, chi square, chi square p-value and chi square df for different distribution functions for EGFR and IRS. The frequency and cumulative distribution curves for different distribution techniques like exponential, log-normal, normal, gamma, chi-square etc are plotted. More generally, these models are flexible, able to incorporate qualitative and noisy data, and powerful enough to produce quantitative predictions and new biological insights about the operation of signaling networks. In future we will design a CAD system that will help radiologists using wavelet transforms.

References

Arteaga C. 2003. Targeting HER1/EGFR: a molecular approach to cancer therapy. Seminars in Oncology, 30: 314

Brockhaus M, Schoenfeld HJ, Schlaeger EJ, Hunziker W, Lesslauer W, and Loetscher H 1990 Identification of two types of tumor necrosis factor receptors on human cell lines by monoclonal antibodies. Proceedings of the National Academy of Sciences of USA, 87: 3127-3131

Jain S, Naik PK, Sharma R 2009 A computational modeling of cell survival/ death using VHDL and MATLAB simulator. Digest Journal of Nanomaterials and Biostructures, 4(4): 863- 879

Jain S, Bhooshan SV, Naik PK 2010 Model of Mitogen Activated Protein Kinases for Cell Survival/Death and its Equivalent Bio-Circuit. Current Research Journal of Biological Sciences, 2(1): 59-71

Jain S, Bhooshan SV, Naik PK. 2011a Mathematical modeling deciphering balance between cell survival and cell death using insulin. Network Biology, 1(1): 46-58

Jain S, Bhooshan SV, Naik PK. 2011b. Mathematical modeling deciphering balance between cell survival and cell death using Tumor Necrosis Factor α. Research Journal of Pharmaceutical, Biological and Chemical Sciences, 2(3): 574-583

Jain S, 2012 Communication of signals and responses leading to cell survival / cell death using Engineered Regulatory Networks. PhD Thesis, Jaypee University of Information Technology, Solan, Himachal Pradesh, India

Jain S. 2014. Implementation of fuzzy system using operational transconductance amplifier for ERK pathway of EGF/ Insulin leading to cell survival/ death. Journal of Pharmaceutical and Biomedical Sciences, 4(8): 701-707

Jain S, Chauhan DS. 2015a. Mathematical analysis of receptors for survival proteins. International Journal of Pharma and Bio Sciences, 6(3): 164-176

Jain S, Chauhan DS. 2015b Implementation of fuzzy system using different voltages of OTA for JNK pathway leading to cell survival/ death. Network Biology, 5(2): 62-70

Jain S. 2015c. Mathematical Analysis and Probability Density Function of FKHR pathway for Cell Survival /Death. Control System and Power Electronics – CSPE 2015. 84-93, Banglore, India

Jain S, Chauhan DS. 2015d, Linear and Non Linear Modeling of Protein Kinase B/ AkT, International Conference on Information and Communication Technology for Sustainable Development (ICT4SD-2015). 81-88, Ahmedabad, India

Jain S. 2016a, Regression analysis on different mitogenic pathways. Network Biology, 6(2): 40-46

Jain S. 2016b. Mathematical analysis using frequency and cumulative distribution functions for mitogenic pathway. Research Journal of Pharmaceutical, Biological and Chemical Sciences, 7(3): 262-272

Janes KA, John AG, Suzanne G, Peter SK, Douglas LA, Michael YB. 2005 A systems model of signaling identifies a molecular basis set for cytokine-induced apoptosis. Science, 310: 1646-1653

Kyriakis JM., Avruch J. 1996. Sounding the alarm: protein kinase cascades activated by stress and inflammation. Journal of Biological Chemistry, 271: 24313-24316

Libermann TA , Razon TA., Bartal AD, Yarden Y, Schlessinger J, Soreq H. 1984 Expression of epidermal growth factor receptors in human brain tumors. Cancer Research, 44: 753-760

Lizcano J. M. Alessi D. R. 2002. The insulin signalling pathway. Current Biology, 12: 236-238

Normanno N, De Luca A, Bianco C, Strizzi L, Mancino M, Maiello MR,, Carotenuto A, De Feo G, Caponiqro F, Salomon DS. 2006. Epidermal growth factor receptor (EGFR) signaling in cancer. Gene, 366: 2-16

Suzanne G, Janes KA, John AG, Emily PA, Douglas LA, Peter SK. 2005. A compendium of signals and responses trigerred by prodeath and prosurvival cytokines. Manuscript M500158-MCP200. MCP,USA

Thoma B, Grell M, Pfizenmaier K, and Scheurich P 1990, Identification of a 60-kD tumor necrosis factor (TNF) receptor as the major signal transducing component in TNF responses. Journal of Experimental Medicine, 172: 1019-1023

Ullrich A, Schlessinger J. 1990. Signal transduction by receptors with tyrosine kinase activity. Cell, 61: 203-211

Weiss R. 2001. Cellular computation and communications using engineered genetic regulatory networks. PhD Thesis, MIT, USA

White MF. 2003. Insulin signaling in health and disease Science, 302(5651): 1710-1711

18

Immunoregulatory network and cancer-associated genes: molecular links and relevance to aging

Robi Tacutu[1], Arie Budovsky[1,2], Hagai Yanai[1], Marina Wolfson[1], Vadim Fraifeld[1]

[1]The Shraga Segal Department of Microbiology and Immunology, Center for Multidisciplinary Research on Aging, Ben-Gurion University of the Negev, POB 653, Beer Sheva 84105, Israel

[2]The Judea Regional R&D Center, Moshav Carmel, Israel

E-mail: vadim.fraifeld@gmail.com

Abstract

Although different aspects of cancer immunity are a subject of intensive investigation, an integrative view on the possible molecular links between immunoregulators and cancer-associated genes has not yet been fully considered. In an attempt to get more insights on the problem, we analyzed these links from a network perspective. We showed that the immunoregulators could be organized into a miRNA-regulated PPI network—the immunoregulatory network. This network has numerous links with cancer, including (i) cancer-associated immunoregulators, (ii) direct and indirect protein-protein interactions (through the common protein partners), and (iii) common miRNAs. These links may largely determine the interactions between the host's immunity and cancer, supporting the possibility for co-expression and post-transcriptional co-regulation of immunoregulatory and cancer genes. In addition, the connection between immunoregulation and cancer may lie within the realm of cancer-predisposing conditions, such as chronic inflammation and fibroproliferative repair. A gradual, age-related deterioration of the integrity and functionality of the immunoregulatory network could contribute to impaired immunity and generation of cancer-predisposing conditions.

Keywords immunoregulators; cancer genes; protein-protein interactions (PPI); PPI network; micro RNAs (miRNAs); aging.

1 Introduction

The interactions between host's immune system and cancer is a field of intensive investigation (for reviews see: Dunn et al., 2002; Croci et al., 2007; Kim et al., 2007; Gregory and Pound, 2011; Ibrahim et al., 2011; Schreiber et al., 2011; Zhang, 2011a). Recognition and elimination of continuously arising transformed cells is an important host protection process aimed at inhibiting carcinogenesis and eventually maintaining regular cellular homeostasis. On the other hand, cancer cells evolved the mechanisms to escape the control from the immune system. Although it seems paradoxical, the immune and cancer cells often use the same "tools" aimed at ensuring their specific activities. Among these "tools" are immunoregulators which comprise a wide range of molecules including cytokines, chemokines, growth factors, adhesion molecules, Toll receptors, eicosanoids and others (reviewed by Clayton, 2010). As mentioned above, many of them are expressed not only by immune but also by tumor cells, resulting in complex and often opposite effects on host's immunity and tumor progression (Kim et al., 2007; Apte and Voronov, 2008). For example, the pro-inflammatory cytokines of IL-1 family are abundant at tumor sites, being produced by cells of the tumor microenvironment or by the tumor cells (Apte and Voronov, 2008). While IL-1α reduces tumorigenicity by inducing antitumor immunity, another IL-1 member, IL-1β, promotes tumor angiogenesis, growth and invasiveness and also suppresses immune responses in the host (Song et al., 2003; Elkabets et al., 2010).

Here we analyze for the first time the possible molecular links between immunoregulators and cancer-associated genes from a network perspective. Since the age-related changes in the immune system ("immunosenescence") are believed to contribute to increased incidence of cancer in the elderly (Pawelec et al., 2010), we also included this aspect in the analysis.

2 Methods

2.1 Data sources and analytic tools

The list of immunoregulatory genes was data mined from the scientific literature and manually curated. They include the genes whose products were reported to activate or suppress immune functions. Cancer-associated genes were collected from scientific literature and from publicly available databases as described previously (for details see: Budovsky et al., 2009; Tacutu et al., 2010, and references therein). They were selected based on (1) mutations associated with a higher frequency of cancer, (2) consistent up- or down-regulation of gene expression, shown to be important for cancer initiation or progression, and (3) gene polymorphisms associated with greater predisposition or susceptibility to cancer. Longevity-associated genes were taken from the Human Aging Genomic Resources—GenAge Database (de Magalhães et al., 2009; http://genomics.senescence.info/genes) and the genes involved in major age-related diseases (ARDs) and associated conditions (e.g., chronic inflammation) were taken from the NetAge database (Tacutu et al., 2010; http://netage-project.org).

Protein-protein interaction (PPI) data was extracted from the BioGRID database (Stark et al., 2011; http://thebiogrid.org; human interactome release 3.1.71). Micro RNAs (miRNAs) and their experimentally verified targets were retrieved from the TarBase database (Papadopoulos et al., 2009; http://diana.cslab.ece.ntua.gr/tarbase/). Annotations of miRNAs involved in cancer were taken from the Human MicroRNA Disease Database (Lu et al., 2008), and their classification as either oncogenic or tumor suppressor miRNAs was taken from Wang et al. (2010).

DAVID Bioinformatics Resources 6.7 (Dennis et al. 2003; Huang et al. 2009; http://david.abcc.ncifcrf.gov) was used for the GO (Ashburner et al., 2000) and KEGG (Kanehisa et al., 2010) pathway analysis. The miRNA-regulated PPI network was constructed using the YABNA software program (Tacutu et al., 2010) and the graphical output of the network was generated using Cytoscape 2.8.0 (Shannon et al. 2003; http://www.cytoscape.org/).

2.2 Statistical evaluation

Chi-test was used for evaluation of the difference between the observed and expected sizes of gene overlaps. Significantly enriched GO categories and KEGG pathways were computed with DAVID Bioinformatics Resources 6.7, using Bonferroni correction. P-values less than 0.05 were considered statistically significant.

3 Results and Discussion

3.1 Immunoregulatory network

Comprehensive data mining revealed 179 genes that have been identified as being important players in the regulation of immune responses (Online Suppl Table 1). Most of them (149 genes) were found in the human interactome. This is important for the analysis as immunoregulators exert their effects via interactions with each other and with their interacting proteins. Together with their partners and miRNAs (Suppl Table 2), the immunoregulators eventually form a miRNA-regulated PPI network which comprises 650 nodes (Fig. 1).

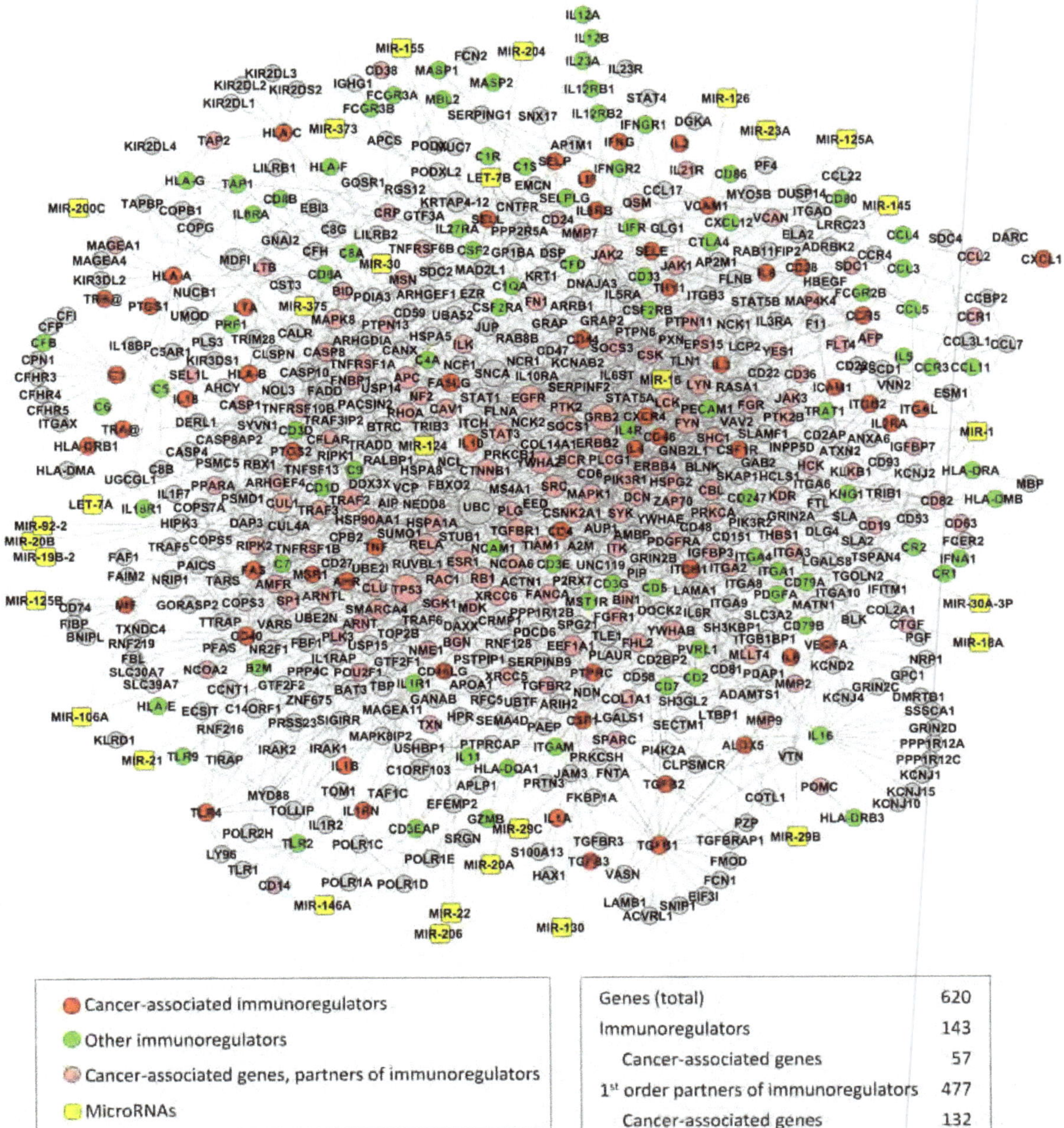

Genes (total)	620
Immunoregulators	143
Cancer-associated genes	57
1st order partners of immunoregulators	477
Cancer-associated genes	132
MicroRNAs	30
MicroRNA targets	50

Fig. 1 The miRNA-regulated PPI immunoregulatory network. The network includes immunoregulators, their first-order partners, and miRNAs with experimentally validated targets. Node size is proportional to the number of PPIs for a given gene. In total, the network comprises 2096 connections.

The very possibility to organize the immunoregulators into an immunoregulatory network indicates that they function in a highly coordinated mode. Another important point is that the immunoregulatory network has a scale-free topology (Fig. 2), which is believed to ensure the network robustness, that is, the ability to resist random node attacks. Despite that, the immunoregulatory network may undergo age-related changes including loss of, or alterations in nodes (e.g, mutations, post-translational modifications) or their interactions and as a result, efficiency of the network decreases. This is in accordance with the numerous observations regarding the impact of aging on the immune system and *vice versa*, which have been reviewed in several papers, where implications for cancer were also addressed (Globerson and Effros, 2000; Kozak and Fraifeld, 2004; Shurin et al., 2007; Agarwal and Busse, 2010; Pawelec et al., 2010).

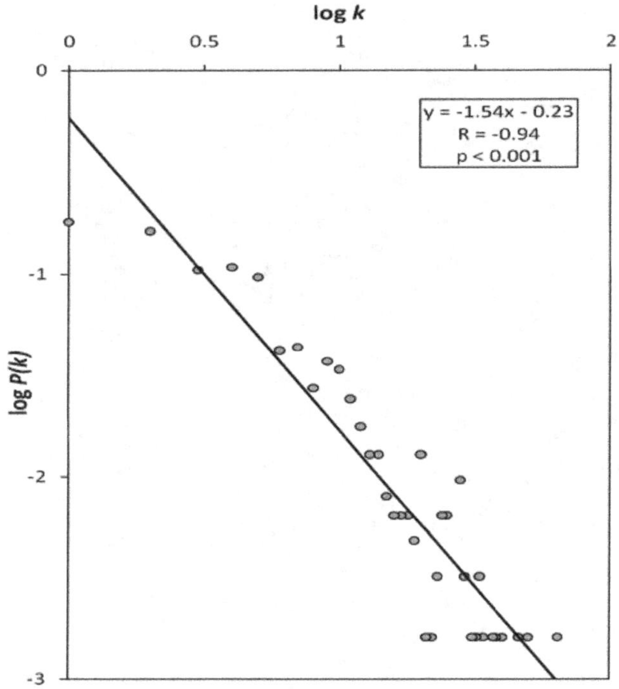

Fig. 2 The log–log plot and dependence of *P(k)* against *k*, illustrating a scale-free topology of the immunoregulatory network, which follows a power-law distribution of connectivity (Barabasi and Oltvai, 2004): $P(k) \sim k^{-\gamma}$, where *P(k)* is the probability that a selected node has exactly *k* connections (degrees) with other nodes; γ is the degree exponent which for most scale-free networks lies within an interval of $2 < \gamma < 3$. The smaller the γ value, the higher contribution of the hubs to the network connectivity.

3.2 Links between immunoregulatory network and cancer-associated genes

As shown in Fig. 3, at least 61 immunoregulators were also identified as cancer-associated genes (cancer-associated immunoregulators). Of them, 57 are found in the immunoregulatory network. The existence of many common genes is the first and most obvious evidence for the connections between the immunoregulatory and cancer-associated genes, suggesting that the same stimuli may induce similar gene responses in the immune and cancer cells. The induction or suppression of a given gene could, however, result in different or even opposite cellular responses. For example, IL-1-induced upregulation of cyclooxygenase 2 (COX-2), with subsequent elevation of PGE_2, acts as a feedback inhibitor of cellular immune responses (Goodwin and Ceuppens, 1983), whereas in tumors, PGE_2 elicits multiple oncogenic signals to promote carcinogenesis (Wang and Dubois, 2010; Wu et al., 2010). Of note, the mitogen-stimulated PGE_2 production is significantly increased in old age, being in part responsible for a reduced proliferative capacity of the immune cells (Fraifeld et al., 1995).

Further strengthening the links between immunoregulators and cancer is that a high portion of immunoregulatory proteins (n = 53) directly interact with proteins involved in cancer (n = 132) and much more may interact via their common protein partners (n = 313) (Fig. 3).

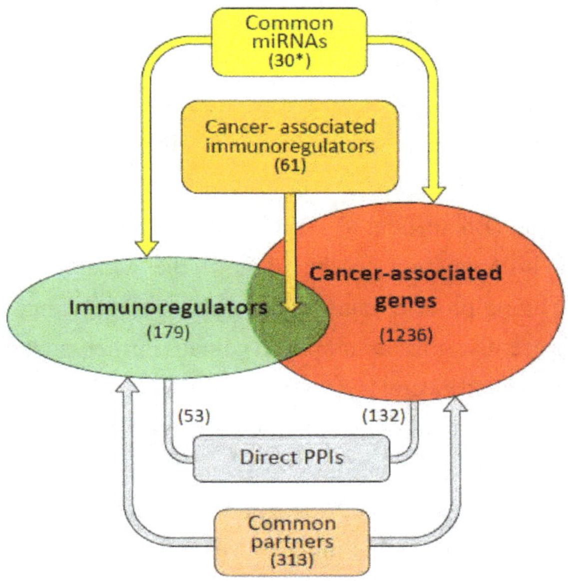

Fig. 3 Schematic representation of molecular links between immunoregulators and cancer-associated genes. *Of the 30 miRNAs which have targets in the immunoregulatory network, 27 miRNAs have also targets among cancer-associated genes (9 miRNAs target immunoregulators and 18 miRNAs target their partners).

In total, almost one third of the genes (n = 189) in the immunoregulatory network overlap with cancer-associated genes (Fig. 1). This value is twofold higher than expected by chance ($p < E{-}25$).

Many genes from the immunoregulatory network are validated targets of miRNAs (Fig. 1, Online Suppl Table 2). Almost all of these miRNAs may have a pleiotropic effect by targeting more than one gene, including cancer-associated genes (n = 99). As such, the common miRNAs could be another line of possible co-regulation of immune responses, cancer development and interaction between them. Important to stress that among the targets of the above miRNAs are also longevity-associated genes (Wolfson et al., 2008; Tacutu et al., 2010). Moreover, the vast majority of the miRNA targets, in addition to cancer, have been reported to be involved in at least one major ARD or aging-associated condition, such as chronic inflammation (Online Suppl Table 3). Thus, the common miRNAs could actually represent one of the bridges connecting immunoregulation, cancer, and aging/longevity.

3.3 GO function and KEGG pathway analysis

To get further insight into the links between the immunoregulatory network and cancer, we undertook a functional analysis of the common genes, that is, cancer-associated genes found in the immunoregulatory network (n=189 genes). The analysis showed more than 350 enriched GO – biological process categories ($p < 0.05$, with Bonferroni correction; Online Suppl Table 4). The most enriched categories were related to regulation of cell proliferation, cell communication, response to wounding, programmed cell death, inflammatory response and as expected, to the various aspects of immune responses.

Enrichment with genes related to regulation of cell proliferation is quite obvious as this process is critical for both immune and cancer cells. Indeed, proliferation of the immune cells ensures a massive attack against cancer cells, while continuous proliferation of cancer cells ensures tumor growth. An immune (inflammatory) component is also essential for the initial phases of wound healing (Gurtner et al., 2008). On the other hand, this fundamental biological process and cancer have much in common and Schäfer and Werner (2008) even considered "cancer as an overhealing wound". With regard to this, it should be noted that the frequency of fibroproliferative conditions dramatically increases with advanced age and this occurs in parallel with an increased incidence of cancer (Cutler and Mattson, 2006). Another important point is a significant overrepresentation of chronic inflammation genes in the immunoregulatory network. The overlap comprises 64 genes, six times higher than expected by chance ($p < E-25$). Of these 64 genes, 47 are also cancer-associated. This coincides well with the notion that chronic inflammation is strongly associated with promotion of malignancy and tumor progression (Elkabets et al., 2010). Moreover, strong links between cancer-associated immunoregulators and chronic inflammation are in favor of the widely accepted assumption that chronic inflammation may underpin the aging process, cancer and other ARDs (Chung et al., 2009; Fulop et al., 2010; Rodier and Campisi, 2011). Recently, this assumption gained further support from the discovery of the phenomenon of Senescence-Associated Secretory Phenotype (SASP) (Coppé et al., 2008; Campisi, 2011). It was shown that senescent cells secrete a wide plethora of pro-inflammatory and pro-cancer molecules (Davalos et al., 2010; Freund et al., 2010). These cells accumulate in aging and are found in tumor microenvironment (fibroblasts, endothelial cells). Nineteen SASP genes are also immunoregulators and almost 40% ($n = 28$ genes) of all currently reported SASP molecules are found in the immunoregulatory network.

The KEGG pathway analysis revealed 42 enriched pathways ($p < 0.05$, with Bonferroni correction), including also several pathways related to various cancers (Online Suppl Table 5). Remarkably, among the topmost enriched pathways are focal adhesion and adherens junction involved in cell-cell and cell-extracellular matrix interactions. As we previously showed, these pathways are considerably overrepresented in major human ARDs including cancer (Wolfson et al., 2009). Thus, any dysfunction in these pathways may have far-reaching consequences not only for immunity and cancer but also for the aging process and associated pathology.

4 Concluding Remarks

It has become increasingly clear that biological systems function as complex networks (Barabasi and Oltvai, 2004; Budovsky et al., 2007; Tacutu et al., 2010; Ibrahim et al., 2011; Zhang, 2011a,b and references therein). The holistic, network-based approach opens new possibilities for understanding the functional organization of processes which include a great number of genes. Here we demonstrate that the immunoregulators, an important part of the immune system, can be organized into a miRNA-regulated PPI network—the immunoregulatory network, highlighting their cooperative mode of action. This network has numerous links with cancer, including (i) cancer-associated immunoregulators, (ii) direct and indirect PPIs (through the common protein partners), and (iii) common miRNAs. These links may largely determine the interactions between the host's immunity and cancer, supporting the possibility for co-expression and post-transcriptional co-regulation of immunoregulatory and cancer genes.

Another line of connections between immunoregulation and cancer lies within the realm of cancer-predisposing conditions, such as chronic inflammation and fibroproliferative repair, in which immune responses play an essential role.

Today it is widely accepted that aging is the major risk factor for the development of ARDs including cancer (Cutler and Mattson, 2006). Even more convincing is that exceptional longevity or longevity-promoting interventions usually lead to the delay of or escape from cancer and also to better preservation of the immune functions (Spaulding et al., 1997; Dixit, 2008; Blagosklonny, 2008; Budovsky et al., 2009; Agarwal and Busse, 2010; Derhovanessian et al., 2010). Age-related changes in the integrity and functionality of the immunoregulatory network could contribute to impaired immunity and generation of ARD-predisposing conditions. The proper maintenance of the network could therefore be the main strategic goal for tackling immunoregulatory disorders and age-associated pathology.

Acknowledgements

This work was funded by the European Union FP7 Health Research Grant number HEALTH-F4-2008-202047 (to V.E.F.) and the Israel Ministry of Science and Technology (to A.B.).

References

Agarwal S, Busse PJ. 2010. Innate and adaptive immunosenescence. Ann Allergy Asthma Immunol, 104(3):183-90; quiz 190-2, 210

Ashburner M, Ball CA, Blake JA, et al. 2000. Gene ontology: tool for the unification of biology. The Gene Ontology Consortium. Nature Genetics, 25(1): 25-29

Apte RN, Voronov E. 2008. Is interleukin-1 a good or bad 'guy' in tumor immunobiology and immunotherapy? Immunological Review, 222: 222-241

Barabasi A, Oltvai ZN. 2004. Network biology: understanding the cell's functional organization. Nature Reviews Genetics, 5(2): 101-113

Blagosklonny MV. 2008. Prevention of cancer by inhibiting aging. Cancer Biology and Therapy, 7(10): 1520-1524

Breitkreutz BJ, Stark C, Reguly T, et al. 2008. The BioGRID Interaction Database: 2008 update. Nucleic Acids Research, 36(Database issue): D637-D640

Budovsky A, Muradian KK, Fraifeld VE. 2006. From disease-oriented to aging/longevity-oriented studies. Rejuvenation Res, 9(2): 207-210

Budovsky A, Abramovich A, Cohen R, et al. 2007. Longevity network: construction and implications. Mechanisms of Ageing and Development, 128(1): 117-124

Budovsky A, Tacutu R, Yanai H, et al. 2009. Common gene signature of cancer and longevity. Mechanisms of Ageing and Development, 130(1-2): 33-39

Campisi J. 2011. Cellular senescence: putting the paradoxes in perspective. Current Opinion in Genetics and Development, 21(1): 107-112

Chung HY, Cesari M, Anton S, et al. 2009. Molecular inflammation: underpinnings of aging and age-related diseases. Ageing Research Reviews, 8(1): 18-30

Clayton T. 2010. Immunoregulators. Encyclopedia of Industrial Biotechnology: Bioprocess, Bioseparation, and Cell Technology. 1-25 (DOI 10.1002/9780470054581.eib609)

Coppé JP, Patil CK, Rodier F, et al. 2008. Senescence-associated secretory phenotypes reveal cell-nonautonomous functions of oncogenic RAS and the p53 tumor suppressor. PLoS Biol, 6(12): 2853-2868

Croci DO, Zacarías Fluck MF, Rico MJ, et al. 2007. Dynamic cross-talk between tumor and immune cells in orchestrating the immunosuppressive network at the tumor microenvironment. Cancer Immunol Immunother, 56(11): 1687-1700

Cutler RG, Mattson MP. 2006. The adversities of aging. Ageing Research Reviews, 5(3): 221-238

Davalos AR, Coppe JP, Campisi J, et al. 2010. Senescent cells as a source of inflammatory factors for tumor progression. Cancer and Metastasis Reviews, 29(2): 273-283 de Magalhães JP, Budovsky A, Lehmann G, et al. 2009. The Human Ageing Genomic Resources: online databases and tools for biogerontologists. Aging Cell, 8(1): 65-72

Dennis G Jr, Sherman BT, Hosack DA, et al. 2003. DAVID: Database for Annotation, Visualization, and Integrated Discovery. Genome Biology, 4(5): P3

Derhovanessian E, Maier AB, Beck R, et al. 2010. Hallmark features of immunosenescence are absent in familial longevity. Journal of Immunology, 185(8): 4618-4624

Dixit VD. 2008. Adipose-immune interactions during obesity and caloric restriction: reciprocal mechanisms regulating immunity and health span. Journal Leukocyte Biology, 84(4): 882-892

Dunn GP, Bruce AT, Ikeda H, et al. 2002. Cancer immunoediting: from immunosurveillance to tumor escape. Nature Immunology, 3(11): 991-998

Elkabets M, Ribeiro VS, Dinarello CA, et al. 2010. IL-1β regulates a novel myeloid-derived suppressor cell subset that impairs NK cell development and function. European Journal of Immunology, 40(12): 3347-3357

Fraifeld V, Kaplanski J, Kukulansky T, et al. 1995. Increased prostaglandin E_2 production by concanavalin A-stimulated splenocytes of old mice. Gerontology, 41(3): 129-133

Freund A, Orjalo AV, Desprez PY, et al. 2010. Inflammatory networks during cellular senescence: causes and consequences. Trends in Molecular Medicine, 16(5): 238-246

Fulop T, Larbi A, Witkowski JM, et al. 2010. Aging, frailty and age-related diseases. Biogerontology, 11(5): 547-563

Globerson A, Effros RB. 2000. Ageing of lymphocytes and lymphocytes in the aged. Immunology Today, 21(10): 515-521

Goodwin JS, Ceuppens J. 1983. Regulation of the immune response by prostaglandins. Journal of Clinical Immunology, 3(4): 295-315

Gregory CD, Pound JD. 2011. Cell death in the neighbourhood: direct microenvironmental effects of apoptosis in normal and neoplastic tissues. Journal of Pathology, 223(2): 177-194

Gurtner GC, Werner S, Barrandon Y, et al. 2008. Wound repair and regeneration. Nature, 453(7193): 314-321

Huang DW, Sherman BT, Lempicki RA. 2009. Systematic and integrative analysis of large gene lists using DAVID bioinformatics resources. Nature Protocols, 4(1): 44-57

Ibrahim SS, Eldeeb MAR, Rady MAH, et al. 2011. The role of protein interaction domains in the human cancer network. Network Biology, 1(1): 59-71

Kanehisa M, Goto S, Furumichi M, et al. 2010. KEGG for representation and analysis of molecular networks involving diseases and drugs. Nucleic Acids Research, 38(Database issue): D355-D360

Kim R, Emi M, Tanabe K. 2007. Cancer immunoediting from immune surveillance to immune escape. Immunology, 121(1): 1-14

Kozak W, Fraifeld V. 2004. Non-prostaglandin eicosanoids in fever and anapyrexia. Frontiers in Bioscience, 9: 3339-3355

Lu M, Zhang Q, Deng M, et al. 2008. An analysis of human microRNA and disease associations. PLoS One, 3(10): e3420

Papadopoulos GL, Reczko M, Simossis VA, et al. 2009. The database of experimentally supported targets: a functional update of TarBase. Nucleic Acids Research, 37(Database issue): D155-D158

Pawelec G, Derhovanessian E, Larbi A. 2010. Immunosenescence and cancer. Critical Reviews in Oncology and Hematology, 75(2): 165-172

Rodier F, Campisi J. 2011. Four faces of cellular senescence. Journal of Cell Biology, 192(4): 547-556

Schäfer M, Werner S. 2008. Cancer as an overhealing wound: an old hypothesis revisited. Nature Reviews Molecular Cell Biology, 9(8): 628-638

Schreiber RD, Old LJ, Smyth MJ. 2011. Cancer immunoediting: integrating immunity's roles in cancer suppression and promotion. Science, 331(6024): 1565-1570

Shannon P, Markiel A, Ozier O, et al. 2003. Cytoscape: a software environment for integrated models of biomolecular interaction networks. Genome Research, 13(11): 2498-2504

Shurin MR, Shurin GV, Chatta GS. 2007. Aging and the dendritic cell system: implications for cancer. Critical Reviews in Oncology and Hematology, 64(2): 90-105

Spaulding CC, Walford RL, Effros RB. 1997. Calorie restriction inhibits the age-related dysregulation of the cytokines TNF-alpha and IL-6 in C3B10RF1 mice. Mechanisms of Ageing and Development, 93(1-3): 87-94

Song X, Voronov E, Dvorkin T, et al. 2003. Differential effects of IL-1 alpha and IL-1 beta on tumorigenicity patterns and invasiveness. Journal of Immunology, 171(12): 6448-6456

Stark C, Breitkreutz BJ, Chatr-Aryamontri A, et al. 2011. The BioGRID Interaction Database: 2011 update. Nucleic Acids Research, 39(Database Issue): D698-D704

Tacutu R, Budovsky A, Fraifeld VE. 2010. The NetAge database: a compendium of networks for longevity, age-related diseases and associated processes. Biogerontology, 11(4): 513-522

Wang D, Dubois RN. 2010. Eicosanoids and cancer. Nature Reviews Cancer, 10(3): 181-193

Wang D, Qiu C, Zhang H, et al. 2010. Human microRNA oncogenes and tumor suppressors show significantly different biological patterns: from functions to targets. PLoS One, 5(9) pii: e13067

Wolfson M, Tacutu R, Budovsky A, et al. 2008. MicroRNAs: relevance to aging and age-related diseases. Open Longevity Science, 2: 66-75

Wolfson M, Budovsky A, Tacutu R, et al. 2009. The signaling hubs at the crossroad of longevity and age-related disease networks. International Journal of Biochemistry and Cell Biology, 41(3): 516-520

Zhang WJ. 2011a. Network Biology: an exciting frontier science. Network Biology, 1(1): 79-80

Zhang WJ. 2011b. Constructing ecological interaction networks by correlation analysis: hints from community sampling. Network Biology, 1(2): 81-98

A Java algorithm for non-parametric statistic comparison of network structure

WenJun Zhang

School of Life Sciences, Sun Yat-sen University, Guangzhou 510275, China; International Academy of Ecology and Environmental Sciences, Hong Kong

E-mail: zhwj@mail.sysu.edu.cn, wjzhang@iaees.org

Abstract

A Java algorithm to statistically compare between-network structure difference was developed. In this algorithm, Euclidean distance, Manhattan distance, Chebyshov distance, and Pearson correlation were available to measure between-network difference. The algorithm was tested and applied for its effectiveness with some arthropod and weed networks.

Keywords network; structure; non-parametric statistic test; comparison; Java algorithm.

1 Introduction

The structure of network refers to the node degree, network connectance, aggregation strength (Dormann, 2011; Zhang, 2011; Zhang and Zhan, 2011), etc. We occasionally need to compare the structure difference between networks. Non-parametric statistics may be used in the difference comparison (Solow, 1993; Manly, 1997; Zhang, 2007).

In this article a Java algorithm, based on previous studies, was presented to statistically compare between-network structure difference.

2 Algorithm

The algorithm is used to compare the difference in structure composition between two networks.

Suppose that a_{ij} is the mass (or degree, etc.) of node j in network i, i=1,2,...,n; j=1,2,...,s. First, define between-network distance measures, i.e., Euclidean distances, Manhattan distances, Chebyshov distance, Pearson correlation (based distance), are as follows:

$$d_{ij}=(\sum^{s}_{k=1}(a_{ik} - a_{jk})^2/s)^{0.5}$$
$$d_{ij}=\sum^{s}_{k=1}|a_{ik} - a_{jk}|/s$$
$$d_{ij}= max_{k} |a_{ik} - a_{jk}|$$
$$d_{ij}=1-\sum^{s}_{k=1}((a_{ik} -a_{ibar})(a_{jk}-a_{jbar}))/(\sum^{s}_{k=1} (a_{ik}-a_{ibar})^2 \sum^{s}_{k=1} (a_{jk}-a_{jbar})^2)^{0.5}$$

where a_{ibar} and a_{jbar} are means of a_{ik} 's and a_{jk} 's.

If $min\ a_{ij}$<0, then let $a_{ij}=a_{ij}-min\ a_{ij}$, i=1,2,...,n; j=1,2,…,s. Suppose z_{ij} is the decimal numbers of a_{ij} if network data contain the decimal value a_{ij}, and calculate $c_{ij}=10^{Z_{ij}}$. Let $a_{ij}=a_{ij}\ max\ c_{kl}$, i=1,2,...,n; j=1,2,…,s. Through these transformations all of the values in network data become integers which are equivalent to numbers of individuals. If no difference exists, then the distribution of individuals in networks i and j will be a result of allocating the mixed network values at random into two networks of size equal to those of the original network (Solow, 1993; Manly, 1997; Zhang, 2007). Assume that the two networks to be tested are i and j, which contain $\sum_{k=1}^{s} a_{ik}$ and $\sum_{k=1}^{s} a_{jk}$ individuals respectively. The $\sum_{k=1}^{s} a_{ik}+\sum_{k=1}^{s} a_{jk}$ individuals of the combined network are randomly reallocated into two randomized networks with $\sum_{k=1}^{s} a_{ik}$ and $\sum_{k=1}^{s} a_{jk}$ labeled individuals. Calculate the expected absolute distance between the two randomized networks and compare whether it is not less than the absolute distance between the true networks i and j. Repeat the simulation many times, calculate the number of the expected are not less than the absolute distance between i and j, and take the percentage as the p value. The p value is used to make statistical test. The threshold p value for test may be defined as 0.05, 0.01, etc. If the calculated p value is less than p threshold, then the structure composition of networks i and j are statistically different.

The algorithm, NetStructComp, is implemented as a Java program based on JDK 1.1.8, in which several classes and an HTML file is included (http://www.iaees.org/publications/software/index.asp). In network data file, the first row is ID numbers of nodes and the first column is ID numbers of networks.

3 Application

I chose the weed data of rice fields in four cities (networks) of Pearl River Delta, China. In total 25 plant families (nodes) were found (Wei, 2010), as indicated in Table 1.

Table 1 Abundance of plants around rice fields in four cities of China

Plant Family	Zhongshan	Zhuhai	Dongguan	Guangzhou
Gramineae	1056.9	184.6	439.3	193.6
Compositae	11.1	95	43.3	63.4
Amaranthaceae	31.1	56	93.4	49
Commelinaceae	0	52.2	14.4	1.4
Onagraceae	0	0.1	2.3	1
Urticaceae	0.3	0	11.9	3.4
Menispermaceae	0	0	0	0.1
Cyperaceae	0	0	0	26.1
Caryophyllaceae	0	0	5.3	6.7
Polygonaceae	0.4	4.2	6.9	3.8
Acanthaceae	0	0	0	0.3
Solanaceae	0.1	0	0.2	0.4
Umbelliferae	0	34.9	0	4.8
Lythraceae	0	0	0	1.6
Scrophulariaceae	0.7	3.9	1.7	2.4
Oxalidaceae	0	0	0	0.2
Chenopodiaceae	1.1	0.4	0.3	0.1
Haloragaceae	0	0	0	4.6
Campanulaceae	0	0	0	0.7
Plantaginaceae	0	0.3	0	0
Rubiaceae	0	0.1	0	0
Euphorbiaceae	0	0	0.1	0
Convolvulaceae	0	0.1	0	0
Pontederiaceae	0	0.1	0	0
Portulacaceae	24.6	0	8	0

Choose Euclidean distance measure, significance level p=0.01, and 1000 randomizations, the results are as follows:

Network pairs with significant statistic difference in structure (with p values):

(1,2)(0.0) (1,3)(0.0) (1,4)(0.0)

(2,3)(0.0) (2,4)(0.0)

(3,4)(0.0)

It is obvious that all network pairs have significant statistic difference.

Another data set is the arthropod data of nine rice fields of Pearl River Delta, China. In total 5 arthropod groups (nodes) were found (Wei, 2010), as indicated in Table 2.

Table 2 Arthropod abundance in nine rice fields

	Herbivorous Insects	Neutral Insects	Predatory Insects	Parasitic Insects	Spiders
1	42.0	0.0	5.3	3.0	8.3
2	66.4	0.0	7.9	4.3	5.7
3	298.8	0.0	10.5	3.2	10.8
4	58.1	0.0	8.9	3.1	6.9
5	50.2	0.0	6.5	3.5	4.5
6	90.6	0.0	19.6	5.6	8.0
7	53.0	0.0	10.0	3.0	7.0
8	36.1	0.1	6.9	3.2	8.2
9	40.3	0.0	8.4	2.4	11.9

Choose Euclidean distance measure, significance level p=0.01, and 1000 randomizations, the results are as follows:

(1,2)(0.0) (1,3)(0.0) (1,5)(0.0010)

(2,3)(0.0) (2,7)(0.0020) (2,8)(0.0) (2,9)(0.0)

(3,4)(0.0) (3,5)(0.0) (3,6)(0.0) (3,7)(0.0) (3,8)(0.0) (3,9)(0.0)

(4,8)(0.0) (4,9)(0.0)

(5,8)(0.0) (5,9)(0.0)

(6,8)(0.0) (6,9)(0.0)

(7,8)(0.0020) (7,9)(0.0)

The results demonstrate that, for example, rice fields 4, 5, 6 and 7 are significantly different from rice fields 8 and 9.

References

Dormann CF. 2011. How to be a specialist? Quantifying specialisation in pollination networks. Network Biology, 1(1): 1-20

Manly BFJ. 1997. Randomization, Bootstrap and Monte Carlo Methods in Biology (2nd Edition). Chapman &

Hall, London, UK

Solow AR. 1993. A simple test for change in community structure. Journal of Animal Ecology, 62: 191-193

Wei W. 2010. Biodiversity Analysis on Arthropod and Weed Communities in Paddy Rice Fields of Pearl River Delta. Master Degree Dissertation. Sun Yat-sen University, China

Zhang WJ. 2007. Computer inference of network of ecological interactions from sampling data. Environmental Monitoring and Assessment, 124: 253-261

Zhang WJ. 2011. Constructing ecological interaction networks by correlation analysis: hints from community sampling. Network Biology, 1(2): 81-98

Zhang WJ, Zhan CY. 2011. An algorithm for calculation of degree distribution and detection of network type: with application in food webs. Network Biology, 1(3-4) (Accepted)

20

Mathematical modeling deciphering balance between cell survival and cell death using insulin

Shruti Jain[1], Sunil V. Bhooshan[1], Pradeep K. Naik[2]*

[1]Department of Electronics and Communication Engineering

[2]Department of Bioinformatics and Biotechnology, Jaypee University of Information Technology, Waknaghat, Solan 173215, Himachal Pradesh, India

*Address for correspondence: Department of Bioinformatics and Biotechnology, Jaypee University of Information Technology, Waknaghat, Distt.- Solan, Himachal Pradesh, India (Pin-173215)

E-mail: jain.shruti15@gmail.com

Abstract

When modelling cell signalling networks, a balance must be struck between mechanistic detail and ease of interpretation. In this paper we apply a deterministic numerical method to the analysis of a large, systematic dataset describing the dynamics of cell signalling downstream of, tumor necrosis factor-α (TNF), epidermal growth factor (EGF), and insulin receptors in human colon carcinoma cells. Deterministic modeling is useful as a means to assemble and test what we know about proteins and networks. We extensively study the space of parameters to show that the model is structurally stable and robust over a broad range of parameter values. We have made the biological view of the main paths of the insulin input showing cell survival and cell death. Than with the help of different parameters relating to that protein present in the model we have designed scheme of the biochemical paths/deterministic model. With those parameters equations were formed which vary with time i.e. differential equation. Thus, our model is suitable for implementation in multi-scale simulation programs that are presently under development to study the behavior of large tumor cell populations.

Keywords tumor necrosis factor-α; epidermal growth factor; insulin; signaling; biological view.

1 Introduction

Large datasets for biological systems are becoming increasingly available due to ongoing improvements in high-throughput measurement methods. However, similarly improved computational methods are needed if we are to gain useful insights from these datasets. Among prominent examples of this current challenge in biology is in understanding how extra cellular cues influence highly interconnected and complex cell signalling pathways to yield cell behavioural responses (Kitano, 2002). In the case of programmed cell death, cytokines such as tumor necrosis factor-alpha (TNF-α) (Wajant et al., 2003) function as a pro-death cue, whereas growth factor such as epidermal growth factor (EGF) and insulin (Avruch, 1998) exert pro-survival effects. The magnitudes of the responses vary with cell type, but the pathways downstream of cytokine receptors are conserved and highly interconnected. It appears that the determination of whether a cell will live or die involves a balance between pro-death and pro-survival signals (Xia et al., 1995). Thus, the intracellular signal transduction network stimulated by TNF, insulin, and other cytokines acts as a signal processor that converts opposing cues into a functional response that controls cell fate.

Although many of the components of cellular signalling systems have now been identified, it is not yet known how binary decisions, such as death–survival, are made. Moreover, the complexity of cell signalling networks precludes a simple protein-by-protein assignment of function. Increasingly, systematic methods are applied to the interpretation and computational analysis of cell signalling (Brent, 2000; Kitano, 2002). These computational methods lie on a spectrum of approaches that vary in their level of abstraction and specificity (Ideker et al., 2003). The most abstract methods, such as multivariate analysis and clustering (Rives et al., 2003), are powerful because they can handle empirical data and prior knowledge about the underlying phenomena in a flexible fashion. At the other hand, differential equation-based models (Schoeberl et al., 2002) are useful for coding existing prior knowledge in pursuit of *in silico* predictions.

Current high-throughput experimental protocols allow us to measure an increasingly large number of variables describing cells or cellular populations. For example, large scale phosphorylation screens (Lund-Johansen et al., 2000), protein–protein binding screens (Cagney et al., 2000) and migration assays (Maliakal, 2002) have demonstrated that it is possible to simultaneously measure the activity of tens to thousands of variables in a biological system. Future advances in micro fluidics (Burns et al., 1998) and high throughput mass spectrometry (Morand et al., 2001), for example, promise even more measurements will be possible at a fraction of their current cost. Analysis of these data has generally focused on identifying a small number of pathway components involved in governing a particular cell behaviour. Growing interest is focusing on how to use these multivariate data to determine integrated pathways in, for instance, signal transduction (Gardner et al., 2003) and metabolic pathways (Price et al., 2003).

In this paper, our purpose is to determine whether numerical model such as deterministic model can be used to uncover important aspects of cellular signals deciphering the fate of cell. Specifically, we examine the TNF-, EGF- and insulin-mediated death-versus-survival response by deterministic approach. We wish to attract the interest of other experimental molecular cell biologists, so we emphasize the conceptual details of deterministic method as a practical modelling technique. A noteworthy feature of deterministic model applied to biological data, is that biological measurements reflect tangible molecular entities with known, mechanistic roles in intracellular processes. Thus, these data-driven models of signalling are empirical, but not phenomenological, and suggest mechanistic dependencies. We show here that numerical and related methods can uncover key contributors to death–survival decisions. These contributions always involve multiple proteins working in concert, but the informative proteins consist of only a fraction of the original protein dataset. Thus, our results suggest that within the entire cellular signalling system lies a reduced set of information-rich protein measurements that together constitute an efficient model of the signalling network state and the relevant signal–response relationships.

2 Theoretical Model

2.1 Binding and internalization of insulin/IRS1 complexes

Current biochemical data show that the IRS receptors rapidly self-trimerized at the cell membrane and interact with insulin homo trimers. Thus, the mechanism of EGF binding to IRS can in principle be viewed as the result of the monomeric interactions between one molecule of insulin and one molecule of receptor. This simplification is further supported by the following considerations:

(1) The mechanism of receptor self-trimerization followed by ligand binding can be modeled by a set of 4

differential equations with 6 parameters. The model can fit experimental data of insulin binding, but one finds that the kinetics of receptor trimerization are much faster than the binding kinetics, and thus the trimerized receptor behaves as an effective monomer. In addition, experimental determinations of the parameter values for intermediate binding reactions are not available;

(2) The model by Bajzer et al. (1989) fitted experimental data well; we use it to describe the early events of TNF interactions with cells. However the original model of Bajzer et al. must be updated to account for some novel aspects of TNF biology. In particular, Bajzer et al. (1989) assumed that internalized ligand/receptor complexes could be recycled back at the cell surface. It is highly probable that TNF/TNF-R1 complexes do not recycle at all but are finally degraded into lysosomes. Therefore we modify the model by Bajzer et al. (1989) and Chignola et al. (2009) model as follows:

$$\begin{aligned}
\frac{d[R]}{dt} &= V_r - k_d[R] - k_{on}[L][R] + k_{off}[N_c] \\
\frac{d[L]}{dt} &= -k_{on}[L][R] + k_{off}[N_c] \\
\frac{d[N_c]}{dt} &= k_{on}[L][R] - (k_{off} + k_{in})[N_c] \\
\frac{d[N_{in}]}{dt} &= k_{in}[N_c] - k_{\deg}[N_c]
\end{aligned} \tag{1}$$

where square brackets denote molar concentrations of free IRS receptors (R), free insulin (L), insulin/IRS complexes bound at the cell membrane (N_C) and internalized complexes (N_{in}). Here k_{on} and k_{off} are the association and dissociation rate constants for insulin binding to IRS, respectively, k_{in} is the internalization rate constant of insulin/IRS complexes and k_{deg} is the rate constant of lysosomal degradation of the complexes.
The two parameters V_r and k_d where introduced by Bajzer et al. although with a slightly different notation, to describe "the zero-order rate of insertion of receptors into the membrane and the turnover (internalization) rate constant of ligand-free receptors (Bajzer et al., 1989; Grell, 1998; Vuk-Pavlovic et al., 1989) respectively.

2.2 Modeling the intracellular signaling pathways triggered by insulin

Cell signaling induced by insulin leading to cell survival and cell death has been represented in Fig. 1*A*. The approximation of this model used for mathematical implementation is shown in Fig.1*B*. We define seven molecular species, *H*, *G*, *F*, *E*, *C*, *B* and *A*, that collectively summarize the various reactions leading to cell survival and death. The molecules *H*, *G*, *F*, *E*, *C*, *B* and *A* can be loosely identified with mTOR, JAK/STAT, FKHR, MEK/ERK, JNK/FLIP, NF-κB/FLIP and caspase-3 respectively. We assuming that after the initial triggering of the pathway, it precedes irreversibly to its endpoint the equations for *A, B, C, E, F, G* and *H* can be written as:

$$\frac{d[A]}{dt}=\alpha[N_{in}]-\gamma[B][A]-\varphi[C][A]-\varepsilon[E][A]+\rho[F][A]-\psi[G][A]-\xi[H][A]-k_{A\deg}$$
$$\frac{d[B]}{dt}=\beta[N_c]-k_{B\deg}[B]$$
$$\frac{d[C]}{dt}=\delta[N_c]-k_{C\deg}[C]$$
$$\frac{d[E]}{dt}=\eta[N_c]-k_{E\deg}[E]$$
$$\frac{d[F]}{dt}=\mu[N_c]-k_{F\deg}[F]\ (Death)$$
$$\frac{d[G]}{dt}=\chi[N_c]-k_{G\deg}[G]$$
$$\frac{d[H]}{dt}=\lambda[N_c]-k_{H\deg}[H] \qquad (2)$$

where the variables N_C and N_{in} are the same as in the differential system (Eq.(1)). We see from the differential system (Eq.(2)), that the cell survival signal, modeled phenomenologically by means of the chemical species *B*, *C*, *E* and *H* depends on the number of insulin/IRS complexes at the cell surface (e.g. N_C), with rate constant parameter β, δ, η, χ and λ. On the other hand, the apoptotic signal, modeled phenomenologically by means of the chemical species *F* that denotes FKHR pathway, with a rate constant μ. While *A*, depends on the number of internalized ligand/receptor complexes (e.g. N_{in}) with rate a constant α. Induction of cell survival pathway inhibits the apoptotic pathway by destroying *A* with rates $\gamma[B]$, $\varphi[C]$, $\varepsilon[E]$, $\psi[G]$, $\xi[H]$. Finally, *H*, *G*, *F*, *E*, *C*, *B* and *A* can be degraded by means of ubiquitination and proteasome cleavage and/or irreversibly inhibited by other molecular species, and these processes are described by the rate constants $k_{H\deg}$,$k_{G\deg}$, $k_{F\deg}$, $k_{E\deg}$, $k_{C\deg}$, $k_{B\deg}$ and $k_{A\deg}$ respectively.

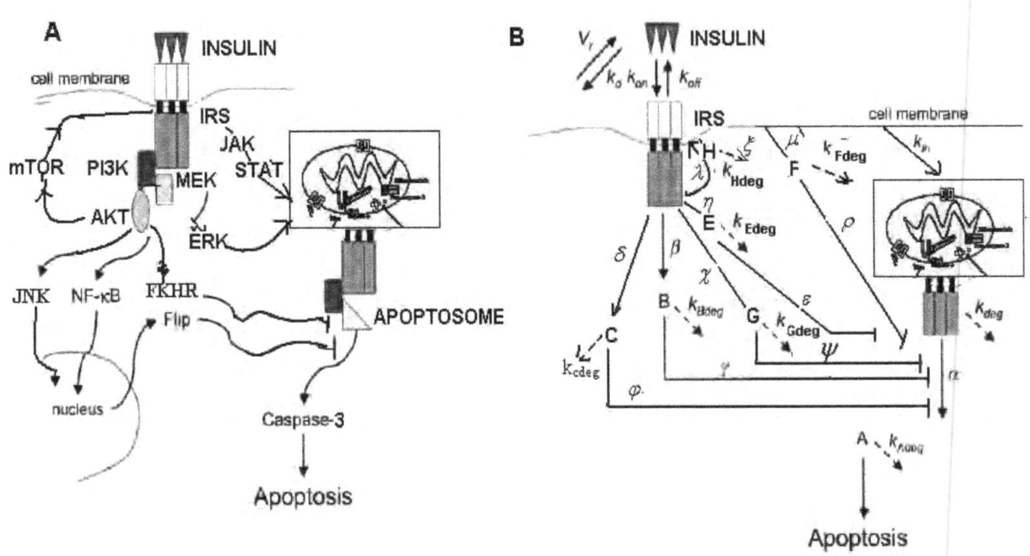

Fig. 1 Modeling insulin mediated signaling. A. Biological view of the main insulin paths. B. Scheme of the biochemical pathways that have been considered in the present model.

3 Numerical Methods

Let us consider a differential equation as

$$\frac{dy}{dx}+Py=Q \tag{3}$$

$\frac{d}{dx}$ can be written as D, replacing in Eq. (3) we get

$$Dy+Py=Q \tag{4}$$

Now, we know the integrating factor is

$$e^{\int P\,dx}$$

Multiply Eq. (4) with integrating factor both sides, we get

$$e^{\int P\,dx}.(Dy+Py)=e^{\int P\,dx}.Q$$

$$Dye^{\int P\,dx}+Pye^{\int P\,dx}=e^{\int P\,dx}.Q$$

$$Dye^{\int P\,dx}=e^{\int P\,dx}.Q$$

Integrating both sides

$$ye^{\int P\,dx}=\int Q.e^{\int P\,dx}\,dx+c$$

Finally the solution of equation is

$$\mathbf{y}=\frac{\int \mathbf{Q.e^{\int Pdx}\,dx+c}}{\mathbf{e^{\int Pdx}}} \tag{5}$$

Now our main equation is $\frac{d[B]}{dt}+k_{B\deg}[B]=\beta[N_c]$ (6)

If we compare the above equation i.e Eq. (6) with Eq. (4) we get

$$y=B;P=k_{B\deg}\ ;\ Q=\beta[N_c]$$

Putting all these values in Eq. (5), we get

$$B=\frac{\int \beta N_c.e^{\int k_{B\deg}\,dt}\,dt+c}{e^{\int k_{B\deg}\,dt}} \tag{7}$$

$$B=e^{-k_{B\deg}t}\left[\left(\int \beta N_c.e^{k_{B\deg}t}\,dt\right)+c\right]$$

After solving the above equation we get

$$B = \frac{\beta N_c}{k_{B\deg}} + c.e^{-k_{B\deg} t} \tag{8}$$

Now applying the initial condition i.e. $t = 0$ and $B = 0$; in the Eq. (8), we get

$$c = -\frac{\beta N_c}{k_{B\deg}}$$

Now putting the value of c in Eq. (7), we get

$$\mathbf{B} = \frac{\boldsymbol{\beta} \mathbf{N_c}}{\mathbf{k_{B\deg}}}\left(1 - \mathbf{e}^{-\mathbf{k_{B\deg} t}}\right) \tag{9}$$

Similarly, we can solve equations of different marker proteins involved in the cell survival and cell death pathway induced by TNF, EGF and insulin.

3.1 Differential equation modeling activity of NF-κB

We have solved the differential equation of NF-κB as

$$\frac{d[B]}{dt} + k_{B\deg}[B] = \beta[N_c]$$

Solution comes out to be

$$[B] = \frac{\beta[N_c]}{k_{B\deg}}\left[1 - e^{-k_{B\deg} t}\right] \tag{10}$$

By considering values of $\beta = 0.33\ \text{min}^{-1}$ and $k_{B\deg} = 0.033\ \text{min}^{-1}$ (Chignola et al., 2009), putting these values in Eq. (10) we get

$$[B] = 10\left[1 - e^{-0.33t}\right] \tag{11}$$

Again by considering the different values of time (Gaudet et al., 2005) and solving Eq. (11); the values of B have been obtained and get a plot between concentration of NF-κB and Time which comes out to be exponential.

3.2 Differential equation modeling activity of JNK

We have solved the differential equation of JNK as

$$\frac{d[C]}{dt} + k_{C\deg}[C] = \delta[N_c]$$

Solution comes out to be

$$[C] = \frac{\delta[N_c]}{k_{C\deg}}\left[1 - e^{-k_{C\deg} t}\right] \tag{12}$$

By assuming the values of $\delta = 0.33/\min$; $k_{C\deg} = 0.018/\min$ and applying these values in Eq. (12) we got

$$[C] = 18.33\left[1 - e^{-0.018t}\right] \tag{13}$$

Considering the different values of time (Gaudet et al., 2005) in Eq. (13) we calculated the corresponding values of C.

3.3 Differential equation modeling activity of MEK/ERK

We have solved the differential equation of MEK/ERK as

$$\frac{d[E]}{dt}+k_{E\deg}[E]=\eta[N_c]$$

Solution comes out to be

$$[E]=\frac{\eta[N_c]}{k_{E\deg}}\left[1-e^{-k_{E\deg}t}\right] \quad (14)$$

By assuming the values of values of $\eta=0.33/\min$; $k_{E\deg}=0.016/\min$ and putting these values in Eq. (14) we solved the equation for calculating E as,

$$[E]=20.62\left[1-e^{-0.016t}\right] \quad (15)$$

By taking the different values of time (Gaudet et al., 2005) and solving the Eq. (15); we calculated the values of E.

3.4 Differential equation modeling activity of JAK/STAT

We have solved the differential equation of JAK/STAT as

$$\frac{d[G]}{dt}+k_{G\deg}[G]=\chi[N_c]$$

Solution comes out to be

$$[G]=\frac{\chi[N_c]}{k_{G\deg}}\left[1-e^{-k_{G\deg}t}\right] \quad (16)$$

Now applying the values of $\chi=0.33/\min$; $k_{G\deg}=0.017/\min$, in Eq. (16) we resolved the final equation for G as,

$$[G]=19.41\left[1-e^{-0.017t}\right] \quad (17)$$

With different values of time (Gaudet et al., 2005) and using Eq. (17) the values of G have been calculated.

3.5 Differential equation modeling activity of mTOR

We have solved the differential equation of mTOR as

$$\frac{d[H]}{dt}+k_{H\deg}[H]=\lambda[N_c]$$

Solution comes out to be

$$[H]=\frac{\lambda[N_c]}{k_{H\deg}}\left[1-e^{-k_{H\deg}t}\right] \quad (18)$$

Now applying the values of $\lambda = 0.33/\min$; $k_{H\deg} = 0.02/\min$ in Eq. (18) we solved the equation for calculating the value of H as:

$$[H] = 16.5\left[1 - e^{-0.02t}\right] \quad (19)$$

Also the values of H with respect to different values of time (Gaudet et al., 2005) have been calculated.

4 Results and Discussions

4.1 Experimental findings

The cultures were stimulated 24 hr later by adding the stimulus diluted in 1/20 of the culture volume of serum-free medium, for final concentrations of 0, 0.2, 5, or 100 ng/ml TNF (Peprotech), 0, 1, or 100 ng/ml EGF (Peprotech), and 0, 1, 5, or 500 ng/ml insulin (Sigma). Triplicate plates were lysed at 0, 5, 15, 30, 60, and 90 min and 2, 4, 8, 12, 16, 20, and 24 hr or prepared for flow cytometry at 12, 24, and 48 hr. To explore systematically relationships between cytokine-receptor interaction, activation of intracellular signaling cascades, and apoptosis-survival cell-fate decisions, cells were exposed to a set of ten cytokine treatments shown in Table 1 and monitored over a 48-hr period. Each treatment consisted of a combination of TNF and either EGF or insulin Cells respond to TNF, EGF, and insulin in a dose-dependent manner and all three cytokines were therefore examined at sub saturating concentrations, designed to mimic physiological conditions, and at saturating concentrations, at which essentially all receptors were ligand-bound. At 13 time points after cytokine addition, three replicate dishes of cells (six for the zero time point) were harvested to measure kinase activities, changes in protein phosphorylation, caspase cleavage, and changes in protein abundance. All together, AkT signals were examined. Kinases such as Akt and ERK were maximally active 5-15 min after cytokine addition whereas caspase cleavage was evident only after 4 hr time points were spaced closely from 0-2 hr (0, 5, 15, 30, 60, 90, 120 min) and then more sparsely from 4-24 hr (4, 8, 12, 16, 20, 24 hr), where $t = 0$ min is the time of cytokine addition taken from Gaudet et al. (2005).

Table 1 Ten cytokine treatments

	(a)	*(b)*	*(c)*	*(d)*	*(e)*	*(f)*	*(g)*	*(h)*	*(i)*	*(j)*
TNF(ng/ml)	-	5	100	-	5	100	-	0.2	5	100
EGF(ng/ml)	-	-	-	100	1	100	-	-	-	-
Insulin(ng/ml)	-	-	-	-	-	-	500	1	5	500

4.2 Deterministic finding

We have presented an integrated theoretical framework for describing the balance between cell survival and cell death regulated by insulin. The theoretical models consider here for mathematical intervention is minimal in the sense it takes into consideration only those reactions that are essential to describe the action of insulin on cell survival and cell death. The results illustrated the rate constants defined in the deterministic models are biologically relevant of key proteins in the signaling pathway which decide the fate of the cell.

Moreover, the rate constant κ parameterizes the triggering kinetics of the survival signal that initiates at the cell membrane level upon binding of TNF to TNF-R1, EGF to EGFR and INSULIN to IRS. A complex of adaptor proteins (consisting of RIP-1 and TRAF-2) is formed at the cell surface transduces signal for activation

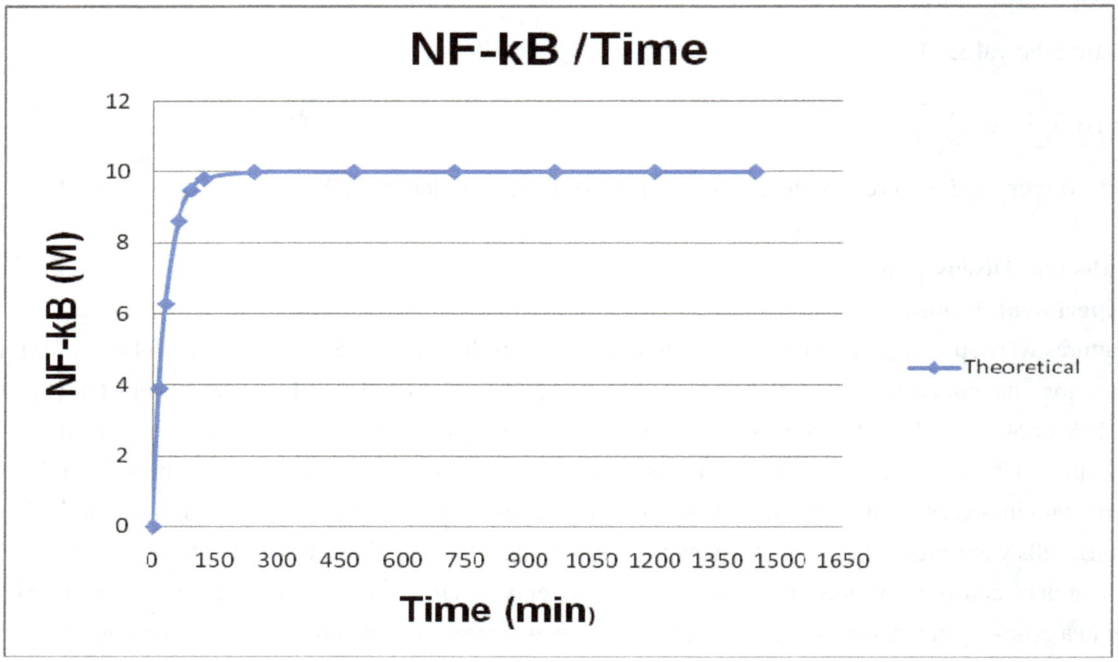

Fig. 2 Theoretical result of NF-κB.

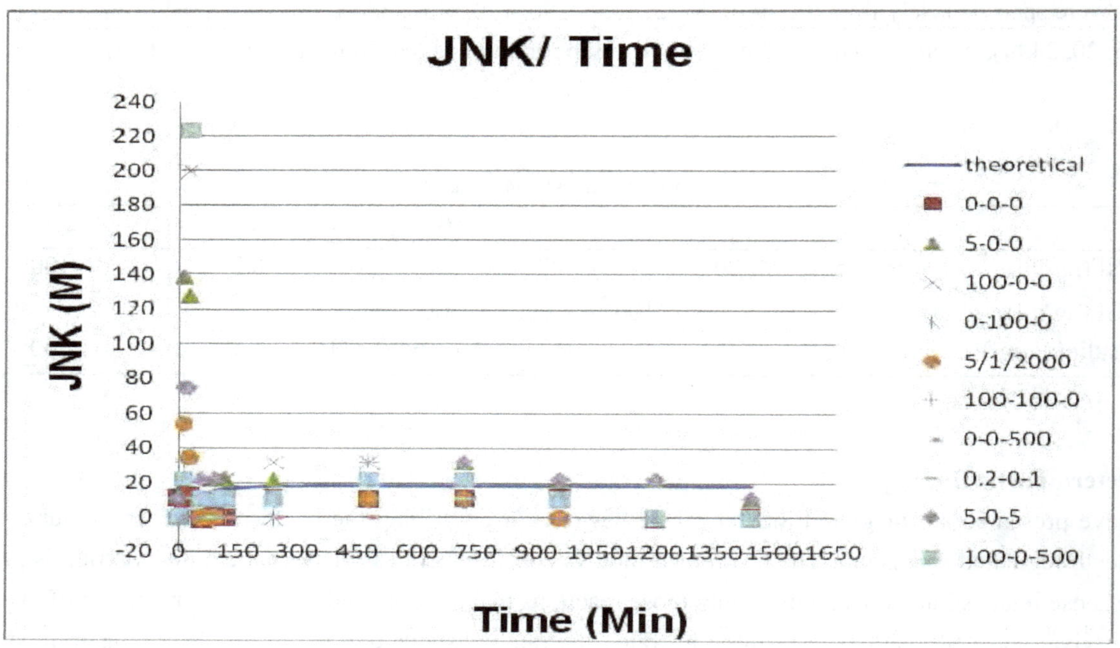

Fig. 3 Theoretical and Practical results of JNK.

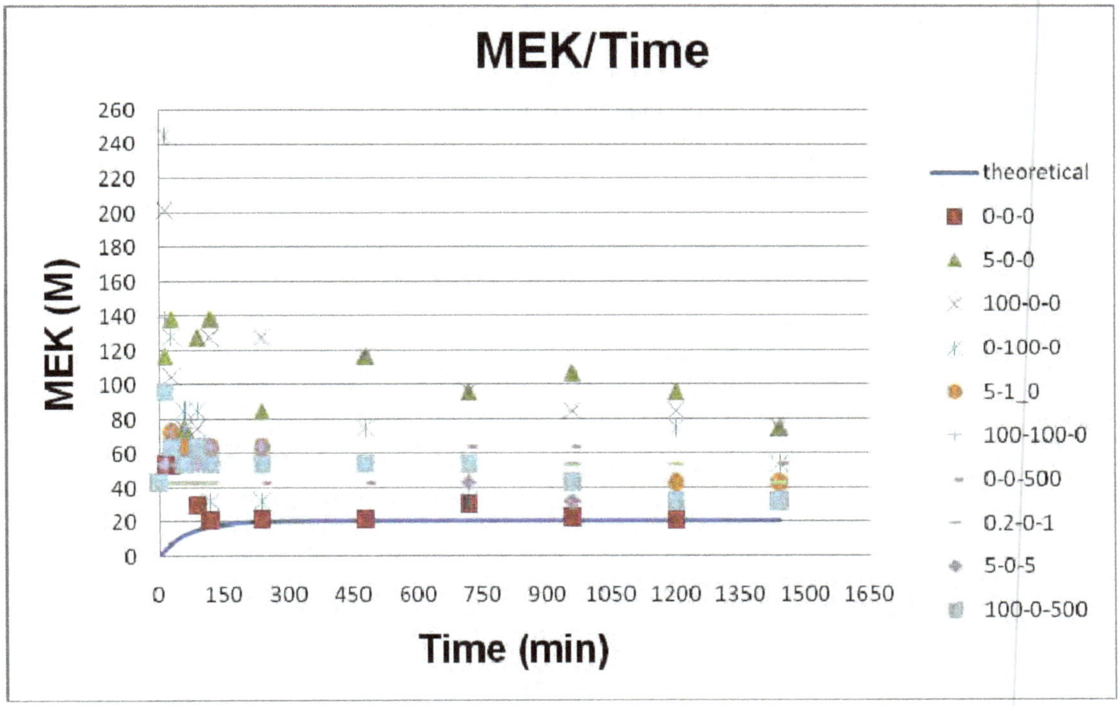

Fig. 4 Theoretical and Practical result of MEK.

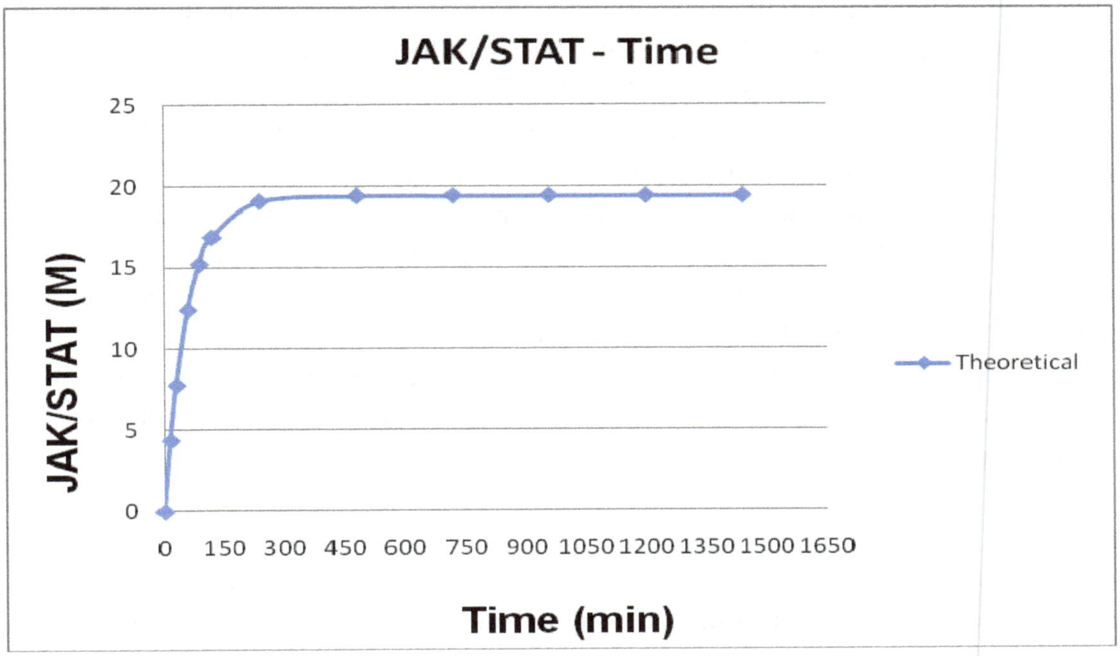

Fig. 5 Theoretical result of JAK/ STAT.

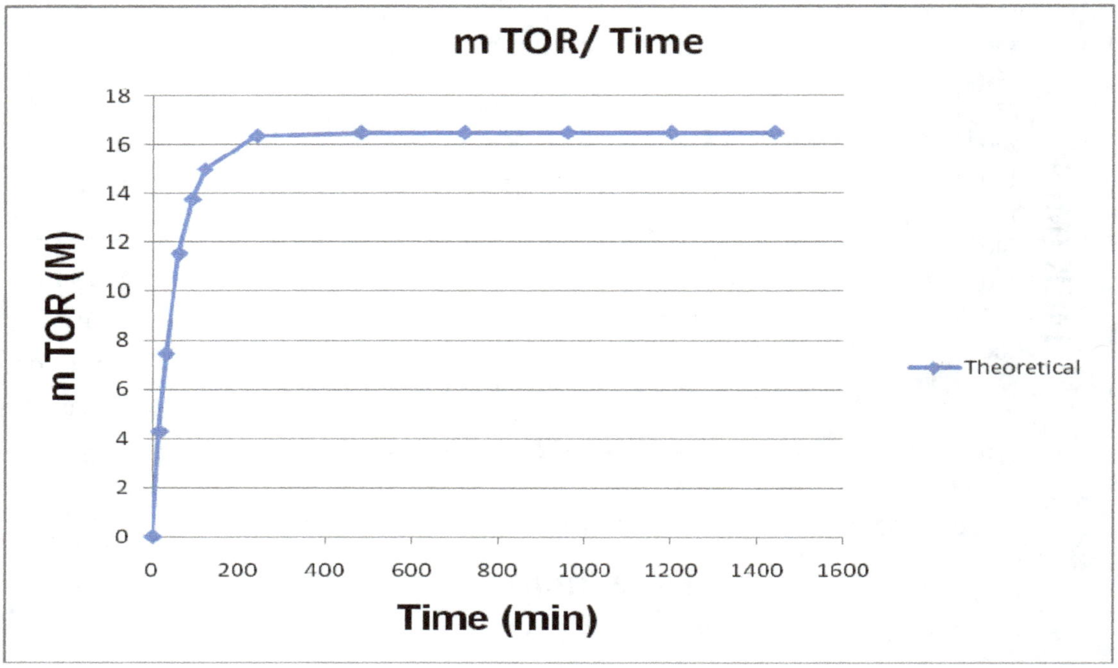

Fig. 6 Theoretical result of mTOR.

of NF-κB and JNK leading to cell survival. Insulin induced activation of proteins such as NF-κB, JNK, MEK/ERK, JAK/STAT and mTOR leading to cell survival, while NF-κB/ FLIP, FKHR, ERK/AP1, JAK/STAT to Bax and JNK/FLIP leads to cell death. Thus it has been revealed that survival and death signals induced by insulin are temporarily separated and this is reflected in our model by the differences between the values of the parameters used. Simulations based on deterministic model recapitulate most features of the data and generate several predictions involving pathway crosstalk and regulation. We uncover a relationship between the key proteins involved in insulin cellular signalling pathways that might account for the cell survival and cell death decision of the cells. More generally, deterministic models are flexible, able to incorporate qualitative and noisy data, and powerful enough to produce quantitative predictions and new biological insights about the operation of signalling networks.

The quality of fit between theoretical and experimental values (ten cytokine combinations) for the proteins such as NFκB, JNK, MEK, JAK-STAT and mTOR respectively revealed good accuracy of the model shown in Figures 2 to 6 respectively.

5 Conclusion

We have made the biological view of the main paths of the insulin input showing cell survival and cell death. We have successfully made a deterministic model for insulin which verifies experimental and theoretical data. Thus, our model is suitable for implementation in multi-scale simulation programs that are presently under development to study the behavior of large tumor cell populations.

Abbreviations

AP-1, Activation Protein 1; ASK1, Apoptosis signal-regulating kinase 1; EGF, epidermal growth factor; EGFR, epidermal growth factor receptor; ERK, extracellular-regulated kinase; FADD, Fas-Associated protein with Death Domain; FKHR, Forkhead transcription factor; Grb2, growth factor receptor-bound 2; IGF, insulin-like growth factor; IκB, I Kappa B (nuclear factor of kappa light polypeptide gene enhancer in B-cells inhibitor); IKK, IκB kinase; IR, insulin receptor; IRS1, insulin receptor substrate 1; JNK1, c-jun NH_2 terminal kinase 1; MAP kinases, mitogen-activated protein kinases; MEK, mitogen-activated protein kinase and extracellular-regulated kinase kinase; MK2, mitogen-activated protein kinase-activated protein kinase 2; mTOR, mammalian target of rapamycin; NF-κB, nuclear factor-κB; PLADD, pre-ligand assembly domain; PDK, Phi Delta Kappa; PI3K, phosphatidylinositol 3-kinase; p38, P38 mitogen-activated protein kinases; Rac, Ras-related C3 botulinum toxin substrate; SAPK/JNK , Stress-activated protein kinase/Jun-amino-terminal kinase; SH2, Src homolgy 2; SODD, Silencer of death domains; SOS, Son of Sevenless; TNF, tumor necrosis factor; TNFR1, tumor necrosis factor receptor 1; TNFR2, tumor necrosis factor receptor 2; TRADD, TNFR associated via death domain; TRAF2, TNF receptor associated factor 2, XIAP, X-linked Inhibitor of Apoptosis Protein.

References

Avruch J. 1998. Insulin signal transduction through protein kinase cascades. Molecular and Cellular Biochemistry, 182: 31–48

Bajzer Z, Myers AC, Vuk-Pavlovic S. 1989. Binding, internalization and intracellular processing of proteins interacting with recycling receptors. The Journal of Biological Chemistry, 264: 13623-13631

Brent R. 2000. Genomic biology. Cell, 100: 169–183

Burns MA, Johnson BN, Brahmasandra SN, et al. 1998. An integrated nanoliter DNA analysis device. Science, 282: 484–487

Cagney G, Uetz P, Fields S. 2000. High-throughput screening for protein–protein interactions using two-hybrid assay. Methods in Enzymology, 328: 3-14

Chignola R, Farina M, Fabrina M, Del Fabbro A, Milotti E. 2009. Balance between cell survival and death: a minimal quantitative model of tumor necrosis factor alpha cytotoxicity. BIOLOGICAL SCIENCES, Biophysics and Computational Biology, 1-32

Gardner TS, di Bernardo D, Lorenz D, Collins JJ. 2003. Inferring genetic networks and identifying compound mode of action via expression profiling. Science, 301: 102-105

Gaudet S, Janes KA, Albeck JG, et al. 2005. A compendium of signals and responses trigerred by prodeath and prosurvival cytokines. Molecular and Cellular Proteomics, 4: 1569-1590

Grell M, Wajant H, Zimmermann G, Scheurich P. 1998. The type 1 receptor (CD120a) is the high-affinity receptor for soluble tumor necrosis factor. Proceedings of the National Academy of Sciences, 95: 570-575

Ideker T, Lauffenburger D. 2003. Building with a scaffold: Emerging strategies for high-to-low-level cellular modeling. Trends in Biotechnology, 21: 255–262

Kitano H. 2002. Systems biology: A brief overview. Science, 295: 1662-1664

Lund-Johansen F, Davis K, Bishop J, et al. 2000. Flow cytometric analysis of immunoprecipitates: high-throughput analysis of protein phosphorylation and protein–protein interactions. Cytometry, 39: 250-259

Maliakal JC. 2002. Quantitative high throughput endothelial cell migration and invasion assay system. Methods in Enzymology, 352:175-182

Morand KL, Burt TM, Regg BT, Tirey DA. 2001. Advances in high-throughput mass spectrometry. Current Opinion in Drug Discovery and Development, 4: 729-735

Price ND, Papin JA, Schilling CH, Palsson BO. 2003. Genome-scale microbial in silico models: the constraints-based approach. Trends in Biotechnology, 21: 162-169

Rives AW, Galitski T. 2003. Modular organization of cellular networks. Proceedings of the National Academy of Sciences, 100: 1128-1133

Schoeberl B, Eichler-Jonsson C, Gilles ED, Muller G. 2002. Computational modeling of the dynamics of the MAP kinase cascade activated by surface and internalized EGF receptors. Nature Biotechnology, 20: 370-375

Vuk-Pavlovic S, Kovach JS. 1989. Recycling of tumor necrosis factor-α receptor in MCF-7 cells. FASEB, J3: 2633-2640

Wajant H, Pfizenmaier K, Scheurich P. 2003. Tumor necrosis factor signaling. Cell Death and Differentiation, 10: 45-65

Xia Z, Dickens M, Raingeaud J, et al. 1995. Opposing effects of ERK and JNK-p38 MAP kinases on apoptosis. Science, 270: 1326-1331

Permissions

All chapters in this book were first published in NB, by International Academy of Ecology and Environmental Sciences; hereby published with permission under the Creative Commons Attribution License or equivalent. Every chapter published in this book has been scrutinized by our experts. Their significance has been extensively debated. The topics covered herein carry significant findings which will fuel the growth of the discipline. They may even be implemented as practical applications or may be referred to as a beginning point for another development.

The contributors of this book come from diverse backgrounds, making this book a truly international effort. This book will bring forth new frontiers with its revolutionizing research information and detailed analysis of the nascent developments around the world.

We would like to thank all the contributing authors for lending their expertise to make the book truly unique. They have played a crucial role in the development of this book. Without their invaluable contributions this book wouldn't have been possible. They have made vital efforts to compile up to date information on the varied aspects of this subject to make this book a valuable addition to the collection of many professionals and students.

This book was conceptualized with the vision of imparting up-to-date information and advanced data in this field. To ensure the same, a matchless editorial board was set up. Every individual on the board went through rigorous rounds of assessment to prove their worth. After which they invested a large part of their time researching and compiling the most relevant data for our readers.

The editorial board has been involved in producing this book since its inception. They have spent rigorous hours researching and exploring the diverse topics which have resulted in the successful publishing of this book. They have passed on their knowledge of decades through this book. To expedite this challenging task, the publisher supported the team at every step. A small team of assistant editors was also appointed to further simplify the editing procedure and attain best results for the readers.

Apart from the editorial board, the designing team has also invested a significant amount of their time in understanding the subject and creating the most relevant covers. They scrutinized every image to scout for the most suitable representation of the subject and create an appropriate cover for the book.

The publishing team has been an ardent support to the editorial, designing and production team. Their endless efforts to recruit the best for this project, has resulted in the accomplishment of this book. They are a veteran in the field of academics and their pool of knowledge is as vast as their experience in printing. Their expertise and guidance has proved useful at every step. Their uncompromising quality standards have made this book an exceptional effort. Their encouragement from time to time has been an inspiration for everyone.

The publisher and the editorial board hope that this book will prove to be a valuable piece of knowledge for researchers, students, practitioners and scholars across the globe.

List of Contributors

Mohammad Khansari, Amin Kaveh, Zainabolhoda Heshmati and Maryam Ashkpoor Motlaq
University of Tehran, Faculty of New Sciences and Technologies, Amir Abad, North Kargar Street 14395-1374, Tehran, Iran

Anindita Bhadra
Behaviour and Ecology Lab, Department of Biological Sciences, Indian Institute of Science Education and Research - Kolkata, Mohanpur, India

Ferenc Jordán
The Microsoft Research – University of Trento Centre for Computational and Systems Biology, Rovereto, Italy

Ranjit Kumar Upadhyay and Argha Mondal
Department of Applied Mathematics, Indian School of Mines, Dhanbad-826004, India

Gaurav and R. S. Anand
Department of Electrical Engineering, Indian Institute of Technology Roorkee, Roorkee 247667, India

Vinod Kumar
Jaypee University of Information Technology, Waknaghat 712100, India

Walter Santos de Araújo
Departamento de Biologia Geral, Universidade Estadual de Montes Claros, 39401-089, Montes Claros, Brazil

Julio Miguel Grandez-Rios
Center for International Forestry Research, 1895, Lima, Peru

Leonardo Lima Bergamini
Departamento de Ecologia, Universidade Federal de Goiás, 74001-970, Goiânia, Brazil

Ján Kollár
Department of Planting Design and Maintenance, Slovak University of Agriculture in Nitra, 949-76, Nitra, Slovakia

Alessandro Ferrarini
Department of Evolutionary and Functional Biology, University of Parma, Via G. Saragat 4, I-43100 Parma, Italy

WenJun Zhang
School of Life Sciences, Sun Yat-sen University, Guangzhou 510275, China; International Academy of Ecology and Environmental Sciences, Hong Kong

Bita Shams and Mohammad Khansari
University of Tehran, Faculty of New Science and Technology, Amir Abad, North Kargar Street 14395-1374, Tehran, Iran

Muhammad Shakil, H. A. Wahab and Muhammad Shahzad
Department of Mathematics, Hazara University, Manshera, Pakistan

Muhammad Naeem
Department of Information Technology, Hazara University, Manshera, Pakistan

Saira Bhatti
Department of Mathematics, COMSATS Institute of Information Technology, Abbottabad, Pakistan

Jasmin Šutković and Mohamed Ragab Abdel Gawwad
Genetics and Bioengineering Department, International University of Sarajevo, Ilidza, 71220 Bosnia and Herzegovina

Navdeep Prashar, Jyotsna Dogra, Meenakshi Sood and Shruti Jain
Department of Electronics and Communication Engineering, Jaypee University of Information Technology, Solan, India

Walter Santos de Araújo
Department of General Biology, Center of Biological Sciences and Health, Universidade Estadual de Minas Gerais, 39401-089, Montes Claros, Brazil

Sudipta Saha
Dalla Lana School of Public Health, University of Toronto, Toronto, ONM5S 2J7, Canada

Munni Begum
Department of Mathematical Sciences, Ball State University, Muncie, IN47306, USA

Shruti Jain
Department of Electronics and Communication Engineering, Jaypee University of Information Technology, Solan-173234, India

Robi Tacutu, Hagai Yanai, Marina Wolfson and Vadim Fraifeld
The Shraga Segal Department of Microbiology and Immunology, Center for Multidisciplinary Research on Aging, Ben-Gurion University of the Negev, Beer Sheva 84105, Israel

Pradeep K. Naik
Department of Bioinformatics and Biotechnology, Jaypee University of Information Technology, Waknaghat, Solan 173215, Himachal Pradesh, India

Vadim E. Fraifeld
The Shraga Segal Department of Microbiology and Immunology, Center for Multidisciplinary Research on Aging, Ben-Gurion University of the Negev, Beer Sheva 84105, Israel

Arie Budovsky
Judea Regional Research & Development Center, Carmel 90404, Israel
The Shraga Segal Department of Microbiology and Immunology, Center for Multidisciplinary Research on Aging, Ben-Gurion University of the Negev, Beer Sheva 84105, Israel

Shruti Jain and Sunil V. Bhooshan
Department of Electronics and Communication Engineering

Index

www.ingramcontent.com/pod-product-compliance
Lightning Source LLC
LaVergne TN
LVHW060518210926
848485LV00005B/1100
* 9 7 8 1 6 3 9 8 9 1 2 1 4 *